# REVIEWS OF PLASMA CHEMISTRY

**Volume 1**

# REVIEWS OF PLASMA CHEMISTRY

## Volume 1


**Edited by**

**B. M. Smirnov**

*Institute of High Temperatures*
*Academy of Sciences of the USSR*
*Moscow, USSR*

Translated from Russian by

**D. H. McNeill**



**CONSULTANTS BUREAU**
**NEW YORK AND LONDON**

**Library of Congress Cataloging-in-Publication Data**

Reviews of plasma chemistry / edited by B.M. Smirnov ; translated from Russian by D.H. McNeill.
v. <1 >
Translated from the Russian.
Includes bibliographical references and index.
ISBN 0-306-11041-5
1. Plasma chemistry. I. Smirnov, B. M. (Boris Mikhaĭlovich), 1938-
QD581.R48 1991
541'.0424--dc20 90-19773
CIP

ISBN 0-306-11041-5

The Russian text was published by Énergoatomizdat in Moscow in 1987 under the title *Khimiya Plazmy, Vyp. 14.*


A Division of Plenum Publishing Corporation
233 Spring Street, New York, N.Y. 10013



Printed in the United States of America

# CONTENTS

# RECOMMENDED DATA ON THE RATE CONSTANTS OF CHEMICAL REACTIONS AMONG MOLECULES CONSISTING OF N AND O ATOMS

**O. É. Krivonosova, S. A. Losev, V. P. Nalivaiko, Yu. K. Mukoseev, and O. P. Shatalov**

The solution of problems involving the motion of gases subject to chemical transformations requires a fairly complete set of quantitative characteristics for the models that are used, including transition probabilities, cross sections, reaction rate constants, relaxation times, etc. A vast amount of data on these quantities have already been published in the literature and one encounters the problems of systematizing, critically analyzing, and selecting these data, developing recommendations, and clarifying what information is lacking. Data bases have been developed for solving this sort of problem. One example is the AVOGADRO data base, which has been developed for research on physicochemical gas dynamics [1]. In this review, which was written in the course of constructing that data base, we analyze published data on the rate constants of chemical reactions in "dry air," i.e., in a system made up of compounds of N and O atoms. This system contains the following 12 species: N, O, NO, $O_2$, $N_2$, $N_3$, $N_2O$, $NO_2$, $O_3$, $NO_3$, $N_2O_4$, and $N_2O_5$. This effort was based on earlier reviews [2–7]. Molecular dissociation and recombination, as well as bimolecular and termolecular reactions, were examined. The recommended values of the rate constants for these reactions are given in tabular form. When experimental or theoretical data are available, these recommendations apply to the widest possible temperature range. New data on the rate constants have been identified with the aid of the abstracts journal *Khimiya* up to May 1986. In addition, the low reliability or absence of certain experimental data meant that, in a number of cases, it was necessary to draw on our own or published theoretical estimates based on the method of single-quantum stepwise excitation, the transition state model, the Troe model, or the modified variational principle of Keck.

The reliability of the experimental and theoretical data was evaluated by an expert analysis of the material. This evaluation was carried out in accordance with the criteria of Baulch et al. [3, 4] and takes into account any information in the original papers on the experimental conditions and on the procedure for processing the raw data, the authors' estimates of the reproducibility and accuracy of their results, and the sensitivity of the results to the model that was chosen. When insufficient information was available, the authority of and the confidence inspired by the author of the paper were taken into account.

The accuracy of the recommended values of the constants is not specified when they have been extrapolated into temperature ranges where there are no experimental (and, usually, theoretical) data.

One characteristic feature of molecular dissociation at high temperatures is an increasing "negative temperature dependence of the preexponential factor" (in the generalized Arrhenius form for the dissociation rate constant) as the temperature is raised. In the last 10–15 years this has been attributed to incomplete vibrational relaxation of molecules in the dissociation process. "Two-temperature dissociation" of diatomic molecules has been extensively investigated theoretically [8–10]. In the approximation of an harmonic oscillator, the rate constant for these processes with $T_v \neq T$ (where $T_v$ and $T$ are the vibrational and translational temperatures, respectively, of a dissociating molecular species) is given by

$$K = K^0 \Phi(T, T_v), \tag{1}$$

where $K^0$ is the equilibrium rate constant for $T_v = T$ and

$$\Phi(T, T_v) = \frac{1 - \exp(-\theta/T_v)}{1 - \exp(-\theta/T)} \exp\left[-\frac{D - \beta kT}{k}\left(\frac{1}{T_v} - \frac{1}{T}\right)\right]. \tag{2}$$

Here $\theta$ is the characteristic temperature of a molecule, $D$ is its dissociation energy, $k$ is Boltzmann's constant, and $\beta$ is a parameter that characterizes the lowering of the dissociation threshold with increasing temperature $T$.

Equation (1) provides a satisfactory description of experiments on dissociation of diatomic molecules at high temperatures, and it is used here. For extrapolation to lower temperatures we have used the single-quantum stepwise excitation model, together with published data obtained from the transition state theory [11], a modification of Keck's variational theory [12], and a single-step deactivation scheme [13]. The proposed formulas for the extrapolation of the dissociation rate constants for $N_2$, NO, and $O_2$ to low temperatures rely on the theoretically predicted temperature dependences of $K$ and have been matched to experimental data for higher temperatures, as well as to the scaled rate constants for the reverse reactions derived from the equilibrium constants.

This analysis allows us to recommend values of the rate constants for dissociation and recombination of $N_2$, NO, and $O_2$ molecules with a certain degree of confidence for temperatures in the range 300–20,000 K. In addition, this analysis revealed the existence of significant gaps in research on some reactions: in the case of the dissociation of $N_2$, the rate of dissociation of nitrogen through interactions with O atoms and NO and $O_2$ molecules is poorly known; there are essentially no reliable data on the efficiency of various partners (O, N, NO, $N_2$, $O_2$) in the dissociation of NO; the efficiency of NO molecules and N atoms in the dissociation of $O_2$ is unknown; there is little experimental data on the dependence of the dissociation rate constants on vibrational excitation; and there are essentially no experimental data on recombination of molecules at high temperatures, not to mention on the efficiency of various collision partners in this process. At the same time, some reactions, especially the dissociation of oxygen ($M$ = Ar, $O_2$) up to T = 10,000 K and nitrogen ($M$ = Ar, $N_2$) for $T \leq 15{,}000$ K, have been studied quite reliably and can serve as reference reactions for describing and studying the processes that take place in air.

In this review we have also analyzed and recommended values for the rates of dissociation and recombination of the molecules $O_3$, $NO_2$, $N_2O$, $NO_3$, $N_2O_4$, and $N_2O_5$, as well as for the rates of bimolecular and termolecular reactions, over a wide range of temperatures at different pressures and for various compositions of the gas mixture. Here, however, even more than in the case of diatomic molecules, there are gaps in the quantitative information. Despite evidence that the vibrational temperature affects the rate of dissociation of polyatomic molecules at high temperatures, there are no systematic data about these effects. For some reactions, such as the dissociation of $NO_3$, the dissociation of $N_2O_4$, and the formation of $N_2O$, $NO_3$, and $N_2O_4$, the data are clearly inadequate or are entirely absent. In a number of cases it is difficult to obtain these data by scaling the rate constants of the reverse reactions, since the data for the reverse reaction are also inadequate (e.g., for the formation and breakup of $NO_3$). Data for some reactions are available only at low temperatures and the scatter in these data is as high as three orders of magnitude at $T = 298$ K.

The relationship among the different channels for the reaction N + $NO_2 \rightarrow$ products is not entirely clear and there are no data on the tempera-

TABLE 1. Rate Constants for Dissociation of Diatomic Molecules

| $M$ | $T$, K | Reaction rate constant, $cm^3/(mole \cdot sec)$ | Error: $\Delta \log K$ | Error: $T$, K | References |
|---|---|---|---|---|---|
| | | 1.1. $N_2+M=N+N+M$ | | | |
| Ar | 300—6000 | $1.08 \cdot 10^{16} \exp(-113\,200/T)\,[1-\exp(-3354/T)]$ | 0.6 | 300—6000 | A |
| | 6000—20 000 | $1.86 \cdot 10^{29}\, T^{-3.6} \exp(-113\,200/T)$ | 0.48<br>0.3<br>0.48<br>— | 6000—7000<br>7000—10 000<br>10 000—15 000<br>15 000—20 000 | A |
| $N_2$ | 300—6000 | $K_{1.1}(M=N_2)=2.96\,K_{1.1}(M=Ar)$ | 0.6 | 300—6000 | A |
| | 6000—20 000 | $2.3 \cdot 10^{29}\, T^{-3.5} \exp(-113\,200/T)$ | 0.48<br>0.3<br>— | 6000—8000<br>8000—14 000<br>14 000—20 000 | [14]<br>[14]<br>A |
| N | 300—6000 | $K_{1.1}(M=N)=6.6\,K_{1.1}(M=Ar)$ | 0.6 | 300—6000 | A |
| | 6000—20 000 | $8.5 \cdot 10^{25}\, T^{-2.5} \exp(-113\,200/T)$ | 0.48<br>— | 6000—14 000<br>14 000—20 000 | [14]<br>A |
| $O_2$, NO | 10 000 | $K_{1.1}(M=O_2) \approx K_{1.1}(M=NO) \approx K_{1.1}(M=N_2)$ | — | 10 000 | A |
| O | 10 000 | $K_{1.1}(M=O) \approx K_{1.1}(M=N)$ | — | 10 000 | A |
| | | $N_2+M=N+N+M$ (for $T \neq T_v$) | | | |

$$K_{1.1}=K^0_{1.1}\,\Phi(T,\ T_v);$$

$$\Phi(T,\ T_v)=\frac{1-\exp(-\theta/T_v)}{1-\exp(-\theta/T)}\exp\left[-\frac{D-\beta kT}{k}\left(\frac{1}{T_v}-\frac{1}{T}\right)\right];$$

$$\theta=3354\ \text{K};\ D/k=113\,200;$$

$$\beta=3$$

| | | | | | |
|---|---|---|---|---|---|
| Ar | 6000—15 000* | $K^0_{1.1}=1.08\cdot10^{16}\exp(-D/kT)[1-\exp(-\theta/T)]$ | 0.48 | 6000—15 000 | [15], A |
| $N_2$ | 6000—15 000* | $K^0_{1.1}\,(M=N_2)=3\,K^0_{1.1}\,M=Ar)$ | 0.48 | 6000—15 000 | [15], A |
| N | 6000—15 000* | $K^0_{1.1}\,(M=N)=6{,}6\,K^0_{1.1}\,(M=Ar)$ | 0.48 | 6000—15 000* | [15], A |
| O | 6000—15 000 | $K^0_{1.1}\,(M=O)=K^0_{1.1}\,(M=N)$ | — | 6000—15 000 | A |
| $O_2$, NO | 6000—15 000 | $K^0_{1.1}\,(M=O_2)=K^0_{1.1}\,(M=NO)=K^0_{1.1}\,(M=N_2)$ | — | 6000—15 000 | A |

1.2. $NO+M=N+O+M$

| | | | | | |
|---|---|---|---|---|---|
| Ar | 300—20 000 | $6.1\cdot10^{15}\exp(-75\,500/T)\,[1-\exp(-2700/T)]$ | 0.48<br>— | 300—7500<br>7500—20 000 | A |
| $N_2$, $O_2$ | 4000—7500 | $K_{1.2}\,(M=N_2)\approx K_{1.2}\,(M=O_2)\approx K_{1.2}\,(M=Ar)$ | — | 4000—7500 | A |
| O, N, NO | 4000—7500 | $K_{1.2}\,(M=NO)\approx K_{1.2}\,(M=O)\approx K_{1.2}\,(M=N)\approx$ $\approx20K_{1.2}\,(M=Ar)$ | — | 4000—7500 | [16, 17] |

$NO+M=N+O+M$ (for $T\neq T_v$)

Table 1. Continued

| $M$ | $T$, K | Reaction rate constant, $cm^3$/(mole·sec) | Error $\Delta \log K$ | Error $T$, K | References |
|---|---|---|---|---|---|
| | | $K_{1.2} = K^0_{1.2}\,\Phi(T,\ T_v)$; $\Phi(T,\ T_v) = \frac{1-\exp(-\theta/T_v)}{1-\exp(-\theta/T)} \exp\left[-\frac{D-\beta kT}{k}\left(\frac{1}{T_v}-\frac{1}{T}\right)\right]$; $\theta=2700$ K; $D/k=75\,994$; $\beta=3$ | | | |
| Ar | 4000—7500† | $K^0_{1.2} = 5.2\cdot10^{15}\exp(-75\,994/T)$ | 0.48 | 4000—7500 | [17], A |
| $N_2$, $O_2$ | 4000—7500 | $K^0_{1.2}(M{=}O_2) = K^0_{1.2}(M{=}N_2) = K^0_{1.2}(M{=}Ar)$ | — | 4000—7500 | A per [3, 16, 17] |
| O, N, NO | 4000—7500 | $K^0_{1.2}(M{=}N) = K^0_{1.2}(M{=}O) = K^0_{1.2}(M{=}NO) = 20\,K^0_{1.2}(M{=}Ar)$ | — | 4000—7500 | A per [3, 16, 17] |
| | | 1.3. $O_2+M=O+O+M$ | | | |
| Ar | 300—3000 | $1.1\cdot10^{15}\exp(-59\,380/T)\,[1-\exp(-2240/T)]$ | 0.48 | 300 | A |
| | 3000—20 000 | $1.8\cdot10^{18}\,T^{-1}\exp(-59\,380/T)$ | 0.3 | 3000—18 000 | [4] |
| | | | — | 18 000—20 000 | A |

| | | | | | |
|---|---|---|---|---|---|
| $O_2$ | 300—4000 | $K_{1.3}(M{=}O_2) = 20\,K_{1.3}(M{=}Ar)$ | 0.48 | 300 | A |
| | 4000—20 000 | $9.8 \cdot 10^{24}\, T^{-2.5} \exp(-59\,380/T)$ | 0.3 | 8000 | [4] |
| | | | — | 10 000—20 000 | A |
| O | 300—4000 | $K_{1.3}(M{=}O) \approx 71\,K_{1.3}(M{=}Ar)$ | 0.6 | 300 | A |
| | 4000—20 000 | $3.5 \cdot 10^{25}\, T^{-2.5} \exp(-59\,380/T)$ | 0.3 | 4000—8000 | [4] |
| | | | — | 8000—20 000 | A |
| N, $N_2$, NO | 2800—7000 | $K_{1.3}(M{=}NO) \approx K_{1.3}(M{=}N_2) \approx K_{1.3}(M{=}N) \approx$ $\approx 0.25\,K_{1.3}(M{=}O_2)$ | — | 2800—7000 | A |

$O_2 + M = O + O + M$ (for $T \neq T_v$)

$$K_{1.3} = K^0_{1.3}\,\Phi(T,\ T_v);$$

$$\Phi(T,\ T_v) = \frac{1-\exp(-\theta/T_v)}{1-\exp(-\theta/T)} \exp\left[-\frac{D-\beta kT}{k}\left(\frac{1}{T_v}-\frac{1}{T}\right)\right];$$

$\theta = 2240$ K; $D/k = 59\,380$ K;

$\beta = 1.5$

**Table 1.** Continued

| $M$ | $T$, K | Reaction rate constant, $cm^3$/(mole·sec) | Error | | References |
|---|---|---|---|---|---|
| | | | $\Delta \log K$ | $T$, K | |
| Ar | 5000—10 000‡ | $K^0_{1.3} = 1.1 \cdot 10^{15} \exp\left(-\frac{59\,380}{T}\right) \times \times \left[1 - \exp\left(-\frac{2240}{T}\right)\right]$ | 0.3 | 5000—10 000 | A |
| $O_2$ | 5000—10 000‡ | $K^0_{1.3}(M{=}O_2) = 20\,K^0_{1.3}(\mathrm{Ar})$ | 0.3 | 5000—10 000 | [18], A |
| O | 5000—10 000‡ | $K^0_{1.3}(M{=}O) = 71\,K^0_{1.3}(M{=}\mathrm{Ar})$ | 0.3 | 5000—10 000 | [18], A |
| N, $N_2$, NO | 5000—10 000 | $K^0_{1.3}(M{=}N_2) = K^0_{1.3}(M{=}NO) = K^0_{1.3}(M{=}N) \approx \approx (1/6)\,K^0_{1.3}(M{=}O_2)$ | — | 5000—10 000 | A per [4] |

*Extrapolation to $T = 300$ K is allowed with an error of $\Delta \log K = 0.6$.
†Extrapolation to $T = 300$ K is allowed with an increase in the error to $\Delta \log K = 1$.
‡Extrapolation to $T = 300$ K is allowed with an increase in the error to $\Delta \log K = 0.6$.

TABLE 2. Rate Constants for Combination of Diatomic Molecules

| $M$ | $T$, K | Reaction rate constant, $cm^6/(mole^2 \cdot sec)$ | Error |  | References |
|---|---|---|---|---|---|
|  |  |  | $\Delta \log K$ | $T$, K |  |
| 2.1. $N+N+M=N_2+M$ |  |  |  |  |  |
| $N_2$ | 100—600 | $3.0 \cdot 10^{14} \exp(500/T)$ | >0.18<br>0.18 | 100—200<br>200—600 | [4]<br>[4] |
|  | 600—6300 | $6.9 \cdot 10^{14}$ | 0.18<br>0.48 | 600<br>6300 | A |
|  | 6300—20 000 | $K_{2.1} = K_{1.1} K_p$ | 0.48<br>— | 6300—14 000<br>14000—20 000 | A |
| Ar | 300—600 | $K_{2.1} (M=Ar) \simeq K_{2.1} (M=N_2)$ | 0.4 | 300—600 | A |
|  | 300—20 000 | $K_{2.1} = K_{1.1} K_p$ | 0.4<br>0.48<br>— | 300<br>6000—15 000<br>15 000—20 000 | A |
| NO, $O_2$ | 100—20 000 | $K_{2.1} (M=NO) = K_{2.1} (M=O_2) = K_{2.1} (M=N_2)$ | — | 100—20 000 | A |
| N, O | 100—6000 | $K_{2.1} (M=O) = K_{2.1} (M=N) \approx 6 K_{2.1} (M=Ar)$ | — | 100—6000 | A |
|  | 6000—20 000 | $K_{2.1} (M=O) \approx K_{2.1} (M=N) \approx K_{1.1} (M=N) K_p$ | 0.48<br>— | 6000—14 000<br>14 000—20 000 | A |

**Table 2.** Continued

| $M$ | $T$, K | Reaction rate constant, $cm^6/(mole^2 \cdot sec)$ | Error | | References |
|---|---|---|---|---|---|
| | | | $\Delta \log K$ | $T$, K | |
| | | 2.2. $N+O+M = NO+M$ | | | |
| $N_2$ | 200—20 000 | $6.4 \cdot 10^{16}\, T^{-0.5}$ | 0.18 | 200—400 | [3] |
| | | | 0.48 | 400—7500 | A per [3] |
| | | | — | 7500—20 000 | A |
| Ar | 200—20 000 | $10^{18}\, T^{-1}$ | 0.3 | 200—400 | A |
| | | | 0.3—0.7 | 400—7500 | A |
| | | | — | 7500—20 000 | A |
| N, O, NO, $O_2$ | 200—20 000 | $K_{2.2}\,(M{=}O_2) = K_{2.2}\,(M{=}NO) = K_{2.2}\,(M{=}N) = K_{2.2}\,(M{=}O) = K_{2.2}\,(M{=}O) = K_{2.2}\,(M{=}N_2)$ | — | 200—20 000 | A |
| | | 2.3. $O+O+M=O_2+M$ | | | |
| Ar | 190—4000 | $1.9 \cdot 10^{13} \exp(900/T)$ | 0.18<br>0.3 | 200<br>4000 | [4] |

| | | | | | |
|---|---|---|---|---|---|
| | 4000—20 000 | $K_{2.3} = K_{1.3} K_p$ | 0.3 | 4000—18 000 | A |
| | | | — | 18 000—20 000 | A |
| $O_2$ | 300—4000 | $8.9 \cdot 10^{16} T^{-0.63}$ | 0.48 | 300—4000 | A |
| | 4000—20 000 | $K_{2.3} = K_{1.3} K_p$ | 0.3 | 4000—10 000 | A |
| | | | — | 10 000—20 000 | A |
| $N_2$ | 190—500 | $1.0 \cdot 10^{14} \exp (720/T)$ | 0.18 | 190—500 | [4] |
| | 500—20 000 | $K_{2.3} (M{=}N_2) = 0.25 K_{2.3} (M{=}O_2)$ | — | 500—20 000 | A |
| N, NO | 300—20 000 | $K_{2.3} (M{=}N) = K_{2.3} (M{=}NO) = 0.25 K_{2.3} (M{=}O_2)$ | — | 300—20 000 | A |
| O | 300—4000 | $K_{2.3} (M{=}O) = 3{,}6 K_{2.3} (M{=}O_2)$ | — | 300—4000 | A |

TABLE 3. Rate Constants for Dissociation of Polyatomic Molecules

| $M$ | $T$, K | Reaction rate constant, $cm^6/(mole^2 \cdot sec)$ | Error: $\Delta \log K$ | Error: $T$, K | References | Notes |
|---|---|---|---|---|---|---|
| **3.1. $N_2O+M=O+N_2+M$** | | | | | | |
| Ar | 500—5000 | $K^0_{Ar}=5.2\cdot10^{15}(T/1324)^{-1.25}\times$ $\times\exp(-32\,000/T)$ | 0.3<br>0.15 | 500<br>900—5000 | [19] | — |
| He, $N_2$, $N_2O$, O, NO | 500—5000 | $K^0_M=K^0_{Ar}(\sum_i R_iX_i)[1+78.3\exp(-2315/T)\times$ $\times X_{NO}/\sum_i R_iX_i]^{-1}$ | 0.3<br>0.1 | 500<br>900—5000 | [20]<br>[21] | $R_{Ar}=1$; $R_{N_2}=1.74$; $R_{N_2O}=6.9$; $R_{O_2}=1.3$; $R_{NO}=3$; $R_{He}=4.2$ |
| Ar | 2500—3600<br>$T_v=$<br>$=2500$—3200 | $K_{3.1}=K^0_{Ar}\frac{T_v}{T}\times$ $\times\exp\left[-32\,000\left(\frac{1}{T_v}-\frac{1}{T}\right)\right]$ | — | — | [19] | For $T>2500$ and $[N_2O]\geq10\%$, vibrational relaxation has a significant effect on $K_{3.1}$ |
| $M$ | 900—2100 | $K_{3.1}=K_\infty(1+K_\infty/(K^0_M[M]))^{-1}$ $sec^{-1}$ | 0.2 | 900—2100 | [3] | For arbitrary pressure and composition of the gas mixture |
| | | $K_\infty=1.3\cdot10^{11}\exp(-30\,000/T)$ $sec^{-1}$ | 0.2 | 900—2100 | [3] | — |
| **3.2. $NO_2+M=O+NO+M$** | | | | | | |

| | | | | | | |
|---|---|---|---|---|---|---|
| Ar | 600—5000 | $K^0_{Ar} = 1.1 \cdot 10^{16} \exp(-33\,000/T)$ | 0.1 | 600—5000 | [20] | — |
| He, $O_2$, $N_2$, NO, $N_2O$ | 600—5000 | $K^0_M = K^0_{Ar}\,[1+706 \times \exp(-2164/T)\, X_{NO_2}/\sum_i R_i X_i]^{-1}$ | 0.15 | 600—5000 | [20] | $R_{Ar} = 1$; $R_{N_2} = 1.6$; $R_{O_2} = 1.3$; $R_{NO_2} = 10$; $R_{He} = 3$; $R_{CO_2} = 4$ |
| $M$ | 600—5000 | $K_\infty = 2 \cdot 10^{14} \exp(-36\,170/T)$ sec$^{-1}$ | 0.4 | 600—5000 | [20] | — |
| $M$ | 600—5000 | $K_{3.2} = K_\infty \left(1 + \frac{K_\infty}{K^0_M [M] \sum_i R_i X_i}\right)^{-1} \times F_c^y$ sec$^{-1}$; $F_c = \exp(-T/1800) + \exp(-7200/T)$; $y = \left\{1 + \left[\log(K^0_M [M] \sum_i R_i X_i/K_\infty)\right]^2\right\}^{-1}$ | 0.4<br>0.1<br>0.2 | 600<br>1400—2500<br>5000 | [20] | For arbitrary pressure and composition of the gas mixture |
| 3.3. $O_3+M = O+O_2+M$ | | | | | | |
| Ar<br>$O_3$<br>$O_2$, $CO_2$<br>$N_2$ | 200—1000<br>300—3000<br>348—433<br>200—1000 | $2.48 \cdot 10^{14} \exp(-11\,430/T)$<br>$4.31 \cdot 10^{14} \exp(-11\,173/T)$<br>$2 \cdot 10^{12} \exp(-11\,726/T)$<br>$(0.3-1.5)\, K_{3.3}\ (M=Ar)$ | 0.1<br>0.15<br>1<br>— | 200—1000<br>300—3000<br>348—433<br>200—1000 | [4]<br>[22]<br>[23]<br>[4] | The reaction has not been studied at high pressures |
| 3.4. $NO_3+M = O+NO_2+M$ | | | | | | |
| $N_2$ | 200—2000 | $K^0_{N_2} = 4.45 \cdot 10^{16} (T/1000)^{-2} \exp(-23\,234/T)$ | 0.2 | 200—2000 | A | — |

Table 3. Continued

| M | T, K | Reaction rate constant, $cm^6/(mole^2 \cdot sec)$ | Error | | References | Notes |
|---|---|---|---|---|---|---|
| | | | $\Delta \log K$ | T, K | | |
| $M$ | 200—2000 | $K_\infty = 2.0 \cdot 10^{14} \exp(-23\,234/T)\ sec^{-1}$ | 0.5 | 200—2000 | A | — |
| $N_2$, $O_2$, Ar, He, CO, $CO_2$ | 200—2000 | $K_{1.4} = K_\infty \left\{1 + K_\infty / (K^0_{N_2} [M] \sum_i R_i X_i)\right\}^{-1} \times$ $\times F_c^y\ sec^{-1}$; $F_c = \exp(-T/1300) + \exp(-5200/T)$; $y = \left\{1 + \left[\log (K^0_{N_2} [M] \sum_i R_i X_i / K_\infty)\right]^2\right\}^{-1}$ | 0.5 | 200—2000 | A | For arbitrary pressure and composition of the gas mixture $R_{N_2} = 1$; $R_{O_2} = 1$; $R_{Ar} = 1.2$; $R_{He} = 1.2$; $R_{CO} = 2$; $R_{CO_2} = 2.9$ |
| 3.5. $NO_3 + M = NO + O_2 + M$ | | | | | | |
| $N_2$ | 3500—3700 | $K^0_{N_2} = 9 \cdot 10^6 (T/1000)^{-2} \exp(-23\,234/T)$ | 0.5 | 3500—3700 | A | — |
| $M$ | 3500—3700 | $K_\infty = 4 \cdot 10^{14} \exp(-23\,234/T)\ sec^{-1}$ | 0.5 | 3500—3700 | A | — |
| $N_2$, $O_2$, Ar, He, CO, $CO_2$ | 3500—3700 | $K_{3.5} = K_\infty \left\{1 + K_\infty / (K^0_{N_2} [M] \sum_i R_i X_i)\right\}^{-1} \times$ $\times F_c^y\ sec^{-1}$; $F_c = \exp(-T/1300) + \exp(-5200/T)$; $y = \left\{1 + \left[\log (K^0_{N_2} [M] \sum_i R_i X_i / K_\infty)\right]^2\right\}^{-1}$ | 0.5 | 3500—3700 | A | For arbitrary pressure and composition of the gas mixture $R_{N_2} = 1$; $R_{O_2} = 1$; $R_{Ar} = 1.2$; $R_{He} = 1.2$; $R_{CO} = 2$; $R_{CO_2} = 2.9$ |

| | | | | | | |
|---|---|---|---|---|---|---|
| **3.6. $N_2O_4+M=NO_2+NO_2+M$** | | | | | | |
| $N_2$ | 200—1000 | $K^0_{N_2}=1.44\cdot10^{18}\,(T/300)^{-5}\exp(-6392/T)$ | 0.3 | 200—1000 | A | — |
| $M$ | 200—1000 | $K_\infty=3\cdot10^{15}\,(T/300)^{-0.8}\times$ $\times\exp(-6392/T)$ sec$^{-1}$ | 0.5 | 200—1000 | A | — |
| $N_2$, Ar, $O_2$, He, CO, $CO_2$ | 200—1000 | $K_{3.6}=K_\infty\left\{1+K_\infty/(K^0_{N_2}[M]\sum_i R_iX_i)\right\}^{-1}\times$ $\times F_c^y$ sec$^{-1}$; $F_c=\exp(-T/250)+\exp(-1050/T)$; $y=\left\{1+\left[\left(\log\frac{K^0_{N_2}[M]\sum_i R_iX_i}{K_\infty}/(0.75-\log F_c)\right)\right]^2\right\}^{-1}$ | 0.5<br>0.3<br>0.5 | 200<br>300—500<br>1000 | A | For arbitrary pressure and composition of the gas mixture $R_{N_2}=1$; $R_{Ar}=0.8$; $R_{CO_2}=2$; $R_{N_2O_4}=2$; $R_{NO_2}=2$; $R_{O_2}=1$ |
| **3.7. $N_2O_5+M=NO_2+NO_3+M$** | | | | | | |
| $N_2$ | 200—1000 | $K^0_{N_2}=1.32\cdot10^{21}\,(T/300)^{-6.1}\times$ $\times\exp(-11\,080/T)$ | 0.2 | 200—1000 | [6] based on [24] | — |
| $M$ | 200—1000 | $K_\infty=9.7\cdot10^{14}\,(T/300)^{-0.9}\times$ $\times\exp(-11\,080/T)$ sec$^{-1}$ | 0.3 | 200—1000 | | — |
| $N_2$, $O_2$, $N_2O_5$, NO, Ar, He, $CO_2$ | 200—1000 | $K_{3.7}=K_\infty\left\{1+K_\infty/(K^0_{N_2}[M]\sum_i R_iX_i)\right\}^{-1}\times$ $\times F_c^y$ sec$^{-1}$; $F_c=\exp(-T/250)+\exp(-1050/T)$; $y=\left\{1+\left[\left(\log\frac{K^0_{N_2}[M]\sum_i R_iX_i}{K_\infty}/(0.75-\log F_c)\right)\right]^2\right\}^{-1}$ | 0.3<br>0.2<br>0.3 | 200<br>300—500<br>1000 | | For arbitrary pressure and composition of the gas mixture $R_{N_2}=R_{O_2}=1$; $R_{NO}=1.3$; $R_{N_2O_5}=4.4$; $R_{CO_2}=1.7$; $R_{He}=0.53$; $R_{Ar}=0.59$ |

TABLE 4. Rate Constants for Recombination Leading to Formation of Polyatomic Molecules

| $M$ | $T$, K | Reaction rate constant, $cm^6/(mole^2 \cdot sec)$ | Error: $\Delta \log K$ | Error: $T$, K | References | Notes |
|---|---|---|---|---|---|---|
| **4.1. $O+N_2+M=N_2O+M$** | | | | | | |
| Ar | 400—5000 | $K^0_{Ar}=1.46\cdot 10^{14}(T/1324)^{-1.25}\times \exp\left(-\frac{13\,400}{T}\right)$ | 0.1 | 400—5000 | A | — |
| He, $N_2$, $O_2$, NO, $N_2O$ | 400—5000 | $K_M=K^0_{Ar}\left(\sum_i R_iX_i\right)\times \left\{1+\left[78.3\exp\left(-\frac{2315}{T}\right)X_{NO}\right]/\sum_i R_iX_i\right\}^{-1}$ | 0.3 | 400—5000 | A | $R_{Ar}=1$; $R_{He}=4.2$; $R_{NO}=3$; $R_{N_2}=1.74$; $R_{N_2O}=6.9$; $R_{O_2}=1.3$ |
| $M$ | 400—5000 | $K_\infty=3.64\cdot 10^9\exp(-11\,400/T)$ $cm^3/(mole\cdot sec)$ | 0.3 | 400—5000 | A | — |
| | | $K_{4.1}=K_\infty(1+K_\infty/(K_M[M]))^{-1}$ $cm^3/(mole\cdot sec)$ | 0.3<br>0.15 | 500<br>900—5000 | A | For arbitrary pressure and composition of the gas mixture |
| **4.2. $O+NO+M=NO_2+M$** | | | | | | |
| Ar | 200—5000 | $K^0=2.3\cdot 10^{16}(T/300)^{-1.8}$ | 0.1 | 200—5000 | A per [25] | — |
| $M$ | 200—5000 | $K_\infty=1.8\cdot 10^{13}(T/300)^{0.3}$ $cm^3/(mole\cdot sec)$ | 0.4 | 200—500 | The same | — |

| | | | | | | |
|---|---|---|---|---|---|---|
| Ar, He, $N_2$, $O_2$, NO, $NO_2$, $N_2O$ | 200—5000 | $K_{4.2}=K_{\infty}\left(\frac{K_{\infty}}{K^0_{N_2}[M]\sum_i R_iX_i}\right)^{-1}\times$ $\times F_c^y$ cm$^3$/(mole·sec); $F_c=\exp(-T/1800)+\exp(-7200/T)$; $y=\left\{1+\left[\log\left(\frac{K^0_{N_2}[M]\sum_i R_iX_i}{K_{\infty}}\right)\right]^2\right\}^{-1}$ | 0.5 | 200—5000 | The same | For arbitrary pressure and composition of the gas mixture $R_{Ar}=1$; $R_{N_2}=1.6$; $R_{O_2}=1.3$; $R_{NO}=2.8$; $R_{NO_2}=10$; $R_{N_2O}=7$ |
| | | 4.3. $O+O_2+M=O_3+M$ | | | | |
| Ar | 219—1000 | $1.4\cdot10^{14}\,(T/300)^{-1.9}$ | 0.1 | 200—300 | [6], A | At low pressures |
| $N_2$ | 219—1000 | $2.25\cdot10^{14}\,(T/300)^{-2}$ | 0.2 | 1000 | [6], A | |
| $O_2$ | 219—1000 | $2.5\cdot10^{14}\,(T/300)^{-1.25}$ | — | — | [6], A | |
| Ar | >1000 | $4.1\cdot10^{12}\exp(1057/T)$ | — | — | [4] | |
| $N_2$, $O_2$ | >1000 | $K(M=Ar):K(M=N_2):K(M=O_2)=1:1.4:1.5$ | — | — | [4] | |
| $N_2$ | 295 | $K_{\infty}=1.7\cdot10^{12}$ cm$^3$/(mole·sec) | 0.18 | 295 | [5] | At high pressures |

Table 4. Continued

| $M$ | $T$, K | Reaction rate constant, $cm^6/(mole^2 \cdot sec)$ | Error | | References | Notes |
|---|---|---|---|---|---|---|
| | | | $\Delta \log K$ | $T$, K | | |
| | | 4.4. $O+NO_2+M = NO_3+M$ | | | | |
| $N_2$ | 200—2500 | $K^0_{N_2} = 2.93 \cdot 10^{15} (T/1000)^{-2}$ | 0.1 | 200—400 | [6] | — |
| $M$ | 200—2500 | $K_\infty = 1.32 \cdot 10^{13}$ $cm^3/(mole \cdot sec)$ | 0.4 | 2500 | [6] | — |
| $N_2$, $O_2$, $CO_2$, Ar, He, CO | 200—2500 | $K_{4.4} = K_\infty \left\{1+K_\infty/(K^0_{N_2}[M]\sum_i R_iX_i)\right\}^{-1} \times$ $\times F_c^y$ $cm^3/(mole \cdot sec)$; $F_c = \exp(-T/1300) + \exp(-5200/T)$; $y = \left\{1+\left[\log\left(K^0_{N_2}[M]\sum_i R_iX_i/K_\infty\right)\right]^2\right\}^{-1}$ | 0.1<br>0.4 | 200—400<br>2500 | A per[26] | In a gas mixture with arbitrary pressure and composition $R_{N_2} = R_{O_2} = 1$; $R_{CO_2} = 2.9$; $R_{Ar} = R_{He} = 1.2$; $R_{CO} = 2$ |
| | | 4.5. $NO+O_2+M = NO_3+M$ | | | | |
| $N_2$ | 300—5000 | $K^0_{N_2} = 5 \cdot 13 \cdot 10^{15} (T/1000)^{-2} \exp(-23\,050/T)$ | 0.4 | 300—5000 | A | — |
| W | 300—5000 | $K_\infty = 2.28 \cdot 10^{13} \exp(-23\,050/T)$ $cm^3/(mole \cdot sec)$ | 0.8 | 300—5000 | A | — |

| | | | | | | |
|---|---|---|---|---|---|---|
| $N_2$, $O_2$ $CO_2$, $CO$, Ar, He | 300—5000 | $K_{4.5} = K_\infty \left\{1+K_\infty/(K^0_{N_2}[M]\sum_i R_iX_i)\right\}^{-1} \times$ $\times F_c^y$ cm$^3$/(mole·sec); $F_c = \exp(-T/1300) + \exp(-5200/T)$; $y = \left\{1+\left[\log\left(K^0_{N_2}[M]\sum_i R_iX_i/K_\infty\right)\right]^2\right\}^{-1}$ | 0.8 | 300—5000 | A | In a gas mixture with arbitrary pressure and composition $R_{N_2} = R_{O_2} = 1$; $R_{CO_2} = 2.9$; $R_{Ar} = R_{He} = 1.2$; $R_{CO} = 2$ |
| | | 4.6. $NO_2+NO_2+M = N_2O_4+M$ | | | | |
| $N_2$ | 200—1000 | $K^0_{N_2} = 2{,}3 \cdot 10^{14}\,(T/300)^{-2.5}$ | 0.3 | 200—1000 | A | — |
| $M$ | 200—1000 | $K_\infty = 5 \cdot 10^{11}\,(T/300)^{1.7}$ cm$^3$/(mole·sec) | 0.5 | 200—1000 | A | — |
| $N_2$, $O_2$, Ar, $CO_2$, $N_2O_4$, $NO_2$ | 200—1000 | $K_{4.6} = K_\infty \left\{1+K_\infty/(K^0_{N_2}[M]\sum_i R_iX_i)\right\}^{-1} \times$ $\times F_c^y$ cm$^3$/(mole·sec); $F_c = \exp(-T/250) + \exp(-1050/T)$; $y = \left\{1+\left[\left(\log\left(K^0_{N_2}[M]\sum_i R_iX_i/K_\infty\right)\right)/N\right]^2\right\}^{-1}$; $N = 0.75 - \log F_c$ | 0.5<br>0.3<br>0.2 | 200<br>300—500<br>1000 | A | In a gas mixture with arbitrary pressure and composition $R_{N_2} = 1$; $R_{Ar} = 0.8$; $R_{O_2} = 1$; $R_{CO_2} = 2$; $R_{NO_2} = 2$; $R_{N_2O_4} = 2$ |

Table 4. Continued

| $M$ | $T$, K | Reaction rate constant, $cm^6/(mole^2 \cdot sec)$ | Error | | References | Notes |
|---|---|---|---|---|---|---|
| | | | $\Delta \log K$ | $T$, K | | |
| 4.7. $NO_2+NO_3+M=N_2O_5+M$ | | | | | | |
| $N_2$ | 200—1000 | $K^0_{N_2}=1.34\cdot10^{18}\,(T/300)^{-5}$ | 0.2 | 200—1000 | A | — |
| $M$ | 200—1000 | $K_\infty=9.6\cdot10^{11}\,(T/300)^{+0.2}$ $cm^3/(mole\cdot sec)$ | 0.3 | 200—1000 | A | — |
| $N_2$, $O_2$, $N_2O_5$, NO, Ar, He, $CO_2$ | 200—1000 | $K_{4.7}=K_\infty\left\{1+K_\infty/(K^0_{N_2}[M]\sum_i R_iX_i)\right\}^{-1}\times$ $\times F_c^y$ $cm^3/(mole\cdot sec)$; $F_c=\exp(-T/250)+\exp(-1050/T)$; $y=\left\{1+\left[\left(\log K^0_{N_2}[M]\sum_i R_iX_i/K_\infty\right)/N\right]^2\right\}^{-1}$; $N=0.75-\log F_c$ | 0.3<br>0.2<br>0.3 | 200<br>300—500<br>1000 | A | In a gas mixture with arbitrary pressure and composition $R_{N_2}=R_{O_2}=1$; $R_{Ar}=0.59$; $R_{N_2O_5}=4.4$; $R_{NO}=1.3$; $R_{CO_2}=1.7$; $R_{He}=0.53$ |

TABLE 5. Rate Constants for Bimolecular Reactions

| Reaction | $T$, K | Reaction rate constant, $cm^3/(mole \cdot sec)$ | Error | | | References | Notes |
|---|---|---|---|---|---|---|---|
| | | | $\Delta K$, % | $\Delta \log K$ | $T$, K | | |
| 5.1 $N+NO=O+N_2$ | 200—400 | $1.87 \cdot 10^{13}$ | — | 0.15 | 200—400 | [6] | — |
| | 200—4000 | $6.3 \cdot 10^{11} T^{0.5}$ | — | — | 200—4000 | [11] | — |
| 5.2 $N+O_2=O+NO$ | 200—300 | $2.7 \cdot 10^{12} \exp\left(-\frac{3220}{T}\right)$ | — | 0.1 | 200—300 | [5] | — |
| | 300—3000 | $6.4 \cdot 10^{9} T \exp\left(-\frac{3150}{T}\right)$ | 30 | — | 300—1500 | [3] | — |
| | | | — | 0.3 | 3000 | [3] | — |
| 5.3 $N+NO_2=O+O+N_2$ | 298 | $5.48 \cdot 10^{11}$ | — | — | 298 | A per [27, 28] | $K_{5.3}=0.1315\ K$; $K=\sum_{i=3}^{6} K_{5.i}$ |
| 5.4 $N+NO_2=O+N_2O$ | 298 | $1.8 \cdot 10^{12}$ | 10 | — | 298 | [6, 7] | $K_{5.4}=0.434\ K$; $K=\sum_{i=3}^{6} K_{5.i}$ |

**Table 5.** Continued

| Reaction | T, K | Reaction rate constant, $cm^3$/(mole·sec) | Error: ΔK, % | Error: Δ log K | Error: T, K | References | Notes |
|---|---|---|---|---|---|---|---|
| 5.5 $N+NO_2=N_2+O_2$ | 298 | $4.2\cdot10^{11}$ | — | — | 298 | A per [27, 28] | $K_{5.5}=0.1015\ K$; $K=\sum_{i=3}^{6}K_{5.i}$ |
| 5.6 $N+NO_2=NO+NO$ | 298 | $1.39\cdot10^{12}$ | — | — | 298 | A per [27, 28] | $K_{5.6}=0.333\ K$; $K=\sum_{i=3}^{6}K_{5.i}$ |
| 5.7 $N+O_3=NO+O_2$ | 300 | $\leqslant1.2\cdot10^{8}$ | — | — | 300 | [29] | — |
| 5.8 $O+N_2=N+NO$ | 2000—5000 | $7.6\cdot10^{13}\ \exp\left(-\frac{38\,000}{T}\right)$ | — | 0.3 | 2000—5000 | [3] | — |
| 5.9 $O+NO=N+O_2$ | 1000—3000 | $1.5\cdot10^{9}T\ \exp\left(-\frac{19\,500}{T}\right)$ | 30<br>— | —<br>0.3 | 1000<br>3000 | [3] | — |
| 5.10 $O+N_2O=N_2+O_2$ | 300—5000 | $3.48\cdot10^{13}\exp\left(-\frac{11\,986}{T}\right)$ | —<br>— | 1<br>0.3 | 300<br>1000—5000 | A | — |
| 5.11 $O+N_2O=NO+NO$ | 300—5000 | $1.09\cdot10^{14}\exp\left(-\frac{14\,494}{T}\right)$ | —<br>— | 0.6<br>0.11 | 300<br>1000—5000 | A | — |

| | | | | | | | |
|---|---|---|---|---|---|---|---|
| 5.12 $O+NO_2=NO+O_2$ | 230—2500 | $6.8\cdot10^{12}(T/1000)^{0.18}$ | — | 0.08 | 230—2500 | A | — |
| 5.13 $O+O_3=O_2+O_2$ | 298 | $4.8\cdot10^{9}$ | — | 0.08 | 298 | [7] | — |
| | 220—440 | $4.8\cdot10^{12}\ \exp\left(-\frac{2060}{T}\right)$ | — | 0.1 | 220—440 | [7] | — |
| | 220—1000 | $1.2\cdot10^{13}\ \exp\left(-\frac{2280}{T}\right)$ | —<br>— | 0.1<br>0.3 | 298<br>1000 | [4]<br>[4] | — |
| 5.14 $O+NO_3=O_2+NO_2$ | 298 | $6.02\cdot10^{12}$ | — | 0.5 | 298 | [6, 30] | — |
| 5.15 $O+N_2O_5$=products | 223—360 | $<1.8\cdot10^{8}$ | — | — | 223—360 | [5, 31] | — |
| 5.16 $N_2+O_2=O+N_2O$ | 300—10 000 | $1.5\cdot10^{14}\ \exp\left(-\frac{50\,391}{T}\right)$ | — | — | 300 | A | — |
| | | | — | 0.6 | 1000—10 000 | A | — |
| 5.17 $N_2+O_2=NO+NO$ | 300—10 000 | $K_{5.17}=K_{5.20}K_p$ | — | — | 300—10 000 | A | No reliable data for this reaction |

**Table 5.** Continued

| Reaction | $T$, K | Reaction rate constant, $cm^3/(mole \cdot sec)$ | Error | | | References | Notes |
|---|---|---|---|---|---|---|---|
| | | | $\Delta K$, % | $\Delta \log K$ | $T$, K | | |
| 5.18 $NO+NO=N+NO_2$ | 298 | $9 \cdot 10^9 \exp\left(-\frac{39\,198}{T}\right)$ | — | — | 298 | A | — |
| 5.19 $NO+NO=O+N_2O$ | 300—10 000 | $3.07 \cdot 10^{12} \exp\left(-\frac{33\,664}{T}\right)$ | — | 1 | 300 | A | — |
| | | | — | 0.2 | 1000—10 000 | | — |
| 5.20 $NO+NO=N_2+O_2$ | 300—10 000 | $3.07 \cdot 10^{11} \exp\left(-\frac{33\,664}{T}\right)$ | — | 1.5 | 300 | A per [33] | — |
| | | | — | 0,5 | 1000—10 000 | | — |
| 5.21 $NO+O_2=O+NO_2$ | 300—550 | $1.7 \cdot 10^{12} \exp\left(-\frac{23\,400}{T}\right)$ | — | — | 300—550 | [3] | — |
| 5.22 $NO+O_3=O_2+NO_2$ | 200—400 | $2.6 \cdot 10^{12} \exp\left(-\frac{1560}{T}\right)$ | 50 | — | 200—400 | [3] | — |
| 5.23 $NO+N_2O=N_2+NO_2$ | 300—5000 | $K_M = K'(1+78{,}3 \times \exp\left(-\frac{2300}{T}\right) \frac{X_{NO}}{\sum_i R_i X_i}\right)^{-1}$; | —<br>— | 0.6<br>0.2 | 300<br>500—5000 | A | $R_{Ar}=1$; $R_{He}=4.2$; $R_{N_2}=1.74$; $R_{N_2O}=6.9$; $R_{O_2}=1.3$; $R_{NO}=3$; $R_{CO_2}=4$ |

| | | | | | | | |
|---|---|---|---|---|---|---|---|
| | | $K' = 5.48 \cdot 10^{15} \exp\times$ $\times\left(-\frac{26\,472}{T}\right)$ | | | | | |
| 5.24 $NO+NO_3=NO_2+NO_2$ | 300—5000 | $10^{13}$ | — | 1 | 300—5000 | A | — |
| 5.25 $O_2+O_2=O+O_3$ | 293—1000 | $4.8 \cdot 10^{12} \exp\left(-\frac{49\,824}{T}\right)$ | — | — | 293—1000 | [3] | — |
| 5.26 $O_2+NO_2=NO+O_3$ | 200—500 | $1.7 \cdot 10^{12} \exp\left(-\frac{25\,400}{T}\right)$ | — | — | 200—500 | [34] | — |
| 5.27 $NO_2+NO_2=NO+NO+$ $+O_2$ | 300—3000 | $2.0 \cdot 10^{12} \exp\left(-\frac{13\,500}{T}\right)$ | — | 0.6 | 300 | [3] | — |
| | | | — | 0.1 | 600—1000 | [3] | — |
| | | | — | 0.3 | 3000 | [3] | — |
| 5.28 $NO_2+NO_2=NO+NO_3$ | 500—2300 | $K_M=K'(1+706\times$ $\times\exp\left(-\frac{2164}{T}\right)\left(\frac{X_{NO_2}}{\sum_i R_i X_i}\right)^{-1}$; $K'=2.4 \cdot 10^{15} \exp\times$ $\times\left(-\frac{19\,628}{T}\right)$ | — | 0.8 | 500 | A | $R_{Ar}=1$; $R_{He}=3$; $R_{NO_2}=10$; $R_{N_2}=1.6$; $R_{O_2}=1.2$; $R_{CO_2}=4.3$; $R_{Ne}=1.3$ |
| | | | — | 0.1 | 1400—2300 | | |

**Table 5.** Continued

| Reaction | $T$, K | Reaction rate constant, $cm^3$/(mole·sec) | Error | | | References | Notes |
|---|---|---|---|---|---|---|---|
| | | | $\Delta K$, % | $\Delta \log K$ | $T$, K | | |
| 5.29 $NO_2+O_3=O_2+NO_3$ | 230—360 | $7.2\cdot10^{10}\exp(-2450/T)$ | — | 0.06 | 298 | [5] | — |
| 5.30 $NO_3+NO_3=O_2+NO_2+$ $+NO_2$ | 293—309 | $3\cdot10^{12}\exp(-3000/T)$ | — | — | — | [34] | — |
| | 600—1100 | $2.6\cdot10^{12}\exp(-3850/T)$ | — | — | — | [34] | — |
| 5.31 $O+N_3=NO+N_2$ | 433—633 | $6.023\cdot10^{12}$ | 40 | — | 433—633 | [35] | — |
| 5.32 $N+N_3=N_2+N_2$ | 298 | $9.6\cdot10^{12}$ | 70 | — | 298 | [36] | — |
| 5.33 $N_3+N_3=N_2+N_2+N_2$ | 298 | $8.4\cdot10^{11}$ | — | — | 298 | [36] | — |
| 5.34 $NO_2+NO_3=NO+$ $+NO_2+O_2$ | 300—850 | $1.4\cdot10^{11}\exp\left(-\frac{1600}{T}\right)$ | — | 0.4 | 300—850 | [3] | — |
| 5.35 $NO_3+O_2=NO_2+O_3$ | 230—360 | $K_{5.35}=K_{5.29}K_p$ | — | — | 230—360 | A | — |

TABLE 6. Rate Constants for Termolecular Reactions

| Reaction | T, K | Reaction rate constant, $cm^6/(mole^2 \cdot sec)$ | Error | | References |
|---|---|---|---|---|---|
| | | | $\Delta \log K$ | T, K | |
| 6.1 $NO+NO+NO=$ $=N_2O+NO_2$ | 753—813 | $1.26 \cdot 10^{10} \exp(-13\ 588/T)$ | — | 753—813 | [37] |
| 6.2 $NO+NO+O_2=$ $=NO_2+NO_2$ | 273—5000 | $1.29 \cdot 10^9 \exp(-530/T)$ | 0.18<br>0.4 | 273—660<br>5000 | [3] |
| 6.3 $NO+NO_2+O_2=$ $=NO_2+NO_3$ | 300—500 | $2.9 \cdot 10^7 \exp(400/T)$ | 0.4 | 300—500 | [3] |

Notation used in these tables: $K_{i,j}$, reaction rate constant; $K_M$, rate constant for a reaction in a gas mixture; $K_\infty$, rate constant for a reaction at high pressures; $X_i$, molar fraction of a mixture component; $R_j$, relative efficiency of a component ($j$) compared to the component ($i$) whose efficiency is listed first ($R_i = 1$) in the notes; $T$, translational temperature; $T_v$, vibrational temperature; A denotes the present review in the "References" column; $K^0$, equilibrium rate constant for dissociation (Table 1); $K^0$, reaction rate constant at low pressures (Tables 3 and 4); $K_p$, equilibrium constant defined as the ratio of the rate constant for a given reaction to the rate constant for the reverse reaction.

ture dependence of the reaction rates. Little information exists about the effect of excited states on the rates of reactions.

It would be inappropriate to continue this list of gaps in the world-wide literature. The shortcomings pointed out here are responsible for the low accuracy of some of the proposed recommendations, but sometimes there is simply no information at all on their accuracy. We might note, however, that the reactions for which information is lacking are mostly of secondary importance for gas dynamics. The information on the basic reactions (dissociation and recombination of diatomic molecules) and on certain exchange reactions is fairly reliable.

## REFERENCES

1. V. P. Varakin, V. G. Gromov, S. A. Losev, et al., "The AVOGADRO research data base for support of computer programs in physicochemical gas dynamics. A

description of the project," Institute of Mechanics, Moscow State University, deposited in VINITI March 19, 1985, No. 2510-85-dep.
2. V. N. Kondrat'ev, *Rate Constants of Gas-Phase Reactions. A Handbook* [in Russian], Nauka, Moscow (1970).
3. D. L. Baulch, D. D. Drysdale, and D. C. Horne, *Evaluated Kinetic Data for High-Temperature Reactions. Vol. 2. Homogeneous Gas-Phase Reactions of the* $H_2$–$N_2$–$O_2$ *System*, Butterworths, London (1973).
4. D. L. Baulch, D. D. Drysdale, J. Duxbary, and S. L. Grant, *Evaluated Kinetic Data for High-Temperature Reactions. Vol. 3. Homogeneous Gas Phase Reactions of* $O_2$–$O_3$ *Systems, the* $CO$–$O_2$–$H_2$ *System, and of Sulphur-Containing Systems*, Butterworths, London (1976).
5. D. L. Baulch, R. A. Cox, R. F. Hampson, et al., "Evaluated kinetic and photochemical data for atmospheric chemistry," *J. Phys. Chem. Ref. Data* **9**, 295–471 (1980).
6. D. L. Baulch, R. A. Cox, P. J. Crutzen, et al., "Evaluated kinetic and photochemical data for atmospheric chemistry: Supplement I. CODATA Task Group on Chemical Kinetics," *J. Phys. Chem. Ref. Data* **11**, 327–496 (1982).
7. D. L. Baulch, R. A. Cox, R. F. Hampson, et al., "Evaluated kinetic and photochemical data for atmospheric chemistry: Supplement II. CODATA Task Group on Chemical Kinetics," *J. Phys. Chem. Ref. Data* **13**, 1259–1379 (1984).
8. S. A. Losev and N. A. Generalov, *Dokl. Akad. Nauk SSSR* **141**, 1072–1075 (1961).
9. N. M. Kuznetsov, *Dokl. Akad. Nauk SSSR* **164**, 1097–1100 (1965).
10. A. I. Osipov, *Teor. Éksp. Khim.* **2**, 649–657 (1966).
11. S. W. Benson, D. M. Golden, R. W. Lawrence, et al., *Int. J. Chem. Kinet. Symp.*, No. 1, pp. 399–440 (1975).
12. V. M. Shui, J. P. Appleton, and J. C. Keck, *J. Chem. Phys.* **53**, 2547–2558 (1970).
13. S. W. Benson and T. Fueno, *J. Chem. Phys.* **36**, 1597–1607 (1962).
14. D. J. Kewley and H. G. Hornung, *Chem. Phys. Lett.* **25**, 531–536 (1974).
15. M. S. Yalovik and S. A. Losev, *Scientific Proceedings of the Institute of Mechanics of Moscow State University* (Nauchn. Tr. Inst. Mekh. Mosk. Gos. Univ.), Izd. MGU, Moscow, No. 18, pp. 4–34 (1972).
16. K. L. Wray and J. D. Teare, *J. Chem. Phys.* **36**, 2582 (1962).
17. M. Koshi, S. Bando, M. Saito, and T. Asaba, 17th Symp. (International) on Combustion, Leeds (1978), Pittsburgh (1979), pp. 553–562.
18. O. P. Shatalov, *Fiz. Goreniya Vzryva* **9**, 699–703 (1973).
19. N. M. Kuznetsov, I. S. Zaslonko, and Yu. P. Petrov, "Thermal dissociation of $N_2O$," *Preprint*, Institute of Physical Chemistry, Academy of Sciences of the USSR, Moscow (1979).
20. I. S. Zaslonko and Yu. K. Mukoseev, *Khim. Fiz.*, No. 11, 1508–1517 (1982).
21. Yu. K. Mukoseev and I. S. Zaslonko, *Khim. Fiz.*, No. 1, 66–73 (1983).
22. J. M. Heimerl and T. P. Coffee, *Combust. Flame* **35**, 117–123 (1979).
23. S. Toby and E. Ullrich, *Int. J. Chem. Kinet.* **12**, 535–546 (1980).
24. M. W. Malko and J. Troe, *Int. J. Chem. Kinet.* **14**, 399–416 (1982).
25. J. Troe, *Ber. Bunsenges. Phys. Chem.* **73**, 906–911 (1969).

26. J. Troe, *J. Chem. Phys.* **66**, 4758–4775 (1977).
27. L. F. Phillips and H. J. Schiff, *J. Chem. Phys.* **42**, 3171–3174 (1965).
28. M. A. Cline and J. Ono, *Chem. Phys.* **69**, 381–388 (1982).
29. L. J. Stief, W. A. Payne, J. H. Lee, and J. V. Michael, *J. Chem. Phys.* **70**, 5241–5243 (1979).
30. R. A. Graham and H. S. Johnston, *J. Chem. Phys.* **82**, 254–268 (1978).
31. E. W. Keiser and S. W. Japar, *Chem. Phys. Lett.* **54**, 265–268 (1978).
32. K. Schofield, *Planet. Space Sci.* **15**, 643–670 (1975).
33. M. Camac and R. M. Feinberg, *11th Combustion Symp.* (1967), p. 137.
34. M. McEwan and L. Phillips, *The Chemistry of the Atmosphere*, Halsted Press, New York (1975).
35. L. G. Piper, R. H. Krech, and R. L. Taylor, *J. Chem. Phys.* **75**, 2099–2104 (1979).
36. K. Yamasaki, T. Fueno, and O. Kajimoto, *Chem. Phys. Lett.* **94**, 425–429 (1983).
37. A. A. Gvozdev, V. B. Nesterenko, G. V. Nichipor, and V. V. Trubnikov, *Vestsyi Akad. Navuk BSSR, Ser. Fyiz. Ehnerg. Navuk,* No. 2, pp. 68–73 (1979).

# METAL DIMERS

**B. M. Smirnov and A. S. Yatsenko**

Dimers are the simplest molecules, made up of two atoms of the same type. Metal dimers are formed during relaxation of the vapor after a metal surface is acted on by laser light, electron or ion beams, gas discharges, or other agents. During these interactions the metal is first converted into a monatomic gas or plasma, and dimers are formed later, as this medium relaxes. Diagnostics of a relaxing metal vapor or plasma based on the detection of dimers yields information on the state of the vapor and its evolution during the relaxation process. Information on dimers is of practical interest for this reason.

A great deal of information on the parameters of diatomic molecules is contained in a book by Huber and Herzberg [1], in which data on diatomic molecules obtained through 1979 have been collected and analyzed. In recent years, however, these data have been revised and supplemented. Improved methods based on modern experimental techniques have been developed. Thus, there is a need to review these data.

This review is concerned with a small fraction of molecules discussed in Huber and Herzberg's book – the metal dimers. We have made some slight changes, compared to that book, in the way the information is discussed, by including potential curves, lifetimes, ionization potentials of the molecules, etc. The observed radiative transitions of the dimers are shown in the figures.*

---

*The location of the energy levels is given in accordance with [1–5]. Each known electronic state corresponds to a horizontal line. The locations of states that are not exactly determined are denoted by wavy lines and the locations of states that have been calculated quantum mechanically are denoted by dashed lines. The numbers in parentheses characterize an absorption or emission band of a given transition in nanometers and the value of $\nu_{00}$ for this band is listed in front of the parentheses. The tabulated values of $T_e$, $\omega_e$, $\omega_e x_e$, $B_e$, and $\alpha_e$ are given in cm$^{-1}$, except for some rare exceptions, and $r_e$ is given in nm.

A few words about the constants of the dimers listed in the tables: since it is a diatomic molecule, a dimer is described by the quantum numbers of an ordinary diatomic molecule,* but it also has an additional symmetry, since it is made up of identical atoms. The additional quantum number is the parity of the state upon reflection of all the electrons with respect to the surface of symmetry – a plane perpendicular to the axis joining the nuclei that divides it in half. If the wave function of the electrons does not change after this reflection, then the state is even ($g$), while if the sign of the wave function changes during this operation, the state of the molecule is odd ($u$).

The total energy of a diatomic molecule is given by

$$E = T_e + E(v, J),$$

where $T_e$ is the electronic energy and $E(v, J)$ is the vibrational–rotational energy, with $v$ and $J$ being the vibrational and rotational quantum numbers, respectively, of the molecule. In this equation,

$$E(v, J) = G(v) + F_v(J),$$

where the vibrational energy is given by

$$G(v) = \omega_e(v + 1/2) - x_e\omega_e(v + 1/2)^2$$

and the rotational energy is given by

$$F_v(J) = B_vJ(J + 1), \; B_v = B_e - \alpha_e(v + 1/2).$$

Here $\omega_e$ is the energy of a vibrational quantum, $\omega_e x_e$ is the anharmonicity constant, $B_e$ is the rotational constant, and $\alpha_e$ is the vibrational–rotational interaction parameter. Sometimes the vibrational–rotational energy is written in the form

*In the case where the spin–orbit interaction in the molecule is relatively small, the quantum numbers of a diatomic molecule include the total spin of the electrons, the projection of the orbital angular momentum of the electrons on the axis joining the nuclei, and the parity upon reflection of the electrons with respect to the plane of symmetry (a plane passing through the axis of the molecules). If, however, the spin–orbit splitting is comparable with the electrostatic interactions within the molecule, then the first two quantum numbers are replaced by the projection of the total angular momentum on the axis of the molecule.

## DIMER $Ag_2$

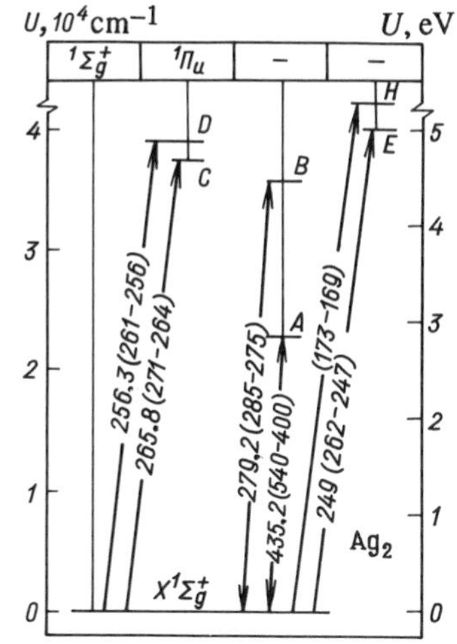


Parameters of states [7–12]

| Term | $T_e$ | $r_e$, nm | $\omega_e$ | $\omega_e x_e$ | $B_e$ | $\alpha_e \cdot 10^{-3}$ |
|---|---|---|---|---|---|---|
| $X^1\Sigma_g^+$ | 0 | 0.27 | 192.4 | 0.643 | 0.496 | 0.19 |
| $A$ | 22 996 | — | 154.6 | 0.587 | — | — |
| $B$ | 35 827 | — | 151.3 | 0.70 | — | — |
| $C(1_u)$ | 37 627 | — | 172.9 | 1.07 | 0.050 | 22 |
| $(1_u)$ | 38 024 | — | 166.7 | 1.13 | — | — |
| $E$ | 40 159 | — | 146.2 | 1.56 | — | — |

I.P. = 7.78 eV; $D_e(X)$ = 1.7 eV.

## DIMER $Al_2$

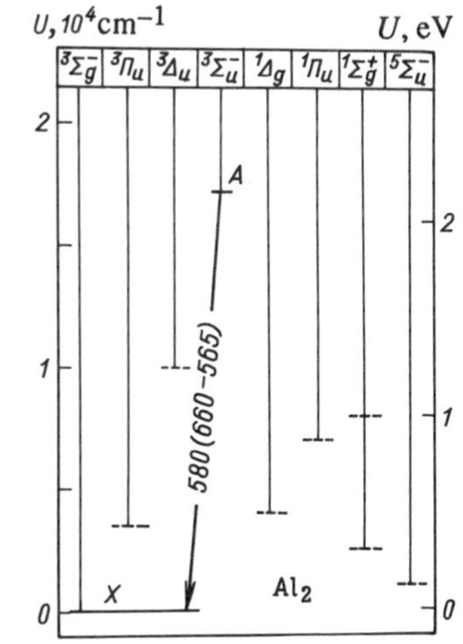


Parameters of states [6, 13, 14]

| Term | $T_e$ | $r_e$ · nm | $\omega_e$ | $\omega_e x_e$ | $B_e$ | $\alpha_e \cdot 10^{-3}$ |
|---|---|---|---|---|---|---|
| $X^3\Sigma_g^-$ | 0 | 0.247 | 350.0 | 2.022 | 0.2054 | 1.2 |
| $A^3\Sigma_u^-$ | 17 270 | 0.256 | 278.8 | 0.831 | 0.1907 | 1.3 |

$D_e(X)$ = 1.03 eV; I.P. = 8.4 eV.

## DIMER $As_2$

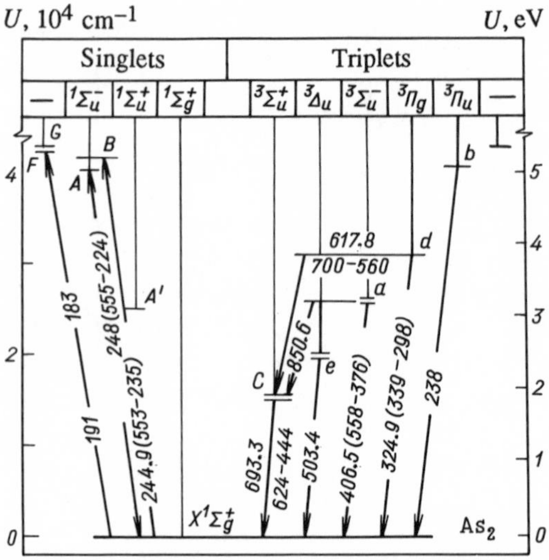


Parameters of states [15, 16]

| Term | $T_e$ | $r_e$, nm | $\omega_e$ | $\omega_e x_e$ | $B_e$ | $\alpha_e \cdot 10^{-3}$ |
|---|---|---|---|---|---|---|
| $X^1\Sigma_g^+$ | 0 | 0.2103 | 429.6 | 1.12 | 0.1018 | 0.333 |
| $c^3\Sigma_u^+$ | 14 495 | 0.2304 | 314.6 | 1,07 | 0.817 | 0.35 |
| $c'^3\Sigma_u^+$ | 14 644 | 0.2302 | 314.3 | 1.17 | 0.849 | 0.35 |
| $e^3\Delta_u$ | 19 801 | — | — | — | — | — |
| $e'^3\Delta_u$ | 19 929 | — | 330.5 | 0.90 | — | — |
| $A'^1\Sigma_u^-$ | 24 559 | — | — | — | — | — |
| $a^3\Sigma_u^-$ | 24 659 | 0.2279 | 336.3 | 0.60 | — | — |
| $a'^3\Sigma_u^-$ | 24 826 | — | 336.8 | 0.69 | — | — |
| $d^3\Pi_g$ | 30 819 | 0.2209 | 336.7 | 1.36 | 0.922 | 0.33 |
| $b0_u^+$ | 38 884 | 0.2358 | 422 | 8.8 | 0.78 | 0.87 |
| $A^1\Sigma_u^+$ | 40 349 | 0.2375 | 260.3 | 5.1 | 0.720 | 0.93 |
| $b'$ | 40 268 | 0.245 | 318.7 | 3.6 | 0.75 | 1.45 |
| $h$ | 41 419 | 0.234 | 318 | 7 | 0.8 | 6.5 |
| $k$ | 42 286 | 0.267 | 246 | 14 | — | — |

I.P. = 12 eV; $D_e(X)$ = 3.9 eV; $\tau(A')$ = 30 μsec, $\tau(c)$ = 12 μsec, $\tau(e)$ = 55 μsec.

## DIMER $Au_2$

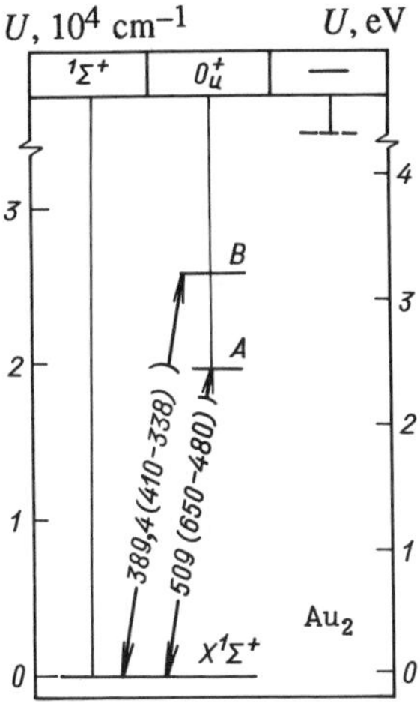


Parameters of states [6, 10, 17, 18]

| Term | $T_e$ | $r_e$, nm | $\omega_e$ | $\omega_e x_e$ | $B_e$ | $\alpha_e \cdot 10^{-3}$ |
|---|---|---|---|---|---|---|
| $X^1\Sigma_g^+$ | 0 | 0.2472 | 190.9 | 0.420 | 0.028 | 0.072 |
| $A(0_u^+)$ | 19 668 | 0.2568 | 142.3 | 0.445 | 0.026 | 0,090 |
| $B(0_u^+)$ | 25 685 | 0.2520 | 179.8 | 0.680 | 0.027 | 0.096 |

I.P. = 10.28 eV; $D_e(X)$ = 2.3 eV; $D_e(A)$ = 1.0 eV; $D_e(B)$ = 1.8 eV.

Computed parameters of state

| Term | $T_e$, eV | $r_e$, nm | $\omega_e$, cm$^{-1}$ | $D_e$, eV |
|---|---|---|---|---|
| 1$u$ | 2.64 | 0.244 | 138 | 0.76 |
| 2$u$ | 2.57 | 0.243 | 143 | 0.94 |
| $0_g^-$ | 2.66 | 0.244 | 140 | 0.85 |
| 1$g$ | 2.67 | 0.244 | 140 | 0.84 |
| 2$g$ | 3.16 | 0.261 | 110 | 0,35 |
| 3$g$ | 3,10 | 0.261 | 112 | 0.41 |

## DIMER $Bi_2$

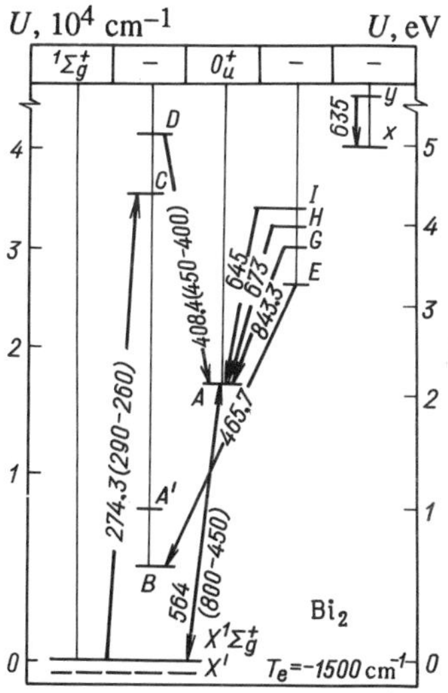


Parameters of states [28–30]

| Term | $T_e$ | $r_e$, nm | $\omega_e$ | $\omega_e x_e$ | $B_e$ | $\alpha_e \cdot 10^{-3}$ |
|---|---|---|---|---|---|---|
| $X^1\Sigma_g^+$ | 0 | 0.2661 | 173.05 | 0.375 | 0.0228 | 0,053 |
| $B^1 0_g^+$ | 10826.4 | 0.3108 | 106.28 | 0.24 | 0.0167 | 0.038 |
| $A0_u^+$ | 17739.3 | 0.2863 | 132.5 | 0.302 | 0.0197 | 0,053 |
| $V$ | 30172.4 | 0.3480 | 32,24 | 0.046 | 0.0133 | 0.057 |

$D_e(X)$ = 2.08 eV = 16,778 cm$^{-1}$; I.P. = 8.9 eV.

## DIMER $B_2$

Parameters of states [6, 19]

| Term | $T_e$ | $r_e$, nm | $\omega_e$ | $\omega_e x_e$ | $B_e$ | $\alpha_e \cdot 10^{-3}$ |
|---|---|---|---|---|---|---|
| $X^3\Sigma_g^-$ | 0 | 0.1589 | 1052.7 | 9.94 | 1.2157 | 14 |
| $a^5\Sigma_u^-$ | 1300 | 0.154 | 1204 | 9.8 | 1.28 | 11 |
| $A^3\Sigma_u^-$ | 1500 | 0.156 | 1104 | 16.0 | 1.25 | 17 |
| $B^3\Sigma_u^-$ | 30 574 | 0.1625 | 937.19 | 2.58 | 1.1648 | 11 |

$D_e(X) = 2.8$ eV, $D_e(a) = 5.6$ eV, $D_e(B) = 7.0$ eV. Observed electronic transition $X \leftrightarrow B$ 327.3 (330–317) nm.

## DIMER $Ba_2$

Parameters of states [20]

| Term | $T_e$ | $r_e$, nm | $\omega_e$ | $\omega_e x_e$ | $B_e$ | $\alpha_e \cdot 10^{-3}$ |
|---|---|---|---|---|---|---|
| $X^1\Sigma_g^+$ | 0 | 0.46 | 84.12 | 0.16 | 0.009 | 8.4 |
| $A$ | — | — | 89.75 | 0.176 | — | — |

Observed absorption bands, nm: (567–571), (583–585), (580–582).

## DIMER $Be_2$

Parameters of states [21–27]

| Term | $T_e$ | $r_e$, nm | $\omega_e$ | $\omega_e x_e$ | $B_e$ | $\alpha_e \cdot 10^{-3}$ |
|---|---|---|---|---|---|---|
| $X^1\Sigma_g^+$ | 0 | 0.245 | 244 | 12.5 | 0.615 | 81 |
| $^3\Pi_g$ | 9557 | 0.213 | 386.7 | 4.24 | 0.828 | 17 |
| $^3\Sigma_u^+$ | 4843 | 0.219 | 517.2 | 5.02 | 0.783 | 52 |
| $A'^1\Pi_g$ | 15 985 | 0.20 | 843.6 | 13.1 | 0.936 | 47 |
| $A^1\Pi_u$ | 26 265 | 0.197 | 1093 | 4.14 | 0.963 | 7 |
| $B^1\Sigma_u^+$ | 30 283 | 0.237 | 421.7 | 15.2 | 0.669 | 11 |

Observed electronic transition $X \rightarrow B$ centered at 350.2 nm; $\tau(A) = 189$ nsec, $\tau(B) = 14.5$ nsec, $D_e(X) = 0.1$ eV $= 790$ cm$^{-1}$.

## DIMER $Be_2^+$

Parameters of states

| Term | $T_e$ | $r_e$, nm | $\omega_e$ | $\omega_e x_e$ |
|---|---|---|---|---|
| $X^2\Sigma_g^+$ | 0 | 0.23 | 368 | 11.5 |
| $^2\Pi_u$ | — | 0.21 | 458 | 13.2 |

## DIMER $Ca_2$

Parameters of states [6, 21, 31, 32]

| Term | $T_e$ | $r_e$, nm | $\omega_e$ | $\omega_e x_e$ | $B_e$ | $\alpha_e \cdot 10^{-3}$ |
|---|---|---|---|---|---|---|
| $X^1\Sigma_g^+$ | 0 | 0.428 | 65.07 | 1.087 | 0.046 | 0.7 |
| $A^1\Sigma_u^+$ | 18 960 | 0.38 | 136.7 | 0.721 | 0.058 | 0.3 |

Observed electron transition $X \leftrightarrow A$, 526.3 nm; $D_e(X) = 1095\ cm^{-1}$; $D_e(A) = 3980\ cm^{-1}$.

Computed parameters

| Term | $T_e$, eV | $r_e$, nm | $\omega_e$, $cm^{-1}$ |
|---|---|---|---|
| $^3\Pi_g$ | 0.67 | 0.33 | 150 |
| $^1\Sigma_u^+$ | 0.78 | 0.35 | 125 |
| $^1\Pi_g$ | 0.78 | 0.33 | 155 |
| $^1\Sigma_u^+$ | 0.93 | 0.35 | 130 |
| $^3\Delta_u$ | 1.15 | 0.36 | 105 |

## DIMER $Ca_2^+$

Parameters of state

| Term | $T_e$ | $r_e$, nm | $\omega_e$ | $B_e$ | $D_e$, eV |
|---|---|---|---|---|---|
| $X^2\Sigma_u^+$ | 0 | 0.37 | 119 | 0.0526 | 1.04 |

## DIMER $Ce_2$ [4]

$D_0 = 2.83$ eV.

## DIMER $Co_2$

Parameters of state [4, 35]

| Term | $T_e$ | $r_e$, nm | $\omega_e$ | $D_e$, eV |
|---|---|---|---|---|
| $X^5\Sigma_g^+$ | 0 | 0.2 | 240 | 1.7 |

## DIMER $Cd_2$

Parameters of states [33–34]

| Term | $T_e$ | $r_e$, nm | $\omega_e$ | $\omega_e x_e$ | $D_e$, cm$^{-1}$ |
|---|---|---|---|---|---|
| $X^1\Sigma_g^+(0_g^+)$ | 0 | 0.33 | 22 | 0.4 | 305 |
| $^3\Pi_u(0_u^+)$ | 30 726 | — | 17.8 | 0.34 | 235 |

Computed parameters

| Term | $T_e$, эВ | $r_e$, nm | $\omega_e$ | $D_e$, eV |
|---|---|---|---|---|
| $^3\Pi_g$ | 2.12 | 0.14 | 116 | 0.8 |
| $^3\Sigma_g^-$ | 2.23 | 0.15 | 104 | 0.7 |
| $^1\Pi_g$ | 3.01 | 0.14 | 137 | 2.17 |
| $^1\Pi_u$ | 4.17 | 0.145 | 119 | 1.02 |
| $^1\Sigma_u^+$ | 4.22 | 0.17 | 78 | 0.96 |
| $2^1\Sigma_g^+$ | 4.47 | 0.165 | 77 | 0.78 |

## DIMER $Cr_2$

Parameters of states [36–41]

| Term | $T_e$ | $r_e$, nm | $\omega_e$ | $\omega_e x_e$ | $B_e$ | $\alpha_e \cdot 10^{-3}$ |
|---|---|---|---|---|---|---|
| $X^1\Sigma_g^+$ | 0 | 0.1678 | 452 | 0.4 | 0.23 | 3.8 |
| $A^1\Sigma_u^+$ | 21751.4 | 0.1685 | 415 | 9.0 | — | — |

Observed electronic transition $X \to A$, 460 nm; $D_e(X) = 1.66$ eV; $D_e(A) = 1.56$ eV; I.P. = 6.76 eV.

## DIMER $Cs_2$

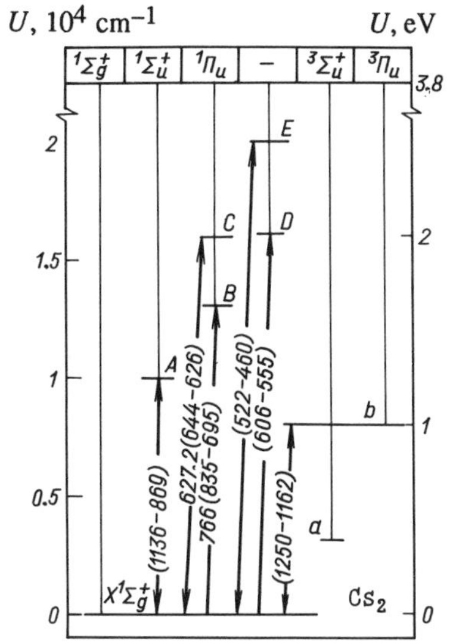


Parameters of states [6, 42–47]

| Term | $T_e$ | $r_e$, nm | $\omega_e$ | $\omega_e x_e$ | $B_e$ | $\alpha_e \cdot 10^{-3}$ |
|---|---|---|---|---|---|---|
| $X^1\Sigma_g^+$ | 0 | 0.465 | 42.02 | 0.082 | 0.0127 | 0.026 |
| $a^3\Sigma_u^+$ | 7850 | 0.43 | 50 | — | 0.013 | — |
| $A^1\Sigma_u^+$ | 9450 | 0.525 | 34 | — | 0.009 | — |
| $b^3\Pi_u$ | 11 000 | — | — | — | — | — |
| $^3\Sigma_g^+$ | 11 602 | 0.554 | — | — | 0.008 | — |
| $^1\Sigma_g^+$ | 12 114 | 0.583 | 23.35 | — | 0.007 | — |
| $B^1\Pi_u$ | 13 044 | — | 34.33 | 0.080 | — | — |
| $^1\Pi_g$ | 13 913 | 0.569 | 18.44 | — | 0.008 | — |
| $C^1\Pi_u$ | 15 948 | 0.434 | 29.7 | 0.058 | 0.0135 | 0.078 |
| $E'$ | 19 410 | 0.775 | 12.69 | — | — | — |
| $E^1\Sigma_u^+$ | 20 195 | 0.534 | 29.1 | — | 0.0081 | — |

$D_e(X) = 0.45$ eV; $D_e(E) = 3648$ cm$^{-1}$; I.P. = 3.76 eV; $D_e(E') = 5550$ cm$^{-1}$; $D_e(^1\Pi_g) = 1467$ cm$^{-1}$.

## DIMER $Cs_2^+$

Parameters of states

| Term | $T_e$ | $r_e$, nm | $\omega_e$ | $D_e$, eV |
|---|---|---|---|---|
| $X^2\Sigma_g^+$ | 0 | 0.444 | 32.4 | 0.67 |
| $^2\Pi_u$ | 12 600 | — | — | — |

## DIMER $Cu_2$

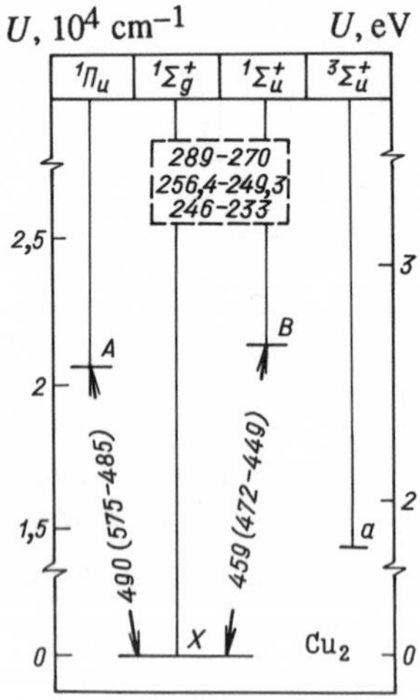


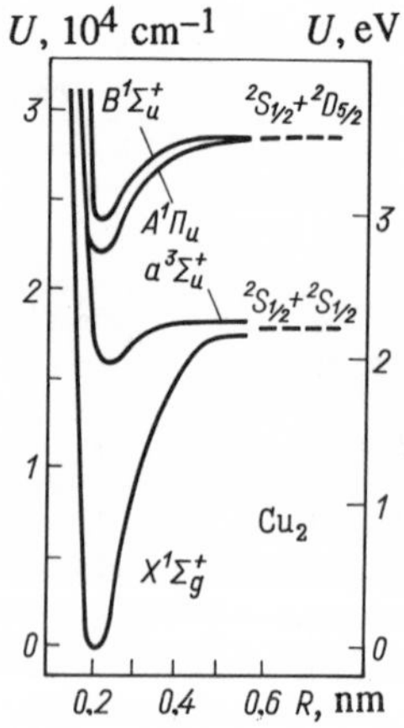


Parameters of states [6, 11, 12, 48–52]

| Term | $T_e$ | $r_e$, nm | $\omega_e$ | $\omega_e x_e$ | $B_e$ | $\alpha_e \cdot 10^{-3}$ |
|---|---|---|---|---|---|---|
| $X^1\Sigma_g^+$ | 0 | 0.222 | 246.5 | 1.085 | 0.1087 | 0.61 |
| $a^3\Sigma_u^+$ | 10 000 | 0.248 | 125 | — | — | — |
| $A^1\Pi_u$ | 20 433 | 0.2558 | 192 | 0.35 | 0.082 | 0.62 |
| $B^1\Sigma_u^+$ | 21 748 | 0.237 | 242.1 | 2.0 | 0.0989 | 0.64 |

$\tau(C)$ = 2.3 nsec; I.P. ≐ 7.4 eV; $\tau(A)$ = 7 nsec; $\tau(B)$ = 3.5 nsec; $D_e(X)$ = 16,760 $cm^{-1}$ = 1.98 eV.

## DIMER $Ga_2$

Parameters of state [4]

| Term | $T_e$ | $\omega_e$ | $\omega_e x_e$ | $D_e$, eV |
|---|---|---|---|---|
| $X^3\Sigma_g^+$ | 0 | 180 | 1.0 | 1.4 |

## DIMER $Ge_2$

Parameters of state [56–58]

| Term | $T_e$ | $r_e$, nm | $\omega_e$ | $\omega_e x_e$ | $B_e$ | $\alpha_e \cdot 10^{-3}$ |
|---|---|---|---|---|---|---|
| $XO_g^+$ | 0 | 0.27 | 270 | 0.8 | 0.081 | 3.31 |
| $^3\Pi_u$ | 501 | — | 195 | — | — | — |
| $^1\Delta_g$ | 3871 | — | 165 | — | — | — |
| $^1\Sigma_g^+$ | 4593 | — | 166 | — | — | — |
| $^1\Pi_u$ | 5166 | — | 199 | — | — | — |
| $^3\Pi_g$ | 8090 | — | 156 | — | — | — |
| $^1\Sigma_g^+$ | 10 969 | — | 220 | — | — | — |

$D_e(X) = 2.8$ eV; I.P. = 7.9 eV.

## DIMER $Gd_2$ [4]

$D_0 = 1.8$ eV.

## DIMER $Fe_2$

Parameters of states [53–55]

| Term | $T_e$ | $\omega_e$ | $\omega_e x_e$ | $D_e$, eV | $r_e$, nm |
|---|---|---|---|---|---|
| $X^7\Delta_u$ | 0 | 299.6 | 1.4 | 1.1 | 0.18 |
| $A$ | 18 355 | — | — | — | — |
| $B$ | 21 095 | — | — | — | — |
| $C$ | — | — | — | — | — |

Observed electronic transitions $X \to A$ 544.8 nm; I.P. = 6.3 eV; $X \to B$ 474.0 nm; $X \to A$ continuum with peak at 414.6 nm.

## DIMER $In_2$

Parameters of states [4]

| Term | $T_e$ | $\omega_e$ | $\omega_e x_e$ | $r_e$, nm |
|---|---|---|---|---|
| $X^3\Sigma_g^-$ | 0 | 143.1 | 0.1 | 0.28 |
| $A$ | 18 025 | 115.0 | — | — |

Observed electronic transitions $X \rightarrow A$ (546–550 nm); $D_e(X) = 1.0$ eV.

## DIMER $Er_2$ [4]

$D_0 = 0.78$ eV.

## DIMER $Eu_2$ [4]

$D_0 = 0.43$ eV.

## DIMER $Hg_2$

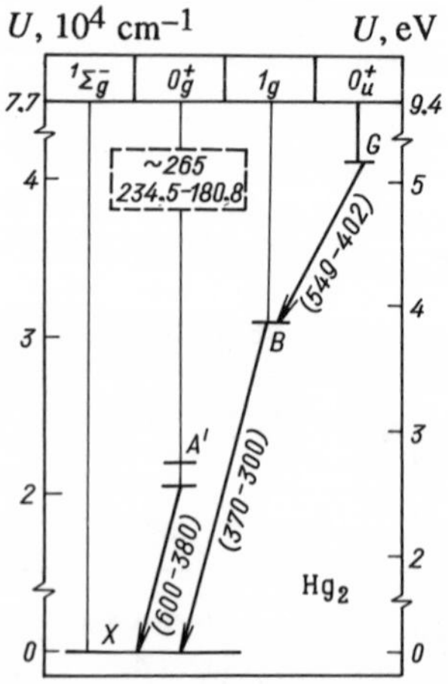

Parameters of states [59–63]

| Term | $T_e$, eV | $r_e$, nm | $\omega_e$ | $\omega_e x_e$ | $B_e$ | $\alpha_e \cdot 10^{-3}$ |
|---|---|---|---|---|---|---|
| $XO_g^+$ | 0 | 0.29 | 44 | 0.5 | 0.02 | 0.073 |
| $AO_g^-$ | 3.6 | — | 144 | 0.5 | — | 0.57 |
| $AO_g^+$ | 3.6 | — | 143 | 0.5 | — | — |
| $B1g$ | 3.8 | — | — | — | — | — |
| $D1u$ | 3.9 | — | — | — | — | — |
| $E2g$ | 4.1 | — | 142 | 0.5 | — | — |
| $F1g$ | 4.5 | — | 152 | 0.9 | — | — |
| $GO_u^+$ | 5.6 | 0.24 | 129 | 0.35 | — | — |
| $1u$ | 6.5 | — | — | — | — | — |

Observed electronic transitions: $D1u \rightarrow X$, continuum max 355 nm: $1u \rightarrow AO_g^+$ 420 nm; $1u \rightarrow E2g$ 509.7 nm; $1u \rightarrow F1g$ 633 nm. I.P. = 9.4 eV.

## DIMER $Hg_2^+$ [4]

$D_0 = 1.4$ eV.

## DIMER $K_2$

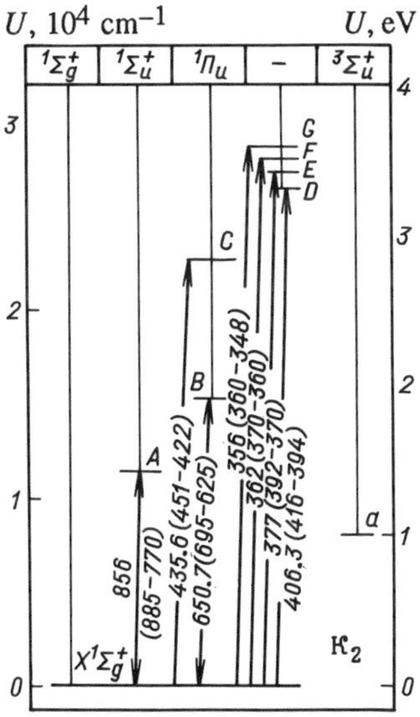


Parameters of states [6, 64–69]

| Term | $T_e$ | $r_e$, nm | $\omega_e$ | $\omega_e x_e$ | $B_e$ | $\alpha_e \cdot 10^{-3}$ |
|---|---|---|---|---|---|---|
| $X^1\Sigma_g^+$ | 0 | 0.39 | 92.09 | 0.283 | 0.0567 | 0.165 |
| $a^3\Sigma_u^+$ | 9600 | 0.39 | 90 | 0.267 | 0.0568 | 0.251 |
| $A^1\Pi_u^+$ | 11 682 | 0.45 | 69.09 | 0.153 | 0.0418 | 0.138 |
| $b^3\Sigma_g^+$ | 13 000 | 0.48 | 70 | 0.294 | 0.0472 | 0.286 |
| $B^1\Pi_u$ | 15 376 | 0.42 | 74.93 | 0.378 | 0.0487 | 0.24 |
| $C^1\Pi_u$ | 22 969 | 0.44 | 61.48 | 0.13 | 0.044 | 0.165 |
| $D^1\Pi_g$ | 19 500 | 0.62 | 20 | 0.079 | 0.0222 | 0.131 |

$D_e(X) = 0.51$ eV; $\tau(B) = 1.1$ nsec; I.P. = 32,775 cm$^{-1}$ = 4.064 eV.

Quantum defects of the dimer $K_2$

| Highly excited state | Defect | Highly excited state | Defect |
|---|---|---|---|
| $ns\sigma^1\Sigma_g^+$ | 0.4 | $nd\sigma^1\Delta_g$ | 0.89 |

## DIMER $K_2^+$

Parameters of states

| Term | $T_e$ | $r_e$, nm | $\omega_e$ | $\omega_e x_e$ | $B_e$ | $\alpha_e \cdot 10^{-3}$ |
|---|---|---|---|---|---|---|
| $X^2\Sigma_g^+$ | 0 | 0.45 | 72.5 | 0.2 | 0.042 | 0.18 |
| $A^2\Pi_u$ | 18 300 | 0.60 | 30.0 | 0.2 | 0.024 | 0.22 |

$D_e(X) = 0.81$ eV.

## DIMER $La_2$

Parameters of states [4]

| Term | $T_e$ | $\omega_e$ | $\omega_e x_e$ | $D_e$, eV |
|---|---|---|---|---|
| $X^2\Sigma$ | 0 | 82.6 | 0.71 | 2.5 |
| $^2\Pi$ | — | 76.9 | 0.57 | — |

## DIMER $Li_2$

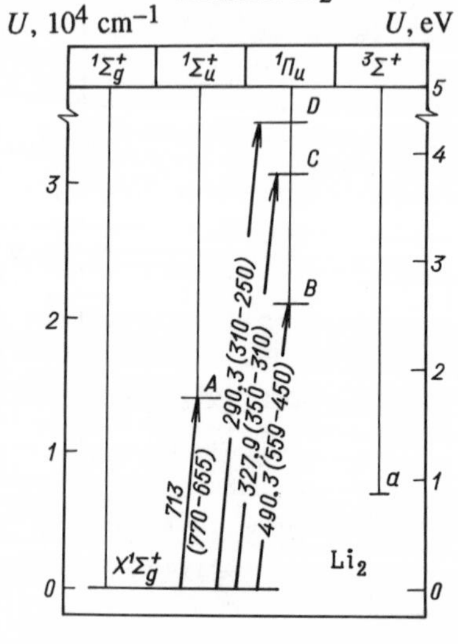

Parameters of states [6, 70–78]

| Term | $T_e$ | $r_e$, nm | $\omega_e$ | $\omega_e x_e$ | $B_e$ | $\alpha_e \cdot 10^{-3}$ |
|---|---|---|---|---|---|---|
| $X^1\Sigma_g^+$ | 0 | 0.267 | 351.4 | 2.59 | 0.672 | 7.04 |
| $a^3\Sigma_u^+$ | 8108 | 0.413 | 73.0 | 3.95 | 0.268 | 2.5 |
| $b^3\Pi_u$ | 11 241 | 0.255 | 340 | 2.5 | 0.695 | 7.7 |
| $A^1\Sigma_u^+$ | 14 068 | 0.311 | 255.5 | 1.58 | 0.497 | 5.4 |
| $c^3\Sigma_g^+$ | 16 700 | 0.31 | 240.0 | 2.1 | 0.500 | 6.6 |
| $2^1\Sigma_g^+$ | 20 102 | 0.365 | 128.7 | 1.7 | 0.360 | 4.17 |
| $B^1\Pi_u$ | 20 436 | 0.294 | 270.6 | 2.92 | 0.557 | 8.3 |
| $E^1\Sigma_g^+$ | 29 652 | — | — | — | — | — |
| $2^3\Pi_g$ | 29 840 | 0.382 | 204.2 | 1.37 | 0.385 | — |
| $F^1\Sigma_g^+$ | 29 975 | — | — | — | — | — |
| $1^3\Delta_g$ | 30 090 | 0.302 | 301.7 | 1.88 | 0.614 | 6.5 |
| $3^3\Sigma_g^+$ | 31 042 | 0.308 | 296.2 | 2.05 | 0.590 | 7.0 |
| $c^1\Pi_u$ | 30 551 | 0.308 | 238 | 3.3 | 0.507 | 9.7 |
| $D^1\Pi_u$ | 34 443 | 0.322 | 201 | — | 0.463 | 7.0 |

$D_e(X) = 8516\ \text{cm}^{-1}$; $D_e(a) = 336\ \text{cm}^{-1}$; $D_e(A) = 1.1$ eV; $D_e(b) = 12{,}172\ \text{cm}^{-1}$; $D_e(B) = 1313\ \text{cm}^{-1}$; $D_e(2'\Sigma_g) = 3319\ \text{cm}^{-1}$; $\tau(A) = 1.81$ nsec; $\tau(B) = 0.8$ nsec; I.P. $= 41{,}496\ \text{cm}^{-1} = 5.14$ eV.

Computed parameters

| Term | $T_e$ | $r_e$, nm | $\omega_e$ | $\omega_e x_e$ | $B_e$ | $\alpha_e \cdot 10^{-3}$ |
|---|---|---|---|---|---|---|
| $^1\Sigma_g^+$ | 20 500 | 0.373 | 140 | 1.6 | 0.345 | 6.0 |
| $^1\Pi_u$ | 27 000 | 0.416 | 80 | 1.0 | 0.278 | 5.5 |

Quantum defects of the dimer $Li_2$

| Highly excited state | Defect |
|---|---|
| $ns\sigma^1\Sigma_g^+$ | 0.44 |
| $nd\sigma^1\Sigma_g^+$ | 0.07 |
| $nd\pi^1\Pi_g$ | 0.15 |

## DIMER $Li_2^+$

Parameters of states

| Term | $T_e$ | $r_e$, nm | $\omega_e$ | $\omega_e x_e$ | $B_e$ | $\alpha_e \cdot 10^{-3}$ |
|---|---|---|---|---|---|---|
| $X^2\Sigma_g^+$ | 0 | 0.312 | 262 | 1.93 | 0.49 | 5.4 |
| $A^2\Pi_u$ | 23 400 | 0.401 | 98 | 0.87 | 0.29 | 3.9 |

$D_e(X) = 10{,}469\ cm^{-1} = 1.45$ eV.

## DIMER $Li_2^-$

Ground state $X^2\Sigma_u^+$.

## DIMER $Mg_2$

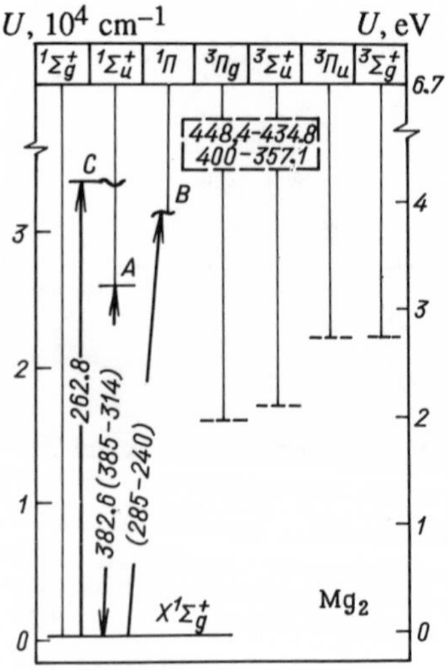


Parameters of states [6, 21]

| Term | $T_e$ | $r_e$, nm | $\omega_e$ | $\omega_e x_e$ | $B_e$ | $\alpha_e \cdot 10^{-3}$ |
|---|---|---|---|---|---|---|
| $X^1\Sigma_g^+$ | 0 | 0.3889 | 51.08 | 1.623 | 0.093 | 3.78 |
| $A^1\Sigma_u^+$ | 26 069 | 0.3082 | 190.7 | 1.164 | 0.1480 | 1.32 |
| $C^1\Sigma_u^+$ | 38 048 | 0.327 | — | — | 0.135 | — |

$D_e(X) = 0.050$ eV = 399 $cm^{-1}$; I.P. = 6.7 eV; $D_e(A) = 1.14$ eV = 9311 $cm^{-1}$.

Computed states

| Term | $T_e$, eV | $r_e$, nm | $\omega_e$, $cm^{-1}$ |
|---|---|---|---|
| $^3\Sigma_u^+$ | 1.95 | 0.3 | 190 |
| $^3\Pi_g$ | 1.96 | 0.265 | 280 |
| $^1\Sigma_u^+$ | 2.38 | 0.30 | 185 |
| $^1\Pi_g$ | 2.28 | 0.26 | 290 |
| $^3\Sigma_g^-$ | 4.12 | 0.24 | 485 |

## DIMER $Mo_2$

Parameters of state [38, 79]

| Term | $T_e$ | $r_e$, nm | $\omega_e x_e$ | $\omega_e$, $cm^{-1}$ | $D_e$, eV |
|---|---|---|---|---|---|
| $X^1\Sigma_g^+$ | 0 | 0.1929 | 1.51 | 477.1 | 4.12 |

Observed electronic transition $X \rightarrow {}^1\Sigma_u{}^+$ 518 nm; I.P. = 6.21 eV.

## DIMER $Mn_2$

Parameters of state [80, 81]

| Term | $T_e$ | $r_e$, nm | $\omega_e$ | $\omega_e x_e$ | $D_e$, eV |
|---|---|---|---|---|---|
| $X^1\Sigma_g^-$ | 0 | 0.288 | 124.7 | 0.24 | 0.79 |

Observed electronic transition $X \rightarrow A$ 693 nm; I.P. = 7.4 eV.

## DIMER $Mn_2{}^+$

$D_0 = 1.39$ eV.

## DIMER $Na_2$

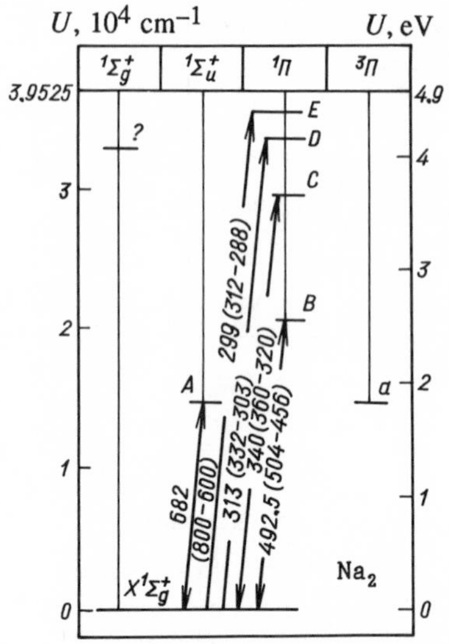


Parameters of states [6, 82–91]

| Term | $T_e$ | $r_e$, nm | $\omega_e$ | $\omega_e x_e$ | $B_e$ | $\alpha_e \cdot 10^{-3}$ |
|---|---|---|---|---|---|---|
| $X^1\Sigma_g^+$ | 0 | 0.308 | 159 | 0.725 | 0.1547 | 0.874 |
| $a^3\Sigma_u^+$ | 5848 | 0.509 | 24.3 | — | — | — |
| $b^3\Pi_u$ | 11 240 | 0.314 | 345 | 2.0 | 0.1718 | 0.597 |
| $^3\Pi_u$ | 13 650 | 0.311 | 152 | 0.498 | 0.152 | 7.2 |
| $A^1\Sigma_u^+$ | 14 681 | 0.364 | 117.3 | 0.358 | 0.1108 | 0.549 |
| $^1\Sigma_g^+$ | 19 338 | 0.445 | 75.2 | — | 0.074 | — |
| $B^1\Pi_u$ | 20 320 | 0.34 | 124.1 | 0.70 | 0.1253 | 0.724 |
| $^1\Pi_g$ | 21 796 | 0.456 | 42.7 | 0.613 | 0.070 | 0.98 |
| $^1\Sigma_g^+$ | 25 692 | 0.356 | 112.7 | 1.047 | 0.115 | 1.125 |
| $2^3\Pi_g$ | 28 399 | 0.429 | — | 0.55 | 0.0796 | 0.329 |
| $2^1\Sigma_u^+$ | 28 453 | — | 106.5 | 1.56 | — | — |
| $1^3\Delta_g$ | 28 523 | 0.339 | 127.7 | 0.54 | 0.127 | 0.887 |
| $C^1\Pi_u$ | 29 621 | — | 116.3 | 0.60 | 0.116 | 0.858 |
| $4^3\Sigma_g^+$ | 32 127 | 0.357 | 112.3 | 0.45 | 0.115 | 0.61 |
| $D^1\Pi_u$ | 33 487 | — | 111.3 | 0.48 | — | — |
| $E^1\Pi_u$ | 35 557 | — | 106.2 | 0.65 | — | — |

$D_e(X) = 6023\ \text{cm}^{-1} = 0.7$ eV; $D_e(A) = 8275\ \text{cm}^{-1}$; $D_e(B) = 520\ \text{cm}^{-1} = 0.06$ eV; $D_e(^1\Pi_g) = 6022\ \text{cm}^{-1}$; $D_e(^3\Sigma_u{}^+) = 174.4\ \text{cm}^{-1}$; $\tau(A) = 12.4$ nsec; I.P. = 39,481 $\text{cm}^{-1}$ = 4.88 eV.

Quantum defects of the dimer

| Highly excited states | Defects |
|---|---|
| $ns\sigma^1\Sigma_g^+$ | 0.6 |
| $nd\sigma^1\Sigma_g^+$ | 0.21±0.02 |
| $nd\sigma^1\Pi_g$ | −0.02±0.02 |
| $nd\sigma^1\Delta_g$ | 0.42±0.01 |

**DIMER** $Na_2^+$

Parameters of state

| Term | $T_e$ | $r_e$, nm | $\omega_e$ | $\omega_e x_e$ | $B_e$ | $\alpha_e \cdot 10^{-3}$ |
|---|---|---|---|---|---|---|
| $X^2\Sigma_g^+$ | 0 | 0.354 | 126 | 0.46 | 0.117 | 0.7 |

$D_e(X) = 0.96$ eV.

**DIMER** $Nb_2$ [4]

$D_0 = 5.21$ eV.

**DIMER** $Nd_2$ [4]

$D_0 = 0.82$ eV.

**DIMER** $Ni_2^+$

Parameters of states [92]

| Term | $T_e$ | $r_e$, nm | $\omega_e$ | $\omega_e x_e$ | $B_e$ |
|---|---|---|---|---|---|
| $X^3\Pi_u$ | 0 | 0.220 | 381 | 1.1 | 0.104 |
| $A$ | 22 246 | 0.2364 | 192 | — | 0.120 |

Observed electronic transition $X \rightarrow A$ 459 nm; $D_e(X) = 2.4$ eV; I.P. = 7.64 eV.

**DIMER** $Ni_2$

Ground state $X^4\Sigma_u^-$.

**DIMER** $Np_2$

I.P. = 6.1 eV [4].

## DIMER $P_2$

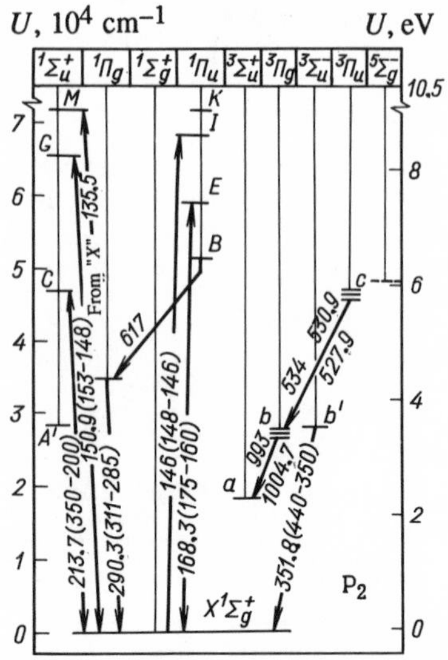


Parameters of states [6, 93]

| Term | $T_e$ | $r_e$, nm | $\omega_e$ | $\omega_e x_e$ | $B_e$ | $\alpha_e \cdot 10^{-3}$ |
|---|---|---|---|---|---|---|
| $X^1\Sigma_g^+$ | 0 | 0.1893 | 780.8 | 2.83 | 0.3036 | 1.5 |
| $a^3\Sigma_u^-$ | 18 790 | 0.209 | 565 | 3 | 0.25 | — |
| $b'^3\Sigma_u^-$ | 28 503 | 0.205 | 604.5 | 2 | 0.258 | 1.4 |
| $b^3\Pi_g$ | 28 069<br>28 197<br>28 330 | —<br>0.197<br>— | —<br>644.7<br>— | —<br>3.21<br>— | —<br>0.28<br>— | —<br>1.8<br>— |
| $A^1\Pi_g$ | 34515.2 | 0.1989 | 618.9 | 3.0 | 0.2752 | 1.7 |
| $C^1\Sigma_u^+$ | 46941.3 | 0.2120 | 473.9 | 2.34 | 0.2421 | 1.8 |
| $c^3\Pi_u$ | 47 139<br>47 159<br>47 177 | —<br>0.223<br>— | —<br>393.7<br>— | —<br>3.85<br>— | —<br>0.219<br>— | —<br>2.4<br>— |
| $B^1\Pi_u$ | 50 846 | 0.219 | 359 | 3 | 0.227 | 6 |
| $E^1\Pi_u$ | 59446.2 | 0.1969 | 700.7 | 2.9 | 0.2807 | — |
| $G^1\Sigma_u^+$ | 66313.4 | 0.1913 | 694.1 | 4.18 | 0.2973 | 1.9 |
| $I^1\Pi_u$ | 68 849 | 0.207 | — | — | 0.2541 | — |
| $^5\Sigma_g^+$ | 31 965 | 0.233 | 395 | 4.0 | 0.20 | — |

I.P. = 10.5 eV.

## DIMER $P_2^-$

Ground state $X^2\Pi_g$.

## DIMER $P_2^+$

Parameters of states

| Term | $T_e$ | $r_e$, nm | $\omega_e$ | $\omega_e x_e$ | $B_e$ | $\alpha_e \cdot 10^{-3}$ |
|---|---|---|---|---|---|---|
| $^2\Pi_u$ $X_1$ | 0 | 0.1985 | 672.2 | 2.74 | 0.276 | 1.51 |
| $^2\Pi_u$ $X_2$ | 260 | — | — | — | — | — |
| $^2\Pi_g$ $D_1$ | 2179 | 0.1893 | 733 | — | 0.303 | 2.1 |
| $^2\Pi_g$ $D_2$ | 18 741 | 0.2226 | 462.2 | 2.45 | 0.219 | 1.42 |
| $B^2\Sigma_u^+$ | 18 832 | — | — | — | — | — |
| $^2\Pi_g$ $C_1$ | 25 566 | 0.2121 | 410.5 | 3.23 | 0.241 | 2.11 |
| $^2\Pi_g$ $C_2$ | 28 686 | 0.2243 | 441.4 | 2.58 | 0.216 | 1.36 |
| $F^2\Sigma_u^+$ | 28 870 | — | — | — | — | — |
|  | 40 180 | — | 810 | — | — | — |

Observed electronic transition { $C \rightarrow X$ (340—450) nm ; $D \rightarrow X$ (560—680) nm .

## DIMER $Pb_2$

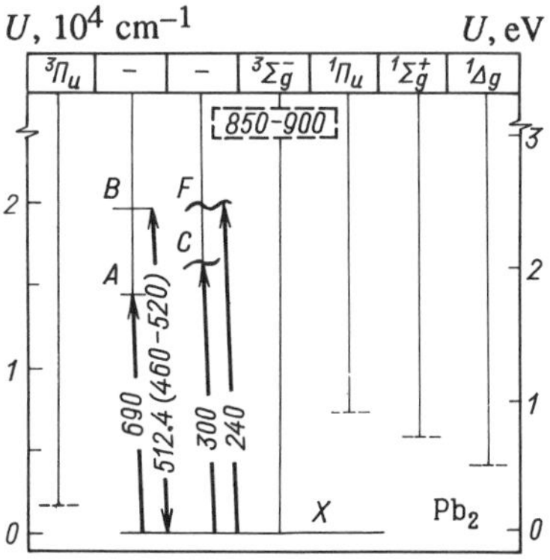

Parameters of states [6, 94–99]

| Term | $T_e$ | $r_e$ | $\omega_e$ | $\omega_e x_e$ | $B_e$ | $\alpha_e \cdot 10^{-3}$ |
|---|---|---|---|---|---|---|
| $X0_g^+$ | 0 | 0.293 | 110.1 | 0.32 | 0.0189 | 0.051 |
| $^3\Pi_u$ | 1584 | 0.27 | 130 | — | — | — |
| $^1\Delta_g$ | 3580 | 0.285 | 123 | — | — | — |
| $^1\Sigma_g^+$ | 5142 | 0.29 | 117 | — | — | — |
| $^1\Pi_u$ | 6240 | 0.27 | 124 | — | — | — |
| $^5\Pi_g$ | 6527 | 0.295 | 103 | — | — | — |
| $A0_u^+$ | 14 465 | — | 162 | 0.4 | — | — |
| $B$ (1 $u$) | 15 311 | 0.277 | 128.1 | 1.13 | 0.0212 | 0.1 |
| $C(0_g^+)$ | 19 490 | — | 48 | 0.13 | — | — |
| $F0_u^+$ | 19 806 | 0.307 | 159.4 | 1.36 | 0.0172 | 0.027 |

$D_e(X) = 0.8$ eV = 6669 cm$^{-1}$.

## DIMER $Pd_2$

Parameters of state [100]

| Term | $T_e$ | $r_e$, nm | $D_e$, eV |
|---|---|---|---|
| $X^1\Gamma_g$ | 0 | 0.2808 | 0.76 |

## DIMER $Po_2$

Parameters of states [1, 4]

| Term | $T_e$ | $\omega_e$ | $\omega_e x_e$ | $D_e$, eV |
|---|---|---|---|---|
| $X0_g^+$ | 0 | 155.7 | 0.335 | 1.8 |
| $A0_u^+$ | 25 149 | 108.5 | 0.4417 | — |

Observed electronic transition $X \rightarrow A$ 398 nm.

## DIMER $Pt_2$

Parameters of states [1, 4]

| Term | $T_e$ | $r_e$, nm | $D_e$, eV |
|---|---|---|---|
| $X^1\Gamma_g$ | 0 | 0.257 | 0.93 |
| $A$ | — | — | — |

Observed electronic transition $X \rightarrow A$ 889 nm.

## DIMER $Rb_2$

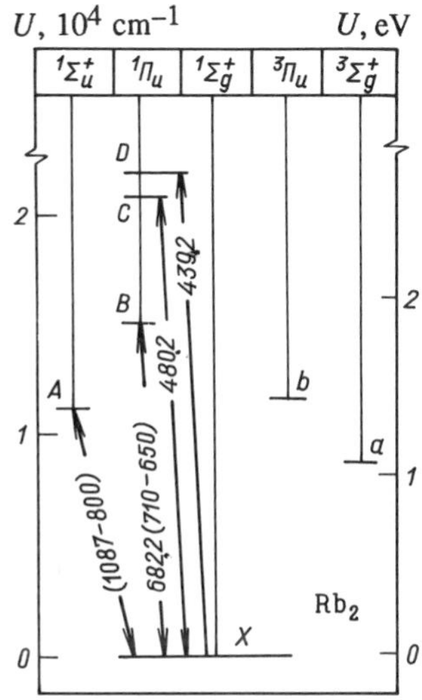


Parameters of states [42, 68, 101]

| Term | $T_e$ | $r_e$, nm | $\omega_e$ | $\omega_e x_e$ | $B_e$ | $\alpha_e \cdot 10^{-3}$ |
|---|---|---|---|---|---|---|
| $X^1\Sigma_g^+$ | 0 | 0.417 | 47.43 | 0.16 | 0.0228 | 0.047 |
| $a^3\Pi_u$ | 9000 | — | — | — | — | — |
| $b^3\Sigma_g^+$ | 12 500 | — | — | — | — | — |
| $A^1\Sigma_u^+$ | 11 000 | — | — | — | — | — |
| $B^1\Pi_u$ | 14 665 | 0.446 | 57.75 | 0.15 | 0.020 | 0.07 |
| $C^1\Pi_u$ | 20 835 | — | 36.46 | 0.124 | — | — |
| $D^1\Pi_u$ | 22 777 | — | 40.42 | 0.074 | — | — |

$D_e(X) = 0.49$ eV $= 4180$ cm$^{-1}$; I.P. $= 3.96$ eV; $\tau(C) = 1.4$ nsec.

## DIMER $Rb_2^+$

Parameters of state

| Term | $T_e$ | $r_e$, nm | $D_e$, eV | $\omega_e$ |
|---|---|---|---|---|
| $X^2\Sigma_g^+$ | 0 | 0.394 | 0.72 | 47 |

## DIMER $Ra_2$

Parameters of state [4]

| Term | $T_e$ | $r_e$, nm | $\omega_e$ |
|---|---|---|---|
| $X^1\Sigma_g^+$ | 0 | 0.49 | 45 |

## DIMER $Rh_2$

$D_0 = 2.9$ eV [4].

## DIMER $Sc_2$

Parameters of state [80, 112–115]

| Term | $T_e$ | $r_e$, nm | $\omega_e$ | $\omega_e x_e$ |
|---|---|---|---|---|
| $X^5\Sigma_u^-$ | 0 | 0.221 | 239.9 | 0.93 |

$D_e(X) = 1.6$ eV

## DIMER $Sb_2$

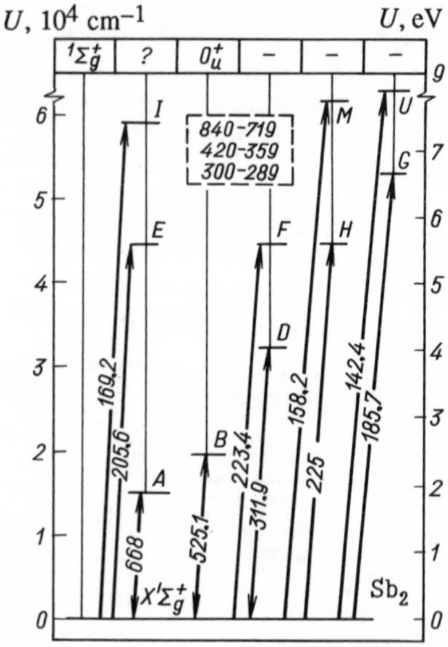

Parameters of states [6, 102, 103]

| Term | $T_e$ | $\omega_e$ | $\omega_e x_e$ | $B_e$ | $\alpha_e \cdot 10^{-3}$ | $r_e$, nm |
|---|---|---|---|---|---|---|
| $X^1\Sigma_g^+$ | 0 | 269.9 | 0.56 | 0.050 | 11.2 | 0.234 |
| $A(1u)$ | 14 991 | 218 | 0.46 | — | — | — |
| $B(0_u^+)$ | 19 067 | 219 | 0.55 | 0.0396 | 10.2 | 0.264 |
| $K$ | 31 397 | 127.4 | 1.076 | 0.0343 | 21 | 0.248 |
| $D$ | 32 087 | 209.6 | — | — | — | — |
| $H$ | 44 329 | 479 | — | — | — | — |
| $F$ | 44 780 | 226 | 1.17 | — | — | — |
| $E$ | 48 645 | 228 | — | — | — | — |
| $G$ | 53 888 | 185 | — | — | — | — |

$D_e(X)$ = 3.1 eV; $D_e(B)$ = 2.76 eV; $D_e(K)$ = 0.46 eV; I.P. = 9.3 eV.

**DIMER $Se_2$**

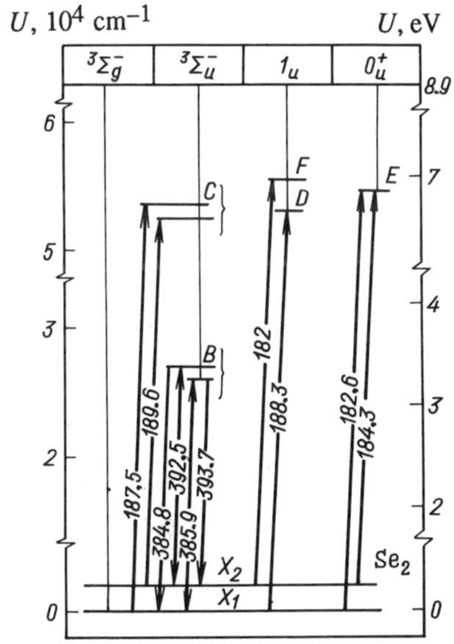

Parameters of states [6, 104–108]

| Term | $T_e$ | $r_e$, nm | $\omega_e$ | $\omega_e x_e$ | $B_e$ | $\alpha_e \cdot 10^{-3}$ |
|---|---|---|---|---|---|---|
| $X^3\Sigma$ $0_g^+$ | 0 | 0.2166 | 385.3 | 0.963 | 0.89 | 0.28 |
| $X^3\Sigma$ $1g$ | 510 | 0.2160 | 387.2 | 0.964 | 0.90 | 0.29 |
| $b^1\Sigma_g^-$ | 7958 | — | 355 | 1.085 | 0.876 | 0.32 |
| $A(0_u^+)$ | 15 131 | — | — | — | — | — |
| $B^3\Sigma$ $0_u^+$ | 25 980 | 0.2446 | 246.3 | 1.016 | 0.704 | 0.34 |
| $B^3\Sigma$ $1_u$ | 26 058 | 0.2440 | 246.4 | 1.225 | 0.708 | 0.55 |
| $C^3\Sigma$ $1_u$ | 53 220 | 0.2089 | 428 | 1.22 | 0.966 | 0.33 |
| $C^3\Sigma$ $0_u^+$ | 53 324 | — | 414 | — | — | — |

$D_e(X) = 3.14$ eV; I.P. = 8.88 eV.

## DIMER $Si_2$

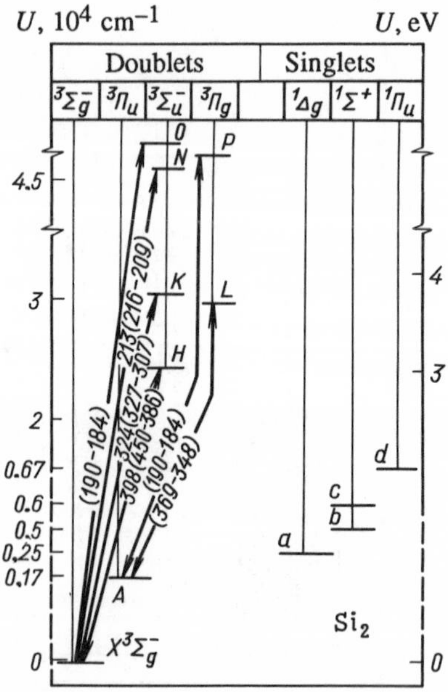

Parameters of states

| Term | $T_e$ | $r_e$, nm | $\omega_e$ | $\omega_e x_e$ | $B_e$ | $\alpha_e \cdot 10^{-3}$ |
|---|---|---|---|---|---|---|
| $X^3\Sigma_g^-$ | 0 | 0.224 | 510.9 | 2.02 | 0.239 | 1.35 |
| $A^3\Pi_u$ | 1700 | 0.215 | 547.9 | 2.43 | 0.259 | 1.55 |
| $a^1\Delta_u$ | 2500 | — | — | — | — | — |
| $b^1\Sigma_g^+$ | 5000 | — | — | — | — | — |
| $c^1\Sigma_g^+$ | 6000 | — | — | — | — | — |
| $d^1\Pi_u$ | 6700 | — | 440 | — | — | — |
| $H^3\Sigma_u^-$ | 24 429 | 0.26 | 275.3 | 1.99 | 0.171 | 1.35 |
| $L^3\Pi_g$ | 29 785 | 0.225 | 494 | — | 0.237 | — |
| $K^3\Sigma_u^-$ | 30 794 | 0.23 | 462.6 | 5.9 | 0.218 | 3.16 |

$D_e(X)$ = 3.28 eV; I.P. = 7.4 eV.

## DIMER $Si_2^+$

Ground state $X^4\Sigma_u^-$.

## DIMER $Sn_2$

Parameters of states [95, 99, 116–118]

| Term | $T_e$ | $r_e$, nm | $\omega_e$ | $\omega_e x_e$ | $B_e$ | $\alpha_e \cdot 10^{-3}$ |
|---|---|---|---|---|---|---|
| $X0_g^+$ | 0 | 0.275 | 189.7 | 0.49 | 0.0385 | 0.01 |
| $^3\Pi_u$ | 1652 | 0.26 | 187 | — | — | — |
| $^1\Delta_g$ | 4139 | 0.27 | 167 | — | — | — |
| $^1\Sigma_g^+$ | 5866 | 0.275 | 162 | — | — | — |
| $^1\Pi_u$ | 6310 | 0.26 | 187 | — | — | — |
| $^5\Pi_g$ | 6727 | 0.28 | 168 | — | — | — |
| $C0_u^+$ | 18 223 | 0.323 | 83.4 | 0.05 | 0.0278 | 0.03 |

$D_e(X)$ = 2.0 eV.

## DIMER $Sr_2$

Parameters of states [119, 120]

| Term | $T_e$ | $r_e$, nm | $\omega_e$ | $\omega_e x_e$ | $B_e$ | $\alpha_e \cdot 10^{-3}$ |
|---|---|---|---|---|---|---|
| $X^1\Sigma_g^+$ | 0 | 0.445 | 40.32 | 0.40 | 0.0189 | 0.2 |
| $A'\Sigma_u^+$ | 17 340 | 0.395 | 8.3 | — | 0.0238 | — |

Observed electronic transition $X \rightarrow A$ (600–450) nm; $D_e(X)$ = 1100 cm$^{-1}$; $D_e(A)$ = 5460 cm$^{-1}$.

## DIMER $Ta_2$

I.P. = 15.5 eV [4].

## DIMER $Te_2$

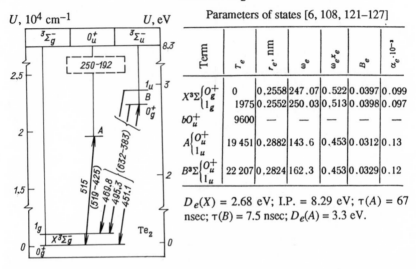

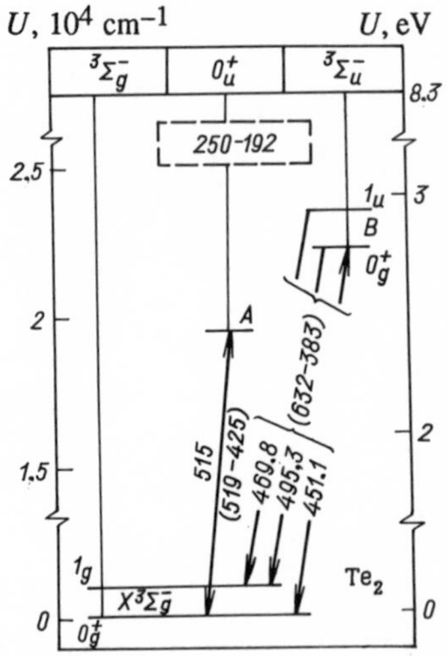


Parameters of states [6, 108, 121–127]

| Term | $T_e$ | $r_e$, nm | $\omega_e$ | $\omega_e x_e$ | $B_e$ | $\alpha_e \cdot 10^{-3}$ |
|---|---|---|---|---|---|---|
| $X^3\Sigma\ 0_g^+$ | 0 | 0.2558 | 247.07 | 0.522 | 0.0397 | 0.099 |
| $X^3\Sigma\ 1_g$ | 1975 | 0.2552 | 250.03 | 0.513 | 0.0398 | 0.097 |
| $b0_u^+$ | 9600 | — | — | — | — | — |
| $A\ \{0_u^+, 1_u\}$ | 19 451 | 0.2882 | 143.6 | 0.453 | 0.0312 | 0.13 |
| $B^3\Sigma\ \{0_u^+, 1_u\}$ | 22 207 | 0.2824 | 162.3 | 0.453 | 0.0329 | 0.12 |

$D_e(X)$ = 2.68 eV; I.P. = 8.29 eV; $\tau(A)$ = 67 nsec; $\tau(B)$ = 7.5 nsec; $D_e(A)$ = 3.3 eV.

## DIMER $Te_2^+$

Parameters of state

| Term | $a^4\Pi_u$ | $A^2\Pi_e$ | $b^4\Sigma_g^-$ | $B^2\Pi_u$ | $C^2\Sigma_g^-$ | $D^2\Pi_u$ | $X^2\Pi_g$ |
|---|---|---|---|---|---|---|---|
| $T_e$ | 9030 | 14 520 | 21 940 | 26 450 | 28 790 | 33 230 | 0 |

## DIMER $Ti_2$

Parameters of states [113, 114, 128]

| Term | $T_e$ | $r_e$, nm | $\omega_e$ | $\omega_e x_e$ | $D_e$, eV |
|---|---|---|---|---|---|
| $X^1\Sigma_g^+$ | 0 | 0.196 | 408 | 1.08 | 1.3 |

I.P. = 6.3 eV.

## DIMER $Tl_2$

Parameters of states [129, 130]

| Term | $T_e$ | $r_e$, nm | $\omega_e$ | $\omega_e x_e$ |
|---|---|---|---|---|
| $X^1\Sigma_u^-$ | 0 | 0.300 | 80 | 0.5 |
| $1u$ | 814 | 0.371 | 30 | 0.4 |
| $0_g^+$ | 860 | 0.397 | 25 | 0.2 |
| $0_u^+$ | 2900 | 0.330 | 85 | 0.7 |
| $2u$ | 6200 | 0.341 | 54 | 0.4 |
| $1g$ | 6770 | 0.30 | 87 | 0.5 |
| $0_g^+$ | 6780 | 0.361 | 54 | 0.2 |
| $2g$ | 8130 | 0.314 | 61 | 0.5 |
| $1u$ | 9280 | 0.349 | 39 | 0.6 |

I.P. = 6.5 eV.

## DIMER $Tm_2$

$D_0 = 0.52$ eV [4].

## DIMER $V_2$

Parameters of states [40, 41, 131]

| Term | $T_e$ | $r_e$, nm | $\omega_e$ | $\omega_e x_e$ | $\alpha_e \cdot 10^{-3}$ |
|---|---|---|---|---|---|
| $X$ | 0 | 0.177 | 508 | 3.3 | 1.4 |
| $A$ | — | — | 537.5 | 4.2 | — |

$D_e(X) = 2.5$ eV; I.P. = 6.74 eV.

## DIMER $W_2$ [132]

$D_0 = 6.9$ eV.

## DIMER $U_2$ [4]

$D_0 = 2.3$ eV.

## DIMER $Y_2$ [4]

$D_0 = 1.6$ eV.

### DIMER $Zn_2$

Parameters of States [133, 134]

| Term | $^1\Sigma_g^+$ | $^3\Pi_g$ | $^3\Sigma_u^+$ | $^1\Pi_g$ | $^1\Sigma_u^+$ | $^1\Pi_u$ | $2'\Sigma_g^+$ |
|---|---|---|---|---|---|---|---|
| $T_e$, eV | 0 | 2.57 | 2.75 | 3.55 | 4.93 | 5.01 | 5.41 |
| $r_e$, nm | 0.31 | 0.12 | 0.13 | 0.12 | 0.14 | 0.13 | 0.145 |
| $\omega_e$, cm$^{-1}$ | 55 | 175 | 154 | 205 | 114 | 175 | 108 |
| $D_e$, eV | 0.17 | 1.05 | 0.8 | 2.4 | 1.1 | 1.0 | 0.6 |

$D_e(X) = 0.17$ eV; I.P. = 8.4 eV.

$$E(v, J) = \sum_{i,k} Y_{ik} (v + 1/2)^i [J(J+1)]^k,$$

where the expansion coefficients $Y_{ik}$ are referred to as the Dunham coefficients. It is clear that $Y_{10} = \omega_e$, $Y_{20} = -x_e\omega_e$, $Y_{01} = B_e$, and $Y_{11} = -\alpha_e$.

In this article we use the following additional notation: $D_0$ is the dissociation energy for the ground vibrational and rotational states; $D_e$ is the depth of the potential well for a given electronic state ($D_e = D_0 + \hbar\omega/2$); I.P. denotes the ionization potential of a dimer (in eV); and $\tau$ is the lifetime of an excited molecule. Usually this refers to the ground vibrational and rotational states of the molecule; $\delta$ represents the quantum defect for highly excited states of the molecule. [The ionization potential of an excited molecule is given by I.P. = Ry $(n - \delta)^2$, where Ry = 13.605 eV and $n$ is the principal quantum number of the excited electron.]

For convenience of use, all the information relating to a given dimer is collected in a single place. In the references, the method by which the information has been obtained is specified by a number in parentheses at the end of each citation. These numbers correspond to the following methods:

1) computation (theoretical),
2) studies of the absorption and fluorescence spectrum of the dimer in a matrix of inert gas,
3) studies of the emission spectrum of dimers in vapor, and
4) laser spectroscopy.

The references are divided into two parts. The first includes monographs and reviews in which a large amount of data is collected and the second part includes original papers containing specific information about certain systems.

We do not list the values of the Franck–Condon factors in this review. In those cases where they are available, a reference is made to the book by Kuz'menko et al. [6], which lists them together with the methods used to calculate them based on the potential curves for the states involved in a transition.

## REFERENCES


1. K. P. Huber and G. Herzberg, *Molecular Spectra and Molecular Structure IV: Constants of Diatomic Molecules*, van Nostrand Reinhold, Princeton (1979).
2. K. P. Huber and G. Herzberg, *Constants of Diatomic Molecules* [Russian translation], Mir, Moscow (1984).
3. B. Rosen, *Spectroscopic Data Relative to Diatomic Molecules*, Pergamon Press, New York, London (1970).
4. S. Suchard and I. E. Meiser, *Spectroscopic Data. Vol. 2. Homonuclear Diatomic Molecules*, IFI/Plenum, New York–Washington–London (1975).
5. *The Thermodynamic Properties of Individual Substances* [in Russian], Nauka, Moscow (1973–1983).
6. N. E. Kuz'menko, L. A. Kuznetsova, and Yu. Ya. Kuzyakov, *Franck–Condon Factors for Diatomic Molecules* [in Russian], Izd. MGU, Moscow (1984).
7. W. Schrittenlacher, H. H. Rotermund, W. Schroeder, and D. M. Kolb, *Surf. Sci.* **156**, 777–784 (1985). (2)
8. J. L. Martin and W. Andreoni, *Phys. Rev. A* **28**, 3637–3639 (1984). (1)
9. H. Stoll, P. Fuentealba, P. Schwerdtfeger, et al., *J. Chem. Phys.* **81**, 2732–2736 (1984). (1)
10. Rabii Sohrab and Cary Y. Yang, *Chem. Phys. Lett.* **105**, 480–483 (1984). (1)
11. C. M. Biagini, D. A. Clemente, and C. Foglia, *Mol. Phys.* **53**, 301–310 (1984). (1)
12. I. Shim and K. A. Gingerich, *J. Chem. Phys.* **79**, 2903–2912 (1983). (1)
13. H. Basch, W. J. Steven, and M. Krauss, *Chem. Phys. Lett.* **109**, 212–216 (1984). (1)
14. T. H. Upton, *J. Phys. Chem.* **90**, 754–759 (1986). (1)
15. Y. Watanabe, Y. Sakai, and H. Kashiwaga, *Chem. Phys. Lett.* **120**, 363–366 (1985). (1)
16. G. Wannous, C. Effantin, F. Martin, and J. D'Incan, *J. Mol. Spectrosc.* **91**, 1–8 (1982). (1)
17. H. Gollisch, *J. Phys. B* **15**, 2569–2578 (1982). (1)
18. W. Ermler, Y. Lee, and K.S. Pitzer, *J. Chem. Phys.* **70**, 288–297 (1979). (1)
19. H. Bredohl, I. Dubois, and P. Nzohabonayo, *J. Mol. Spectrosc.* **93**, 281–285 (1982). (3)
20. R. M. Clements and R. F. Barrow, *J. Chem. Soc. Faraday Trans.* **81**, Pt. 2, 625–727 (1985). (3)
21. C. Malinowska-Adamska, "Orbital energies for the $^1\Sigma_g{}^+$ states of $Be_2$, $Mg_2$, and $Ca_2$," *Z. Nauk. Plodz.*, No. 365, 73–78 (1981). (1)

22. S. L. Richardson, M. Y. Chou, and M. L. Cohen, *Phys. Rev. A* **31**, 3444–3446 (1985). (1)
23. V. E. Bondybey, *Chem. Phys. Lett.* **109**, 436–441 (1984). (4)
24. C. W. Bauschlicher and H. Partridge, *J. Chem. Phys.* **80**, 334–337 (1984). (1)
25. S. Wilson, *Mol. Phys.* **49**, 1489–1493 (1983). (1)
26. B. H. Lengsfield, A. D. McLean, B. Lin, and M. Yoshimine, *J. Chem. Phys.* **79**, 1891–1895 (1983). (2)
27. Y. S. Lee and R. J. Bartlett, *J. Chem. Phys.* **80**, 4371–4377 (1984).
28. B. Bühler, C. Gremer, J. Janes, and G. Gerber, *Ber. Bunsenges. Phys. Chem.* **89**, 284–286 (1985). (4)
29. P. Eberle, H. Sontag, and R. Weber, *Chem. Phys.* **92**, 417–422 (1985). (2)
30. G. Ehret and G. Gerber, *Opt. Commun.* **51**, 145–150 (1984). (4)
31. V. E. Bondybey and J. H. Englisch, *Chem. Phys. Lett.* **111**, 195–200 (1984). (4)
32. R. T. Hofmann and D. O. Harris, *J. Chem. Phys.* **81**, 1047–1048 (1984). (3)
33. A. Kowalski, M. Czajkowski, and W. H. Breckenridge, *Chem. Phys. Lett.* **119**, 368–370 (1985). (4)
34. M. W. McGeoch, *J. Chem. Phys.* **72**, 140–146 (1980). (4)
35. I. Shim and K. A. Gingerich, *J. Chem. Phys.* **78**, 5693–5698 (1983). (1)
36. M. M. Goodgame and W. A. Goodard, *Phys. Rev. Lett.* **54**, 661–664 (1985). (1)
37. M. Moskovits, W. Limm, and T. Meican, *J. Phys. Chem.* **89**, 3886–3890 (1985). (2, 4)
38. A. P. Klyagina, G. L. Gutsev, V. D. Fursova, and A. A. Levin, *Zh. Neorg. Khim.* **29**, 2765–2770 (1984). (1)
39. S. J. Riley, E. K. Parks, L. G. Pobe, and S. Wexier, *J. Chem. Phys.* **79**, 2577–2582 (1983). (4)
40. S. P. Walch, C. W. Bauschlicher, B. O. Roos, and C. J. Nalin, *Chem. Phys. Lett.* **103**, 175–179 (1983). (1)
41. N. A. Baykara, B. N. McMaster, and D. R. Salahub, *Mol. Phys.* **52**, 891–905 (1984). (1)
42. G. S. Wagner and N. R. Isenor, *Can. J. Phys.* **63**, 976–982 (1985). (3)
43. C. Amiot and J. Verges, *Chem. Phys. Lett.* **116**, 273–278 (1985). (4)
44. C. Amiot, C. Crepin, and J. Verges, *J. Mol. Spectrosc.* **107**, 28–47 (1984). (4)
45. C. Amiot, C. Crepin, and J. Verges, *Chem. Phys. Lett.* **106**, 162–165 (1984). (4)
46. H. Helm and R. Müller, *Phys. Rev. A* **27**, 2493–2502 (1983). (4)
47. B. C. Laskowski and S. R. Langhoff, *Chem. Phys. Lett.* **92**, 49–53 (1982). (1)
48. P. Scharf, S. Brode, and R. Ahlrichs, *Chem. Phys. Lett.* **113**, 447–450 (1985). (1)
49. M. Pellisir, *J. Chem. Phys.* **79**, 2099–2100 (1983). (1)
50. D. M. Kolb, H. H. Rotermund, W. Schroeder, and W. Schrittenlacher, *J. Chem. Phys.* **80**, 695–700 (1984). (2)
51. R. L. Martin, *J. Chem. Phys.* **78**, 5840–5842 (1983). (1)
52. L. Pauling, *J. Chem. Phys.* **78**, 3346–3350 (1983).
53. H. M. Nagarathna, P. A. Montano, and V. M. Naik, *J. Am. Chem. Soc.* **105**, 2938–2943 (1983). (1)
54. I. Shim and K. A. Gingerich, *J. Chem. Phys.* **77**, 2490–2497 (1982). (1)
55. D. Guenzburger, E. Saitovitch, and M. Baggio, *Phys. Rev. B* **24**, 2368–2379 (1981). (1)
56. G. Pacchioni, *Mol. Phys.* **49**, 727–736 (1983). (1)
57. F. M. Frohen and W. Schulze, *Surf. Sci.* **156**, 765–769 (1983). (1)

58. J. E. Northrup and M. L. Cohen, *Chem. Phys. Lett.* **102**, 440–441 (1983). (1)
59. J. Supronowicz, R. J. Niefer, J. B. Atkinson, and L. Krause, *J. Phys. B* **19**, 1153–1164 (1986). (4)
60. K. C. Calestino and W. C. Ermler, *J. Chem. Phys.* **81**, 1872–1881 (1984). (1)
61. R. Niefer, J. B. Atkinson, and L. Krause, *J. Phys. B* **16**, 3531–3541 (1983). (4)
62. R. J. Niefer, J. B. Atkinson, and L. Krause, *J. Phys. B* **16**, 3767–3773 (1983). (4)
63. S. H. Linn, C. L. Liao, C. X. Liao, J. M. Brom, and C. Y. Ng, *Chem. Phys. Lett.* **105**, 645–650 (1984). (4)
64. D. E. Johnson and J. G. Eden, *J. Opt. Soc. Am.* **32**, 721–728 (1985). (3)
65. C. Brechignase and Ph. Cahuzac, *Surf. Sci.* **156**, 183–188 (1985). (4)
66. M. Broyer, J. Chevaleyre, G. Delacretaz, P. Fayet, and L. Wöste, *Chem. Phys. Lett.* **114**, 477–482 (1986). (4)
67. C. Brechignac and Ph. Cahuzac, *Chem. Phys. Lett.* **112**, 20–24 (1984). (3, 4).
68. H. Partridge, D. A. Dixon, S. P. Walch, et al., *J. Chem. Phys.* **79**, 1859–1865 (1983). (1)
69. M. P. Anzin'sh, R. S. Ferber, and I. Y. Pirags, *J. Phys. B* **16**, 2759–2771 (1983). (4)
70. B. Barakat, R. Bacis, S. Churassy, et al., *J. Mol. Spectrosc.* **116**, 271–285 (1986). (4)
71. H. H. Michels, R. H. Holbs, and L. A. Wright, *Chem. Phys. Lett.* **118**, 67–71 (1985). (1)
72. A. R. Rajaei-Rizi, F. B. Orth, J. T. Bahns, and W. C. Stwalley, *J. Mol. Spectrosc.* **109**, 123–133 (1985). (4)
73. J. M. Wadehra and H. H. Michels, *Chem. Phys. Lett.* **114**, 380–383 (1985). (4)
74. D. D. Konowalow, R. M. Regan, and M. E. Rosenkrantz, *J. Chem. Phys.* **81**, 4534–4536 (1984). (1)
75. B. Hemmerling, S. B. Rai, and W. Demtröder, *Z. Phys.* **A320**, 135–140 (1985). (4)
76. W. Preuss and G. Baumgartner, *Z. Phys.* **A320**, 125–133 (1985). (4)
77. D. D. Konowalow and L. B Ratcliff, *Chem. Phys. Lett.* **111**, 413–415 (1984). (1)
78. D. Eisel, W. Demtröder, W. Miller, and P. Botschwina, *Chem. Phys.* **80**, 329–344 (1983). (4)
79. J. B. Hopkins, P. R. R. Langridge-Smith, M. D. Morse, and R. E. Smalley, *J. Chem. Phys.* **78**, 1627–1637 (1983). (4)
80. M. Moskovits, D. P. Di Lella, and W. Limm, *J. Chem. Phys.* **80**, 626–633 (1984). (3)
81. K. Ervin, S. K. Loh, N. Aristov, and P. B. Armentront, *J. Phys. Chem.* **87**, 3593–3596 (1983). (1)
82. C. Effantin, O. Babaky, K. Hussein, et al., *J. Phys. B* **18**, 4077–4087 (1985). (3)
83. L. Li, S. F. Race, and R. W. Field, *J. Chem. Phys.* **82**, 1178–1182 (1985). (4)
84. D. L. Cooper, R. F. Barrow, J. Verges, et al., *Chem. Phys. Lett.* **114**, 483–485 (1985). (4)
85. A. M. Bonch-Bruevich, T. A. Vartanyan, Yu. N. Maksimov, and V. V. Khromov, *Opt. Spektrosk.* **58**, 546–550 (1985).
86. M. Ch. Bordas, M. Broyer, J. Chevaleyre, et al., *J. Phys.* **46**, 27–38 (1985). (1)
87. G. Gerber and R. Möller, *Chem. Phys. Lett.* **113**, 546–553 (1985). (4)
88. R. N. Ahmad-Bitar, A. S. Al-Ayash, *J. Mol. Spectrosc.* **106**, 299–306 (1984). (4)

89. J. Verges, C. Effantin, J. D'Incan, et al., *Phys. Rev. Lett.* **53**, 46–47 (1984). (4)
90. R. Montagnani, P. Riani, and O. Salvetti, *Chem. Phys. Lett.* **102**, 571–573 (1983). (1)
91. S. Martin, J. Chevaleyre, C. M. Bordas, et al., *J. Chem. Phys.* **79**, 4132–4141 (1983). (4)
92. A. V. Zaitsevskii, I. A. Topol', B. É. Dzevitskii, and G. N. Zviadadze, *Zh. Fiz. Khim.* **56**, 769–776 (1982). (1)
93. J. T. Snodgrass, J. V. Coe, C. B. Friedhoff, et al., *Chem. Phys. Lett.* **112**, 352–355 (1985). (3)
94. H. Sontag and R. Weber, *J. Mol. Spectrosc.* **100**, 75–81 (1983). (4)
95. G. Pacchioni, *Mol. Phys.* **55**, 211–223 (1985). (1)
96. H. Sontag, B. Eberle, and R. Weber, *Chem. Phys.* **80**, 279–288 (1983). (2)
97. K.S. Pitzer and K. Balasubramanian, *J. Phys. Chem.* **86**, 3068–3070 (1982). (1)
98. V. E. Bondybey and J. H. Englisch, *J. Chem. Phys.* **74**, 6978–6979 (1981). (4)
99. K. Balasubramanian and K. S. Pitzer, *J. Chem. Phys.* **78**, 321–327 (1983). (1)
100. I. Shim and K.A. Gingerich, *J. Chem. Phys.* **80**, 5107–5119 (1984). (1)
101. G. Pichler, S. Milosevich, D. Vezä, and D. Vikicevic, *J. Phys. B* **16**, 4633–4642 (1983). (4)
102. H. Sontag and R. Weber, *Chem. Phys.* **70**, 23–28 (1982). (2)
103. H. Sontag and R. Weber, *J. Mol. Spectrosc.* **91**, 72–79 (1982). (4)
104. M. Heaven, T. A. Miller, J. H. Englisch, and V. E. Bondybey, *Chem. Phys. Lett.* **91**, 251–257 (1982). (4)
105. R. P. Sambasiva, P. T. V. Ramakrishna, and R. R. Ramakrishna, *Acta. Phys. Hung.* **56**, 3–8 (1984). (1)
106. A. Jenouvrier, *Can. J. Phys.* **61**, 1531–1544 (1983). (3)
107. S. J. Prosser, R. F. Barrow, C. Effantin, and J. D'Incan, *J. Phys. B* **15**, 4151–4160 (1982). (4)
108. V. E. Bondybey and J. H. Englisch, *J. Chem. Phys.* **72**, 6479–6484 (1980). (2)
109. J. E. Northrup, M. T. Yin, and M. L. Cohen, *Phys. Rev. A* **28**, 1945–1950 (1983). (1)
110. I. Dubois and H. Leclerq, *J. Phys. B* **14**, 2807–2812 (1981). (3)
111. M. Weinert, E. Wimmer, A. J. Freeman, and H. Krakauer, *Phys. Rev. Lett.* **47**, 705–708 (1981).
112. S. P. Walch and C. W. Bauschlicher, *Chem. Phys. Lett.* **94**, 290–205 (1983). (1)
113. V. D. Fursova, A. P. Klyagina, A. A. Levin, and G. L. Gutsev, *Chem. Phys. Lett.* **116**, 317–322 (1985). (1)
114. V. D. Fursova, A. P. Klyagina, A. A. Levin, and G. L. Gutsev, *Dokl. Akad. Nauk SSSR* **280**, 146–149 (1985). (1)
115. L. B. Knight, R. J. von Zee, and W. Weltner, *Chem. Phys. Lett.* **94**, 296–299 (1983). (2)
116. V. E. Bondybey, M. Heaven, and T. A. Miller, *J. Chem. Phys.* **78**, 3593–3598 (1983). (4)
117. M. A. Epting, M. T. McKenzie, and E. R. Nixon, *J. Chem. Phys.* **73**, 134–136 (1980). (2)
118. V. E. Bondybey and J. H. Englisch, *J. Mol. Spectrosc.* **84**, 383–390 (1980). (2)
119. G. Gerber, R. Möller, and H. Schneider, *J. Chem. Phys.* **81**, 1538–1551 (1984). (4)
120. T. Bergeman and P. F. Liao, *Chem. Phys.* **72**, 886–898 (1980). (4)
121. M. Ya. Tamanis, *Izv. Akad. Nauk Latv. SSR, Ser. Fiz. Tekh. Nauk*, No. 1, 13–21 (1983). (3)

122. H. Schnöckel, *Z. Anorg. Allg. Chem.* **510**, No. 3, pp. 72–78 (1984). (2)
123. J. Carion, Y. Gueru, J. Lotrian, and P. Luc, *J. Phys. B* **15**, 2841–2844 (1982). (4)
124. R. S. Ferber, O. A. Shim, and M. Ya. Taminis, *Chem. Phys. Lett.* **92**, 393–397 (1982). (4)
125. A. Pardo, J. M. Poyato, and J. Basulio, *J. Mol. Spectrosc.* **93**, 245–247 (1982). (4)
126. R. S. Ferber and M. Ya. Tamanis, *Chem. Phys. Lett.* **98**, 577–579 (1983). (4)
127. N. E. Kuz'menko, A. V. Stolyarov, and Yu. Ya. Kuzyakov, *Vestn Mosk. Gos. Univ., Khimiya*, Moscow (1985). (1)
128. C. Cosse, M. Fouassier, T. McJean, M. Tranguille, D. P. DiLella, and M. Moskovits, *J. Chem. Phys.* **79**, 6076–6085 (1980). (3)
129. P. A. Christiansen, *J. Chem. Phys.* **79**, 2928–2931 (1983). (1)
130. F. W. Froben, W. Schulze, and U. Kloss, *Chem. Phys. Lett.* **99**, 500–502 (1983). (3)
131. V. D. Fursova, A. P. Klyagina, A. A. Levin, and G. A. Gutsev, *Izv. Akad. Nauk SSSR, Ser. Khim.*, No. 9, pp. 2032–2035 (1985). (1)
132. R. Hague, M. Pelino, and K. A. Gingerich, *J. Chem. Phys.* **71**, 2929–2933 (1979). (1)
133. C. H. Su, Y. Huang, and R. F. Brebrick, *J. Phys. B* **18**, 3187–3195 (1985). (3)
134. T. Hiroshi, T. Mutsumi, and N. Takashi, *J. Chem. Phys.* **82**, 5608–5615 (1985). (1)

# PHOTODISSOCIATION AND RADIATIVE RECOMBINATION OF DIATOMIC MOLECULES CONTAINED IN THE ATMOSPHERE

A. M. Pravilov

## 1. INTRODUCTION

Calculations of the kinetics of processes in plasmas, flames, shock tubes, gas lasers, and the atmospheres of the Earth and other planets require quantitative information on the properties of the excited states of the molecules involved in these processes, and on the kinetics and mechanisms of the reactions by which these states are populated. Data of these types for the molecules which exist in the atmosphere are of special interest, primarily because of the requirements of aeronomy and gas dynamics.

The stream of published data on the radiative decay of atmospheric molecules, as well as on the mechanism and kinetics of processes which occur during collisions of their fragments, has grown continuously over the last 5–10 years. A review of the photochemistry of these molecules by Okabe [1] was not critical in character, as it absolutely should have been, given the large volume of literature that was examined (1085 references). Much of the data given in that and in earlier books and reviews [2–7] should now be considered erroneous or incomplete, since new data have been published. The latest review of data on chemiluminescence during collisions of fragments of small molecules was published in 1973 [8] and is completely out of date. This is also true of the earlier reviews [9–11].

In this review we offer a critical analysis of published data on:

1) the absolute quantum yields of radiative decay processes (dissociation, predissociation) of diatomic molecules found in the atmosphere ($O_2$, CO, NO, $N_2$, $H_2$). Luminescence, collisional deexcitation, and reactions of excited states of these molecules are examined very briefly. Information on these processes can be found elsewhere [1, 4,

12]. The spectral range extends from the threshold for a given process to the first ionization potential (IP) of these molecules. Information on the photoionization of simple molecules can be found in [13, 14].

2) the spectral distributions of the rate constants for chemiluminescence and the total rate constants for chemiluminescence during collisions of fragments of these molecules.

We discuss the mechanism for radiative decay of these molecules and for radiative recombination of their fragments, as well as the properties of the excited states of these molecules that have been established through combined analyses of photochemical, chemiluminescence, spectroscopic, and computational data. All the quantitative information is tabulated.

## 2. DEFINITIONS

Although research on photochemical and chemiluminescence processes in gases and vapors has a long history, the terminology in this branch of chemical physics has not stabilized. This often leads to confusion and methodological errors. Thus, we must introduce some definitions to be used in the following discussion.

### 2.1. Photochemical Processes [4, 15]

*1. A radiative process or photoprocess* involves the absorption of a photon and the subsequent processes caused by the excited particle that has absorbed the photon.

If the particle that absorbs the photon is a free molecule A, i.e., a molecule in the gaseous phase at moderate pressures which will permit the free escape of its photofragments without occurrence of the "cage effect" [16], then radiative processes can be described in terms of the following scheme (recall that $h\nu <$ IP):

$$A + h\nu \rightarrow \{A^*\} \rightarrow B_1 + C_1, \quad (1)$$

$$\rightarrow B_2 + C_2, \quad (2)$$

. . . . . . . . . .

$$B_i + C_i, \quad (3)$$

. . . . . . . . . .

$$A^{**} + h\nu_L^k ; \quad (4)$$

. . . . . . . . . . .

$$\rightleftarrows A^{**} \tag{5}$$

$$A^* + M \to A^{**} + M, \tag{6}$$

$$\rightleftarrows A^{**} + M, \tag{7}$$

$$\to A^*_{v-\Delta v, J-\Delta J} + M, \tag{8}$$

$$\to \text{products}, \tag{9}$$

$$\to B_l + C_l, \tag{10}$$

$$A^{**},\ A^*_{v-\Delta v, J-\Delta J} \to \cdot\ \ \cdot \cdot \cdot, \tag{11}$$

$$B^*_k \to C_l + D, \tag{12}$$

$$B_k + M \to \text{products}, \tag{13}$$

$$B^*_k \to B_h + h\nu_L, \tag{14}$$

$$B^*_k + M \to B^{**}_k + M, \tag{15}$$

$$\to B^*_{kv-\Delta v, J-\Delta J} + M. \tag{16}$$

Here {A*} denotes electronic–vibrational–rotational (rovibronic) states of the molecule A formed during absorption of a photon and A** denotes other rovibronic states.

Radiative processes are usually subdivided into primary and secondary processes.

2. *Primary radiative processes* include the absorption of a photon by a particle and the processes whereby the excitation energy of the resulting particle degrades without involvement of particles – third bodies (any of the particles in the medium or the vessel wall). If the particle that absorbs a photon is a molecule, then these include processes leading to the formation of A*, dissociation (the lifetime of A* is $\tau_{A^*} \leq 10^{-12}$ sec), predissociation ($\tau_{A^*} \geq 10^{-13}$ sec) of A* into products $B_i$ and $C_i$ which may be in excited states [reactions (1)–(3)], luminescence of A* molecules which have not undergone electron-impact deexcitation and vibrational–rotational relaxation (see below) [$\tau_{A^*} \geq 10^{-9}$ sec, reaction (4)], and spontaneous internal conversion into another electronic state, including intercombination conversion [$\tau_{A^*} \geq 10^{-12}$ sec, reaction (5)].

3. *Photodissociation processes for a molecule* A are primary radiative processes that lead to a reduction in the concentration of A ([A]), i.e., absorption of a photon plus dissociation or predissociation of A*.

*4. Secondary photoprocesses* are those which occur during and immediately after a collision between a particle that has absorbed a photon and a particle or third body, as well as processes which involve the radiative decay products of A. For molecules this includes the following: collisional deexcitation of the excited states of A into another electronic state, including the ground state, with a large loss of energy by the A* molecule, while the reverse process is excluded [reaction (6)]; collisionally induced internal conversion into another electronic state with a small loss of excitation energy by A*, which allows the reverse transition [reaction (7)]; vibrational–rotational relaxation of A* within a single electronic state [reaction (8)]; a collisionally induced reaction involving A* [reaction (9)]; predissociation of A* [reaction (10)] into products, including some of the same type obtained from the radiative decay of A; degradation of the excitation energy of A* and $A^*_{v-\Delta v,J-\Delta J}$ [dissociation, predissociation, luminescent reactions, etc. – reaction (11)]; monomolecular decay of one of the radiative decay products (in other words, of the primary photolysis products [4]), $A - B_k^*$, into products, some of which are of the same type obtained by radiative decay [reaction (12)]; reactions of the photodissociation products [reaction (13)]; their luminescence [reaction (14)]; collisional deexcitation [reaction (15)]; vibrational–rotational relaxation [reaction (16)]; etc.

Noyes and Leighton [17] and, following them, Okabe [1] propose that the primary radiative decay processes should also include processes involving A* and other particles [the third bodies in reactions (6)–(10)], while processes involving only the products of radiative decay [reactions (12)–(16)] should be regarded as secondary. In our opinion, this definition of primary processes is methodologically incorrect in the gaseous phase for the following reason: in this case the absolute quantum yield of the primary photoprocesses [reactions (1)–(4)] depends on the type and pressure of the gases or third bodies and on the surface to volume ratio of the reactor, but is not a molecular constant that characterizes the properties of a molecule.

Reactions (6)–(10) are conventional bi- or termolecular reactions involving excited particles that are not in thermodynamic equilibrium with the surrounding medium and must be described in terms of reaction rate constants which depend on the state of A*, rather than in terms of the concept of an absolute quantum yield.

Our definition of a primary radiative process may cause some methodological complications in practical applications because the primary and secondary photolysis products of reactions (1)–(3) and (12) may have the same form (but, as a rule, not the same excitation energies). For example, during photolysis of $C_2H_5$, the ethylene molecule may be formed by two paths [18]:

$$C_2H_5I + h\nu \rightarrow C_2H_5I^* \rightarrow C_2H_4 + HI, \tag{17}$$

$$\rightarrow C_2H_5^{\#} + I(^2P), \tag{18}$$

$$C_2H_5^{\#} \rightarrow C_2H_4 + H \tag{19}$$

(where # denotes vibrational–rotational excitation).

If reaction (17) is a dissociation ($\tau_{C_2H_5I^*} \sim 10^{-13}$ sec) or clearly allowed predissociation ($\tau_{C_2H_5I^*} \sim 10^{-12}$-$10^{-11}$ sec) process for $C_2H_5I$, then it is possible to establish whether $C_2H_4$ is a primary or secondary photolysis product by studying the dependence of the absolute quantum yield for formation of $C_2H_4$, $\Phi_{C_2H_4}^{C_2H_5I}(\lambda)$, on the pressure of the thermalizer for the vibrational energy or of the acceptor of $C_2H_5^{\#}$ radicals (the rate of reaction (19) should be much lower than that of reaction (17) [18]). If, however, reaction (17) is a strongly forbidden predissociation process, then in order to establish the mechanism for formation of $C_2H_4$ it may be useful to study the dynamics of formation and the angular distribution of the $C_2H_4$ and other photolysis products.

5. *The absolute quantum yield for formation of a primary photoproduct* $B_i$ is the ratio of the rate of change in the concentration of $B_i$ to the density of the absorbed radiation $n(\lambda)$ (photon/cm$^3$·sec), i.e., to the intensity of the radiation absorbed per unit volume:

$$\Phi_{B_i}^{A}(\lambda) = \frac{d[B_i]/dt}{n(\lambda)}. \tag{20}$$

6. *The absolute quantum yield of the* i*th photodissociation process* is the ratio (taken with the sign) of the rate of change in the concentration of the irradiated particles [A] through the *i*th photoprocess to $n(\lambda)$

$$\varphi_i^{A}(\lambda) = -\frac{\{d[A]/dt\}_i}{n(\lambda)}. \tag{21}$$

Evidently, for a photodissociation process, $\Phi_{B_i}{}^{A}(\lambda) = \Phi_{C_i}{}^{A}(\lambda) = \varphi_i{}^{A}(\lambda)$.

The absolute quantum yield for photodissociation of a molecule A is given by

$$\varphi_A^F(\lambda) = \sum_i \varphi_i^A(\lambda). \tag{22}$$

The only significant difference between the definitions (20) and (21) and those given by Calvert and Pitts [4] is that here the absolute quantum yield of primary products or of a process refers only to the yield measured for monochromatic radiation within an interval $\lambda \pm \Delta\lambda$, in which $\Phi_{B_i}{}^A(\lambda \pm \Delta\lambda) \neq f(\Delta\lambda)$. It has been shown [15] that these refinements are essential. We note also that the definitions of $\varphi_i{}^A(\lambda)$ and $\Phi_{B_i}{}^A(\lambda)$ given by Calvert and Pitts [4] are more correct than those given by Okabe [1], since for the secondary photolysis products, it is more correct to define $\Phi^A(\lambda)$ as the yield of the photoprocesses taking place within a unit volume, where $n(\lambda) =$ const and, therefore, the concentration of radicals is also constant, rather than within the entire volume of a cuvette [1].

7. *The absolute integral quantum yield of the* ith *primary photoprocess in the absorption band* $\lambda_i - \lambda_{i+1}$ of A is given by the ratio of the area under the partial absorption cross section $\sigma_i{}^A(\lambda)$ of A resulting in the $i$th process to the total area under the absorption spectrum $\sigma_A(\lambda)$ of A over the entire band:

$$\varphi_i^A = \frac{\int_{\lambda_1}^{\lambda_2} \varphi_i^A(\lambda)\,\sigma_A(\lambda)\,d\lambda}{\int_{\lambda_1}^{\lambda_2} \sigma_A(\lambda)\,d\lambda} = \frac{\int_{\lambda_1}^{\lambda_2} \sigma_i^A(\lambda)\,d\lambda}{\int_{\lambda_1}^{\lambda_2} \sigma_A(\lambda)\,d\lambda}. \tag{23}$$

This is actually the ratio of the oscillator strength of the optical transition leading to breakup of A through the $i$th channel to the sum of the oscillator strengths for all the transitions that occur within a given band. The quantity $\varphi_i{}^A$ characterizes the probability of an optical transition leading to breakup of A through the $i$th channel and can be used in theoretical calculations [15].

8. *The absolute integral quantum yield for formation of the primary photoproduct* $B_i$ is given by

$$\Phi_{B_i}^A = \frac{\int_{\lambda_1}^{\lambda_2} \Phi_{B_i}^A(\lambda)\,\sigma_A(\lambda)\,d\lambda}{\int_{\lambda_1}^{\lambda_2} \sigma_A(\lambda)\,d\lambda} = \frac{\int_{\lambda_1}^{\lambda_2} \sigma_i^A(\lambda)\,d\lambda}{\int_{\lambda_1}^{\lambda_2} \sigma_A(\lambda)\,d\lambda}. \tag{24}$$

*9. The observed quantum yield of the* i*th primary photoprocess over the band* $\lambda_j - \lambda_k$ is given by

$$ {}^{H}\varphi_i^A = \frac{\int_{\lambda_j}^{\lambda_k} \varphi_i^A(\lambda) J_0(\lambda) \{1 - \exp\{-\sigma_A(\lambda)[A]l\}\} d\lambda}{\int_{\lambda_j}^{\lambda_k} J_0(\lambda) \{1 - \exp\{-\sigma_A(\lambda)[A]l\}\} d\lambda}. \quad (25) $$

Here $J_0(\lambda)$ (photon/nm) is the emission spectrum from the light source that enters the photochemical cell and $l$ is the length of the cell.

*10. The observed quantum yield for formation of the photoproduct* $B_i$ is given by

$$ {}^{H}\Phi_{B_i}^A = \frac{\int_{\lambda_j}^{\lambda_k} \Phi_{B_i}^A(\lambda) J_0(\lambda) \{1 - \exp\{-\sigma_A(\lambda)[A]l\}\} d\lambda}{\int_{\lambda_j}^{\lambda_k} J_0(\lambda) \{1 - \exp\{-\sigma_A(\lambda)[A]l\}\} d\lambda}. \quad (26) $$

In general, the values of ${}^{H}\varphi_i{}^A$ and ${}^{H}\Phi_{B_i}{}^A$ are determined by the mutual positions of the curves $J_0(\lambda)$, $\sigma_A(\lambda)$, and $\varphi_i{}^A(\lambda)$ [or $\Phi_{B_i}{}^A(\lambda)$], as well as by the values of [A] and $l$; therefore, the observed quantum yields depend on the experimental conditions, i.e., they are not comparable quantities. (For more detail, see [15].)

## 2.2. Chemiluminescent Processes

*1. Chemiluminescent reactions* include particle collisions accompanied by a change in the chemical composition of the particles and the emission of a photon, as well as the recombination of colliding particles into a short-lived complex (quasimolecule) which loses part or all of its excitation energy through emission of a photon.

Without the second part, this definition of chemiluminescent reactions would exclude, for purely formal reasons, collisions of particles, such as excimer fragments (XeO, $He_2$, HeH, etc.), accompanied by the formation of excited (including bound) states of complexes of these particles and the emission of photons over a wide range of the spectrum in an optical transition to a repulsive (so there is no change in the chemical composition) ground state. Examples include the typical chemiluminescent processes

$$\mathrm{Xe}\,(^1S_0) + \mathrm{O}\,(^1D) \rightleftharpoons \mathrm{Xe\,O}\,(b^1\Pi) \rightarrow \mathrm{XeO}\,(X^3\Pi) + h\nu; \quad \downarrow \quad \mathrm{Xe}\,(^1S_0) + \mathrm{O}\,(^3P) \quad [19] \tag{27}$$

and

$$2\mathrm{O}_2\,(a\,^1\Delta g) \rightarrow \mathrm{O}_2\left(X\,^3\Sigma_g^-\right) + h\nu, \; \lambda = 634, \; 703 \text{ nm } [8]. \tag{28}$$

The last reaction seems to proceed through the formation of an electronically excited $O_4$* complex.

This definition of chemiluminescent reactions does exclude energy transfer processes leading to deactivation of one of the particles and the appearance of excitation in the others, such as

$$\mathrm{O}\,(^1D) + \mathrm{CO}\left(X\,^1\Sigma^+\right)_{v'=0} \rightarrow \mathrm{O}\,(^3P) + \mathrm{CO}\left(X\,^1\Sigma^+\right)_{v>0} \;[20], \tag{29}$$

$$\mathrm{O}\,(^1D) + \mathrm{O}_2\left(X\,^3\Sigma_g^-\right) \rightarrow \mathrm{O}\,(^3P) + \mathrm{O}_2\left(b\,^1\Sigma_g^+\right)_{v'=0,\,1} \;[21], \tag{30}$$

and

$$\mathrm{N}_2\left(A\,^3\Sigma_u^+\right) + \mathrm{CO}\left(X\,'\Sigma^+\right) \rightarrow \mathrm{N}_2\left(X\,'\Sigma_g^+\right) + \mathrm{CO}\,(A^1\Pi) \;[22], \tag{31}$$

which are accompanied by emission of photons in the IR [reaction (29)], visible [reaction (30)], or UV.

Chemiluminescent reactions accompanied by emission of a photon in the visible or UV can be divided quite distinctly into recombination with radiation and exchange chemiluminescence. The first group can be subdivided into radiative recombination and three-body recombination with radiation, while radiative recombination, in turn, can be subdivided further into inverse dissociation and inverse predissociation.

2. *Radiative recombination* is the recombination of two particles into a third, whose electronic–vibrational–rotational excitation degrades through emission of a photon:

$$\mathrm{A} + \mathrm{B} \rightarrow \mathrm{AB}^* \rightarrow \mathrm{AB} + h\nu. \tag{32}$$

*Inverse dissociation* (the reverse of dissociation) refers to radiative recombination during which the excited particle (molecule, excimer) is formed in an adiabatic process, i.e., along a single potential surface (curve) (Fig. 1a). Examples include reaction (27) and the reaction

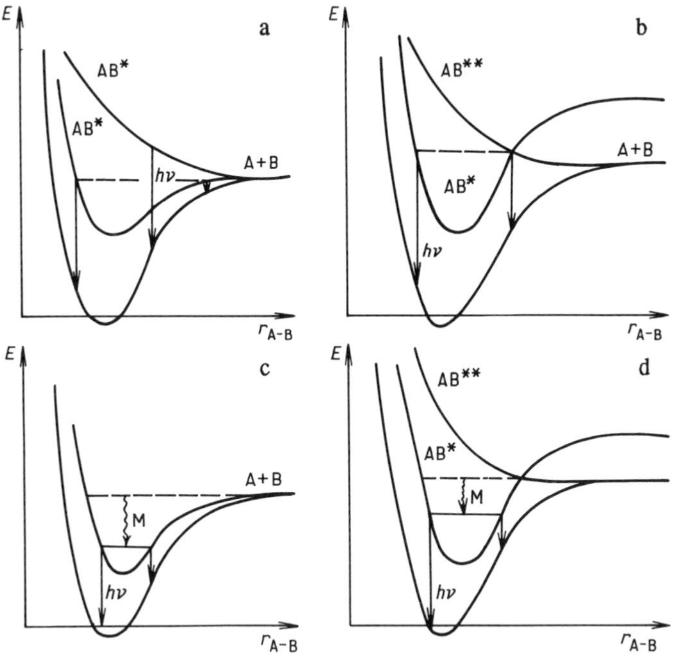


**Fig. 1.** Mechanisms for recombination with radiation: a) inverse dissociation; b) inverse predissociation; c, d) three-body recombination with radiation.

$$O\,(^1D) + CO\,(X\,^1\Sigma^+) \rightarrow CO_2\,(^1B_2) \rightarrow CO_2\,(\tilde{X}\,^1\Sigma_g^+) + h\nu \;[23], \quad \lambda \simeq 300 - 200\,\text{nm}. \tag{33}$$

*Inverse predissociation* (the reverse of predissociation) refers to radiative recombination in which a nonadiabatic transition occurs (Fig. 1b). Examples include

$$O\,(^3P) + N\,(^4S) \rightleftarrows NO\,(a^4\Pi) \rightleftarrows NO\,(C^2\Pi)_{v'=0} + h\nu, \tag{34}$$

with $\lambda$ = 199.5–250 nm, $\delta$ bands of NO, $p_{N_2,O_2} \leq 2.6$ kPa [24], and

$$O\,(^3P) + CO\,(X\,^1\Sigma^+) \rightleftarrows CO_2\,(^3A',\ ^3A'') \rightleftarrows CO_2\,(^1B_2) \rightarrow CO_2\,(\tilde{X}\,^1\Sigma_g^+) + h\nu, \tag{35}$$

with $\lambda$ = 270–800 nm, $p_M < 1.3$ kPa; M = He, Ar, $N_2$ [25–29].

3. *Three-body recombination with radiation* is the recombination of two particles with the participation of a third (a third body), which carries away part of the excitation energy of the molecule (excimer) formed during recombination, and is accompanied by radiation from this molecule (excimer) (Fig. 1c, d). The kinetics of this process may be of third,

$$A + B \xrightarrow{M} AB^* \rightarrow AB + h\nu, \quad (36)$$

second, or even first order, as well as of an intermediate order,

$$A + B \xrightarrow{(M)} AB^* \rightarrow AB + h\nu. \quad (37)$$

Examples include the reaction

$$2N\,(^4S) \xrightarrow{(M)} \ldots \rightarrow N_2\,(B^3\Pi_g)_{v'} \rightarrow N_2\left(A\,^3\Sigma_u^+\right)_{v''} + h\nu, \quad (38)$$

with $\lambda > 500$ nm, the first positive system of $N_2$. For M = He and $p_{He} \leq 2$ kPa, the kinetics are termolecular [30, 31]; for M = He, $N_2$, $p_{He} = 450$–1300 kPa, and $p_{N2} \leq 130$ kPa, the kinetics are intermediate between bi- and termolecular [30, 32]; and, for M = $N_2$ and $p_{N2} > 130$ Pa, the kinetics, are bimolecular [24, 30]. Another example is the reaction

$$2N\,(^4S) \xrightarrow{(M)} \ldots \rightarrow N_2\left(A\,^3\Sigma^+_{u\;v'=0,\;1}\right) \rightarrow N_2\left(X^1\,\Sigma_g^+\right)_{v''} + h\nu; \quad (39)$$

which is a first-order reaction with respect to $[N(^4S)]$ and [M] when $[N(^4S)] \geq 10^{12}$ cm$^{-3}$ and $p_{N2} \leq 0.8$ kPa and a reaction of first order with respect to $[N(^4S)]$ and zeroth order with respect to $[N_2]$ when $[N(^4S)] = 10^{12}$–$10^{13}$ cm$^{-3}$ and $p_{N2} > 1.3 \cdot 10^5$ Pa [33].

4. *Exchange chemiluminescence* includes bimolecular reactions of the types

$$A + BC \rightarrow AB + C^*, \quad (40)$$

$$\rightarrow AB^* + C, \quad (41)$$

$$\rightarrow AB^{\#} + C, \quad (42)$$

and

$$AB + CD \rightarrow AC + B^* + D, \quad (43)$$

$$\rightarrow AC + BD^* \tag{44}$$

accompanied by electronic, electronic-vibrational–rotational, or vibrational–rotational excitation of one of the products [11, 34]. The following are examples of such reactions [34]:

$$Sn\,(^3P_J) + N_2O \rightarrow SnO\,(a\,^3\Sigma^+,\ b^3\Pi) + N_2, \tag{45}$$

$$H + Cl_2 \rightarrow HCl^{\#} + Cl, \tag{46}$$

$$I_2\,(BO_u^+) + F_2 \rightarrow IF\,(BO^+) + IF, \tag{47}$$

and

$$O_3\,(X\,^1A_1) + NO\,(X\,^2\Pi) \rightarrow NO_2\,(^1B_2,\ ^2B_2) + O_2\,(X\,^3\Sigma_g^-). \tag{48}$$

*The spectral distribution of the rate constant for chemiluminescence in reactions*

• *with monomolecular kinetics*, $k_{CL}{}^1(\lambda)$ [photon/(mole·sec·nm)], is given by

$$k_{CL}^1(\lambda) = \frac{J_{CL}(\lambda)}{[A]}. \tag{49}$$

Here and in the following, $J_{CL}(\lambda)$ [photon/(cm$^3$·sec·nm)] represents the intensity of chemiluminescent emission per cm$^3$ of the reaction zone per sec over a solid angle of $4\pi$ sr within a spectral interval of $\Delta\lambda = 1$ nm;

• *with kinetics between bi- and monomolecular*, $k_{CL}{}^1(\lambda, [M])$ [photon/(mole·sec·nm)], is given by

$$k_{CL}^1(\lambda, [M]) = \frac{J_{CL}(\lambda, [M])}{[A]}; \tag{50}$$

• *with bimolecular kinetics*, $k_{CL}{}^2(\lambda)$ [photon·cm$^3$/(mole$^2$·sec·nm)], is given by

$$k_{CL}^2 = \frac{J_{CL}(\lambda)}{[A]\,[B]}; \tag{51}$$

• *with kinetics intermediate between bi- and termolecular*, $k_{CL}{}^2(\lambda, [M])$ [photon·cm$^3$/(mole$^2$·sec·nm)], is given by

$$k_{CL}^2(\lambda, [M]) = \frac{J_{CL}(\lambda, [M])}{[A]\,[B]}; \tag{52}$$

• *with termolecular kinetics*, $k_{CL}{}^3(\lambda)$ [photon·cm$^6$/(mole$^3$·sec·nm)], is given by

$$k^3_{CL}(\lambda) = \frac{J_{CL}(\lambda)}{[A]\,[B]\,[M]}\,. \tag{53}$$

*The rate constant for chemiluminescence in reactions with monomolecular kinetics*, $k_{CL}{}^1$ [photon/(mole·sec)], is given by

$$k^1_{CL} = \frac{J_{CL}}{[A]} = \int_{\lambda_1}^{\lambda_2} k_{CL}(\lambda)\, d\lambda. \tag{54}$$

Here and in the following, $J$ [photon/(cm$^3$·sec)] denotes the intensity of chemiluminescent emission per cm$^3$ of the reaction zone per sec over a solid angle of $4\pi$ sr within the spectral interval $\lambda_1 - \lambda_2$ over which chemiluminescence occurs.

The rate constants for chemiluminescence in reactions with different kinetic orders are defined analogously as the integral over the spectral interval $\lambda_1 - \lambda_2$ of the spectral distribution.

## 3. PHOTODISSOCIATION AND RADIATIVE RECOMBINATION IN DIATOMIC MOLECULES

Published data on photodissociation of these molecules are, as a rule, extremely fragmentary; the same is true of data on chemiluminescence associated with recombination of their fragments. We discuss these data in the following order: the discussion proceeds by individual molecules. First a brief general review is given, then photoprocesses in isolated absorption bands of the molecule are examined, beginning with the lowest photon energies, and, finally, recombination processes with radiation leading to population of the same states of the molecules. Computational, spectroscopic, photochemical, and chemiluminescence data are invoked in the discussion. A brief, combined analysis of all these data is made to the extent possible.

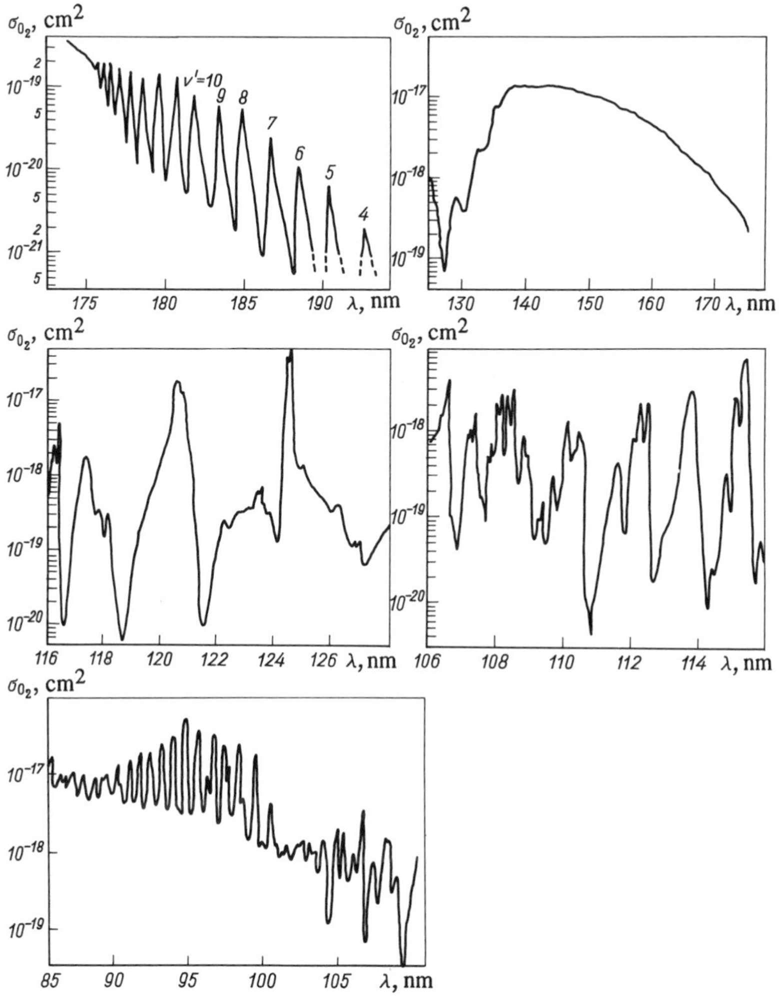


Fig. 2. The absorption spectrum of $O_2$ [35].

### 3.1. Molecular Oxygen [1, 4, 6, 9, 24, 35–60]*

Absorption in molecular oxygen becomes noticeable beginning at roughly 200 nm, which corresponds to the 0–0-band of the $B^3\Sigma_u^- \leftarrow X^3\Sigma_g^-$ transition (Fig. 2).† In this spectral region, i.e., at $\lambda$ = 200 nm, fairly precise measurements have been made of the spectral dependence of the absorption cross section of $O_2$. (See [1, 4] and the literature cited there.) At lower photon energies, transitions into the $a^1\Delta_g$, $b_1\Sigma_g^+$, $c^1\Sigma_u^-$, $C^3\Delta_u$, $A^3\Sigma_u^+$, and, possibly, $^3\Pi_u$ states may occur. Since optical transitions among these states and from them into the ground state are forbidden, information on the assignment of bands, spectroscopic constants, and the lifetimes of these states is incomplete [36–39]. The most exact potential energy curves for the various states of $O_2$ have been calculated by Gilmore [40] and Michels [41] on the basis of experimental data. Several of these curves that are of interest for photodissociation and radiative recombination processes are shown in Fig. 3. Nonempirical calculations of a large number of excited states of $O_2$ have been reported [43, 44, 47–49]; their accuracy, unfortunately, is far from what might be desired.

*Photodissociation of* $O_2$ [1, 4, 42, 51–55]. For photon energies below the first ionization potential of $O_2$ (12.04 eV, $\lambda$ = 102.7 nm [13]), the following photodissociation processes are energetically possible for $O_2$:

$$O_2(X^3\Sigma_g^-) + h\nu \rightarrow \quad \lambda_{thr}\ (\text{nm})$$

$$\rightarrow 2O(^3P) \quad 242.4, \tag{55}$$

$$\rightarrow O(^3P) + O(^1D) \quad 175.9, \tag{56}$$

$$\rightarrow O(^1D) + O(^1D) \quad 136.6^*, \tag{57}$$

$$\rightarrow O(^3P) + O(^1S) \quad 133.2, \tag{58}$$

$$\rightarrow O(^1D) + O(^1S) \quad 110.0^*, \tag{59}$$

where an asterisk denotes a transition forbidden by the Wigner rule and the values of $\lambda_{thr}$ are taken from [1, 4, 13].

*Here and in the following the section headings include references to publications in which data on photodissociation and radiative recombination processes in the given molecules are discussed.

†In Fig. 2 and subsequently in the absorption spectra, $k_A(\lambda)$ (cm$^{-1}$) represents the absorption coefficient in the formula

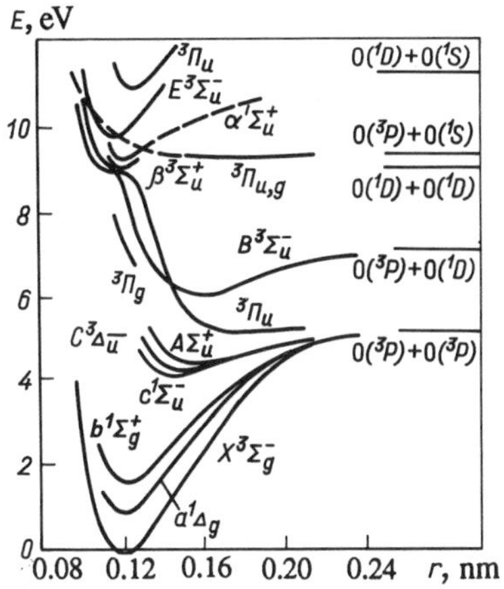


**Fig. 3.** Potential energy curves of the $O_2$ molecule constructed from data in [40-48].

*The spectral region* $\lambda$ = 242.4–175.9 nm [51]. Very little work has been done on photochemistry in the spectral range from the threshold for dissociation of $O_2$ to 184.9 nm and the resulting data cannot be regarded as reliable. Ogawa [50] has measured the cross section for absorption between rotational lines of the Schumann–Runge absorption bands for $\lambda$ = 235–181.4 nm, where the Schumann–Runge bands and the continuum apparently overlap, which corresponds to the transition (55) under the assumption that the values of $\sigma_{O_2}(\lambda)$ obtained in this way refer to the latter. The spectral dependence of the absorption cross section of $O_2$ in the Schumann–Runge bands as a function of the pressure of $O_2$ and the ratios of the lifetimes of the rotational levels with respect to the radiative transition and predissociation through states that converge on $2O(^3P)$ were determined there [50]. Thus, the variation in the absolute quantum yield for formation of $O(^3P)$ by photolysis of $O_2$, $\Phi_{^3P}^{O_2}(\lambda)$, which should depend

---

$$J = J_0 \exp\left[-\frac{k_A(\lambda)\, p_A l \cdot 273}{760T}\right],$$

where $p_A$ is the pressure of the molecules (Torr), $l$ is the length (cm), and $T$ is the temperature (K); $\sigma_A(\lambda)$ (cm$^2$) is the absorption cross section in the formula

$$J = J_0 \exp\left[-\sigma_A(\lambda)\,[A]\,l\right];$$

$k_A(\lambda) = \sigma_A(\lambda)\cdot 2.69\cdot 10^{-19}$, and $\sigma_A(\lambda) = k_A(\lambda)\cdot 3.72\cdot 10^{-20}$ [1].

**TABLE 1.** Photodecay Processes in Diatomic Molecules Contained in the Atmosphere

| Reaction No. | Photoprocess | Notation for quantum yield | λ, нм | Quantum yield | Limitations (pressure, kPa) | Category of data | References |
|---|---|---|---|---|---|---|---|
| (55) | $O_2(X^3\Sigma_g^-)+h\nu\rightarrow 2O(^3P)$ | $\varphi^{O_2}(\lambda)$ | 184.9 | 1.0±? | $P_{O_2}<10^2$ | RD | [51] |
| | | | 193.1—175.9 | 1.0 | $P_{O_2}<10^2$ | RD | [1] |
| (56) | $O_2(X^3\Sigma_g^-)+h\nu\rightarrow O(^3P)+O(^1D)$ | $\Phi^{O_2}_{^1D}(\lambda)$ | 177—116 | Fig. 4 | ? | RD, error ±10% | [52] |
| | | $\Phi^{O_2}_{^1D}(\lambda)$ | 175.9—162.6 | 1.00—0.03 | $P_{O_2}\leqslant 10^2$ | RD | [53] |
| | | $\Phi^{O_2}_{^1D}(\lambda)$ | 162.6—158.5 | >0.97 | $P_{O_2}<10^2$ | RD | [53] |
| | | $^H\Phi^{O_2}_{^1D}$ | 175.9—140.0 | $0.98^{+0.02}_{-0.03}$ | $P_{O_2}<10^2$ | | [54] |
| (55)—(59) | $O_2(X^3\Sigma_g^-)+h\nu\rightarrow O+O$ | $\Phi^{O_2}_{O}(\lambda)$ | 165—115 | Fig. 4 | ? | RD, error ±10% | [52] |
| (58), (59) | $O_2(X^3\Sigma_g^-)+h\nu\rightarrow O(^1S)+O$ | $\Phi^{O_2}_{^1S}(\lambda)$ | 121—80 | Fig. 5 | $p<1.3$ | RD | [42, 55] |

**Note.** All data were obtained for $T = 300$ K. RD denotes recommended data.

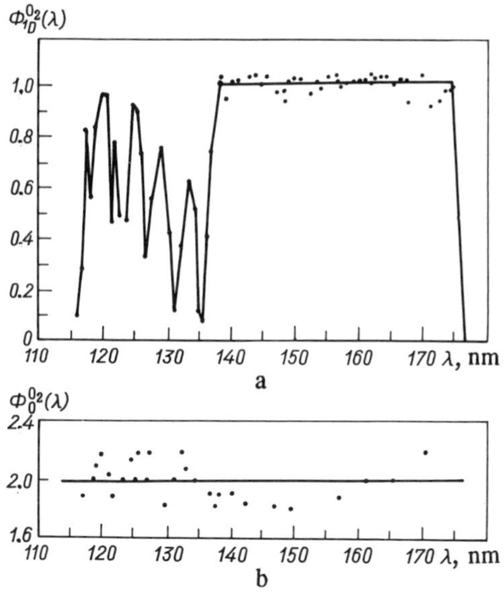


**Fig. 4.** Spectral dependences of the absolute quantum yields for formation of $O(^1D)$, $\Phi_{^1D}(\lambda)$ (a), and of oxygen atoms $\Phi_O(\lambda)$ $\Phi_{^3P}(\lambda) + \Phi_{^1D}(\lambda) + \Phi_{^1S}(\lambda)$ (b), during photolysis of $O_2$ [52] ($T = 293$ K).

on the pressure and may have a very sharp structure in this spectral range, has not been established in the range $\lambda = 242.4$–$193.1$ nm. For $\lambda \leq 193.1$ nm, which corresponds to optical transitions in $O_2(B^3\Sigma_u^-)_{v' \geq 4}$, the spectroscopic data [1] show that predissociation is observed and that $\Phi_{^3P}{}^{O_2}(\lambda) = 2$ (Table 1). Photochemical data for $\lambda = 193.1$ nm do not confirm this [51], although they appear to be in error.

*The Schumann–Runge continuum*, $\lambda = 175.9$–$133$ nm [52–54]. Recent photochemical studies [52–54] show that in this spectral range the major transitions are into the $B^3\Sigma_u^-$ states and into Rydberg states which predissociate into the $B^3\Sigma_u^-$ state (Figs. 3 and 4). The idea [46] that transitions into the valence states converging to $2O(^3P)$, in particular to the $^3\Pi_g$ and $^3\Pi_u$ states, has not been confirmed either experimentally or computationally [53, 54]. (Note that the transition $^3\Pi_g \leftarrow {}^3\Sigma_g$ is strictly forbidden [2].) Thus, in this spectral range, the absolute quantum yield for formation of $O(^1D)$, $\Phi_{^1D}{}^{O_2}(\lambda)$, is extremely close to unity (Fig. 4; Table 1). More details on the photochemistry and spectroscopy of $O_2$ in this spectral range are given elsewhere [53, 54].

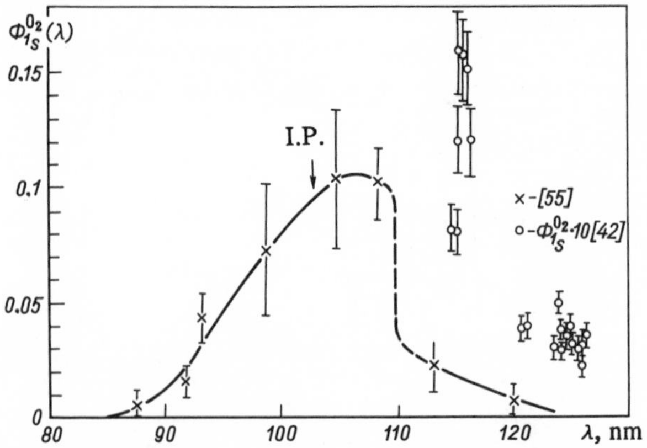


**Fig. 5.** Spectral dependence of the absolute quantum yield for formation of $O(^1S)$ during photolysis of $O_2$ [42, 55] ($T$ = 293 K). (IP is the ionization potential.)

*The spectral range* $\lambda$ = 133–102.7 nm [42, 55]. Absorption by molecular oxygen in this range corresponds to transitions into Rydberg states which converge to the ground state of the ion, $O_2^+(X^2\Pi_g)$; in the interval $\lambda$ = 133–113 nm, these are the $\beta^3\Sigma_u^+$, $\alpha^1\Sigma_u^+$, $E^3\Sigma_u^-$, and $^3\Pi_u$ states [42]. The major channel for degradation of the excitation energy of these states is predissociation through the valence states converging to $2O(^3P)$ and $O(^3P) + O(^1D)$, for example, $^3\Pi_u$ and $B^3\Sigma_u^-$, and for the $E^3\Sigma_u^-$ state, the nonadiabatic transition $E^3\Sigma_u^- \rightarrow B^3\Sigma_u^-$ [42] (see Fig. 3). As a consequence, $\Phi_{1D}^{O2}(\lambda)$, and $\Phi_{3P}^{O2}(\lambda)$ are considerably higher than the absolute quantum yield for formation of $O(^1S)$, $\Phi_{1S}^{O2}(\lambda)$, (Figs. 4 and 5). The absolute quantum yield for fluorescence of $O_2$ is negligibly small: $\Phi_L^{O2}(\lambda) < 10^{-2}$ [52]. The large difference in the rate of predissociation of the Rydberg states through channels (55) and (56), on the one hand, and channels (58) and (59), on the other, is caused by the existence of a strong selection rule for predissociation, $g \leftarrow / \rightarrow u$, and by the heterogeneity of predissociation of the above-mentioned Rydberg states through the valence states $^3\Pi_{u,g}$ which converge to $O(^3P) + O(^1S)$. The restriction on predissociation of these Rydberg states through channels (55) and (56) is considerably weaker [42].

*Chemiluminescence during recombination of atomic oxygen* [24, 56–58, 60]. Reports dealing with chemiluminescence during recombination of oxygen atoms can be divided into two groups. The first (more numerous) group is devoted to studies of afterglow spectra of $O_2$, which are assumed to be caused by chemiluminescence during recombination of oxygen atoms [59], while the second group is devoted to studies of the kinetics of chemiluminescence. Here we shall be interested only in papers from the second group [56–58, 60].

Up to now, chemiluminescence in reactions of two $O(^3P)$ atoms has been observed and, to some extent, studied. A chemiluminescent reaction should also occur at a fairly high rate with a small negative activation energy during collisions of $O(^1D)$ and $O(^3P)$ atoms:

$$O(^1D) + O(^3P) \xrightarrow{(M)} O_2(B^3\Sigma_u^-) \rightarrow O_2(X^3\Sigma_g^-) + h\nu, \qquad (60)$$

with $\lambda \geq 175.9$ nm (Schumann–Runge bands) and $k_{60} \simeq 10^{-35}$–$10^{-34}$ photons·cm$^6$/(mole·sec) ($T = 300$ K) for [M] $\rightarrow$ 0, which is close to the rate constant for chemiluminescence during collisions of $N(^4S)$ atoms (see below). Eight triplet states of $O(^3P) + O(^1D)$ converge to the products $O_2$ and only one of them, the $B^3\Sigma_u^-$ state, is bound. Thus, the value of $k_{60}$ may be close to 1/8 times the recombination rate for two atoms, i.e., ~$1/8 \cdot n \cdot 10^{-33}$ cm$^6$/sec ($n \leq 10$)[61]. Some of the states which converge to $O(^1D) + O('D)$ are bound [49]; thus, radiative transitions into the lower singlet states of $O_2$ may also be observed during collisions of these particles. The states converging to $O(^3P) + O(^1S)$ are repulsive [47, 48]; thus, the rate constant for photorecombination of these particles should have a large activation energy [62]. None of these processes have been observed experimentally.

*Chemiluminescence during recombination* of $O(^3P)$ [24, 56–58, 60]. These processes are among the least studied. The kinetic data obtained for these processes can be considered semiquantitative at best, since the spectral distributions of their rate constants have not been measured (except by Myers and Bartle [58]). The effect of the experimental conditions on these rates has also not been studied.

We observe that $6 \times 3 = 18$ singlet, triplet, and quintet states converge to the limit $2O(^3P)$: at least five (not including the ground $X^3\Sigma_g^-$ state) of these, $a^1\Delta_g$, $b^1\Sigma_g^+$, $C^3\Delta_u$, $A^3\Sigma_u^+$, and $c^1\Sigma_u^-$, are bound and at temperatures close to 300 K they can be populated with a substantial probability during collisions of two $O(^3P)$ atoms. Kinetic data are available for only two processes:

$$2O(^3P) \xrightarrow{(M?)} O_2(b^1\Sigma_g^+) \rightarrow O_2(X^3\Sigma_g^-) + h\nu \qquad (61)$$

the atmospheric bands of $O_2$ at $\lambda \simeq 762$ nm, and

$$2O(^3P) \overset{(M?)}{\rightarrow} O_2(A^3\Sigma_u^+) \rightarrow O_2(X^3\Sigma_g^-) + h\nu \quad (62)$$

the Herzberg I bands at $\lambda = 400$–$254$ nm.

The transitions $c^1\Sigma_u^- \rightarrow X^3\Sigma_g^-$, $C^3\Delta_u \rightarrow X^3\Sigma_g^-$, and $C^3\Delta_u \rightarrow a'\Delta_g$ [59] have also been observed in an oxygen afterglow; however, since no kinetic studies were carried out, it is unclear whether the upper states of these transitions are products of the recombination of $O(^3P)$ or excited $O_2$ molecules were transported out of the discharge.

According to the data of Young et al. [24, 56], at temperatures close to 300 K the mechanism for process (61) is described by the reactions

$$2O(^3P) \overset{M}{\rightarrow} O_2(b^1\Sigma_g^+), \quad (63)$$

$$O_2(b^1\Sigma_g^+) \rightarrow O_2(X^3\Sigma_g^-) + h\nu, \quad (64)$$

and

$$O_2(b^1\Sigma_g^+) + M \rightarrow O_2(?) + M, \quad (65)$$

with $k_{63} \simeq 1.1 \cdot 10^{-37}$ cm$^6$/sec, M = $N_2$ [24], $k_{64} \simeq 11$ sec$^{-1}$ [5, 39], $k_{65} \simeq 2 \cdot 10^{-15}$ cm$^3$/sec, M = $N_2$, and $k_{65} \cong 1.5 \cdot 10^{-16}$ cm$^3$/sec, M = $O_2$ [5], while that for process (62) is described by the reactions

$$2O(^3P) \overset{(M?)}{\rightarrow} O_2(A^3\Sigma_u^+), \quad (66)$$

$$O_2(A_3\Sigma_u^+) \rightarrow O_2(X^3\Sigma_g^-) + h\nu, \quad (67)$$

and

$$O_2(A^3\Sigma_u^+) + M \rightarrow O_2(?) + M, \quad (68)$$

with $k_{66} \simeq 2 \cdot 10^{-37}$ cm$^6$/sec, M = $N_2$ [56], $k_{67} \simeq 1$ sec$^{-1}$ [5], and $k_{68}$ = ?. Thus, we have

$$k_{61} = \frac{k_{63}k_{64}}{k_{64} + k_{65}[M]} \simeq \begin{cases} 1.1 \cdot 10^{-37} \text{ photons·cm}^6/\text{(mole·s) for } k_{65}[N_2] \ll k_{64}, \\ 0.6 \cdot 10^{-21} \text{ photons·cm}^3/\text{(mole·s) for } k_{65}[N] \gg k_{64}. \end{cases}$$

The data from [24] and [56] yield

$$k_{62} \simeq \begin{cases} 2\cdot 10^{-37} \text{ photons}\cdot\text{cm}^6/(\text{mole}\cdot\text{s}) \text{ for } p_{N_2} < 40 \text{ Pa}, \\ 2\cdot 10^{-21} \text{ photons}\cdot\text{cm}^3/(\text{mole}\cdot\text{s}) \text{ for } p_{N_2} > 40 \text{ Pa}. \end{cases}$$

The above data are in need of verification, since:

1) $k_{63}$ and $k_{66}$ are $10^4$ times smaller than $k_{69} = 10^{-33}$ cm$^3$/sec for M = $O_2$ [5, 61] in the reaction

$$2O(^3P) \xrightarrow{M} O_2(X^3\Sigma_g^-)\,; \tag{69}$$

2) the chemiluminescence spectrum in reaction (62) depends [57] on the composition of the He(Ar) + $O_2$ mixture (see [60], as well), and

3) the kinetic data pertaining to the influence of $O_2$ on $k_{62}(\lambda)$ are contradictory [56, 57, 60].

Chemiluminescence in the reaction

$$2O(^3P) \xrightarrow{M} O_2(B^3\Sigma_u^-) \rightarrow O_2(X^3\Sigma_g^-) + h\nu \tag{70}$$

over $\lambda$ = 230–450 nm in the Schumann–Runge bands (Table 2, Fig. 6) has been observed in a shock tube according to Myers and Bartle [58]. They assumed that the $B^3\Sigma_u^-$ state is populated by inverse predissociation induced by collisions of two $O(^3P)$ atoms through

$$2O(^3P) \rightleftarrows O_2^{\bullet}\,, \tag{71}$$

with

$$O_2^{\bullet} + M \rightarrow O_2(B^3\Sigma_u^-) + M, \tag{72}$$

$$O_2(B^3\Sigma_u^-) + M \rightarrow O_2(X^3\Sigma_g^-) + M, \tag{73}$$

and

$$O_2(B^3\Sigma_u^-) \rightarrow O_2(X^3\Sigma_g^-) + h\nu. \tag{74}$$

We consider the mechanism for populating $O_2(B^3\Sigma_u^-)$ to be unproven, since they [58] did not observe the competition among the reactions (72), (73), and (74).

**TABLE 2.** Recombination Processes with Radiation Involving Fragments of Diatomic Molecules Contained in the Atmosphere

| Reaction No. | Reaction | Notation for rate constants | M | $k^3_{CL}$ | $k^2_{CL}$ | Limits on pressure (Pa), temperature (K) | Category of data | References |
|---|---|---|---|---|---|---|---|---|
| (61) | $2O(^3P) \xrightarrow{(M?)} O_2(b^1\Sigma_g^+) \rightarrow O_2(X^3\Sigma_g^-) + h\nu$ | $k^2_{CL}$, $k^3_{CL}$ | $N_2$ | $2 \cdot 10^{-37}$ | — | $p_{N_2} < 1.3 \cdot 10^{-2}$, $T = 293$ | 1 | [24] |
| | | | $N_2$ | — | $10^{-21}$ | $p_{N_2} > 1.3$, $T = 293$ | 1 | Calculated (see text) |
| (62) | $2O(^3P) \xrightarrow{(M?)} O_2(A^3\Sigma_u^+) \rightarrow O_2(X^3\Sigma_g^-) + h\nu$ | $k^2_{CL}$, $k^3_{CL}$ | $N_2$ | $2.1 \cdot 10^{-37}$ | — | $p_{N_2} \ll 40$, $T = 293$ | 1 | [56] |
| | | | $N_2$ | — | $2.3 \cdot 10^{-21}$ | $p_{N_2} \gg 40$, $T = 293$ | 1 | [56] |
| | | | $O_2$ | — | $2.5 \cdot 10^{-21}$ | $p_{O_2} = 133 - 1330$, $T = 293$ | 1 | [57] |
| | | | Ar<br>He | — | $3.7 \cdot 10^{-21}$<br>$1.8 \cdot 10^{-21}$ | $p_M \gg p_{O_2}$, $T = 293$ | 1 | [57] |
| (70) | $2O(^3P) \xrightarrow{(M?)} O_2(B^3\Sigma_u^-) \rightarrow O_2(X^3\Sigma_g^-) + h\nu$ | $k^2_{CL}(\lambda)$ (Fig. 6), $k^2_{CL}$ | Ar, $O_3$, $O_2$ | — | $1.8 \cdot 10^{-18} \times \exp\left(-\frac{28.9}{RT}\ \text{kcal}\right)$ | $p_M = (5-20) \cdot 10^4$, $T = 2500-3800$ | 2 | [58] |

| | | | | | | | | |
|---|---|---|---|---|---|---|---|---|
| (81) | $N(^4S)+O(^3P)\to$ $\to NO(C^2\Pi)_{v'=0}\to$ $\to NO(X^3\Pi)_{v''}+h\nu$ | $k^2_{CL}$ | $O_2$, $N_2$ | — | $1.1\cdot10^{-17}$ | $p_M\leqslant2600$, $T=293$ | 1 | [24] |
| | | | $O_2$, $N_2$ | — | $0.7\cdot10^{-17}\times$ $\times(T/300)^{-0.35}$ | $p_M=?$, $T=300-1500$ | 1 | [65] |
| (82) | $N(^4S)+O(^3P)\to NO(C^2\Pi)_{v'=0}\to$ $\to NO(A^2\Sigma^+)_{v''=0}+h\nu$ | $k^2_{CL}$ | $O_2$, $N_2$ | — | $0.6\cdot10^{-17}$ | $p_M\leqslant2600$, $T=293$ | 2 | Calculated (see text) |
| (83) + (84) | $N(^4S)+O(^3P)\xrightarrow{(M?)}$ $\to NO(A^2\Sigma^+)_{v'=0-2}\to$ $\to NO(X^2\Pi)_{v''}+h\nu$ | $k^2_{CL}$, $k^3_{CL}$ | $O_2$, $N_2$ | $1.2\cdot10^{-17}(T/300)^{-0.35}+$ $+2.1\cdot10^{-34}(T/300)^{-1.24}[N_2]$ | | $T=300-1500$ | 1 | [65] |
| (85) | $N(^4S)+O(^3P)\xrightarrow{(M?)}$ $\to NO(B^2\Pi)_{v'=0-3}\to$ $\to NO(X^2\Pi)+h\nu$ | $k^3_{CL}$ | $O_2$ $N_2$ | $3.1\cdot10^{-34}\times$ $\times(T/300)^{-1.4}$ | — | $T=300-1500$ | 1 | [65] |
| (86) | $N(^4S)+O(^3P)\xrightarrow{(M?)}NO(b^4\Sigma^-)\to$ $\to NO(a^4\Pi)+h\nu$ | $k^2_{CL}$ | $O_2$, $N_2$ | — | $2.3\cdot10^{-17}$ | $T=293$ | 1 | [81] |
| (109) + (110) | $2N(^4S)\xrightarrow{(M?)}N_2(B^3\Pi_g)_{v'=0-11}\to$ $\to N_2(A^3\Sigma_u^+)_{v''=0\div8}+h\nu$ | $k^2_{CL}(\lambda[M])$ (Fig. 11a) | $N_2$, He | — | — | $p_{N_2}=14.6$, $p_{He}=1380$, $T=293$ | ±30 %, RD | [30] |
| | $2N(^4S)\xrightarrow{(M)}N_2(B'^3\Sigma_u^-)_{v'=3-7}\to$ $\to N_2(B^3\Pi_g)+h\nu$ | $k^2_{CL}(\lambda)$ (Fig. 11b) | $N_2$ | — | — | $p_{N_2}>133$, $T=293$ | ±30 %, RD | [30] |
| (109) | | $k^{2,v}_{CL}([M])$ (Fig. 12a, b) | He | — | — | $p_{He}=133-1600$, $T=293$ | ±30 % | [31] See text |

**Table 2.** Continued

| Reaction No. | Reaction | Notation for rate constants | M | $k^3_{CL}$ | $k^2_{CL}$ | Limits on pressure (Pa), temperature (K) | Category of data | References |
|---|---|---|---|---|---|---|---|---|
| (110) | | $k^{2,v}_{CL}([M])$ (Fig. 12c) | He, $N_2$ | — | — | $p_{He}=133\div1600$, $T=293$ | ±30 % | [31] See text |
| (109) + (110) | | $k^2_{CL}([M])$ | He, $N_2$ | — | $2.9\cdot10^{-17}\cdot p_{He}$ Pressure in mm Hg | $p_{N_2}=14,6$, $p_{He}=200\div1600$, $T=293$, $\lambda=500-1050$ nm | RD, ±30 % | [30] |
| | | $k^2_{CL}$ | $N_2$ | — | $(7.0\pm1.4)\cdot10^{-17}$ | $p_{N_2}>260$, $T=293$, $\lambda=500-1050$ nm | RD | [30] |
| (109) + (110) | | $k^2_{CL}$ | $N_2$ | — | $3.1\cdot10^{-17}$ | $p_{N_2}>226$, $T=293$, $\lambda=500-1500$ nm | 1 | [24] |
| (109) | | $k^2_{CL}$ | $N_2$ | — | $1.06\cdot10^{-17}$ | $p_{N_2}>266$, $T=293$, $\lambda=500-1500$ nm | 1 | [85] See text |
| | | $k^2_{CL}([M])$ | Ar, $N_2$ | — | $7.3\cdot10^{-17}$ | $p_{N_2}=9.3$, $p_{Ar}=730$ | 1 | |

| | | | | | | | | |
|---|---|---|---|---|---|---|---|---|
| (110) | | $k^2_{CL}$ | $N_2$ | — | $0.18 \cdot 10^{-17}$ | $T=293$, $\lambda=500$–1500 nm, $p_{N_2}=266$ | 1 | [85] See text |
| | | $k^2_{CL}([M])$ | Ar, $N_2$ | — | $2.9 \cdot 10^{-17}$ | $p_{N_2}=9.3$, $p_{Ar}=740$, $T=293$, $\lambda=500$–1500 nm | 1 | [85] See text |
| (111) | $2N(^4S) \xrightarrow{(M)} N_2(A^3\Sigma_u^+)_{v'=0,1} \rightarrow N_2(X'\Sigma_g^+)+h\nu$ | $k^2_{CL}$ | He | — | $8.2 \cdot 10^{-25}$ | $p_{N_2} \leqslant 13.3$, $p_{He}=266$–1600, $T=293$ | ±30 % RD | [33] |
| | | | $N_2$ | — | $2 \cdot 10^{-24}$ | $p_{He} < 800=$const, $p_{N_2}=130$–800, $T=293$ | ±30 % RD | [33] |
| | | | $N_2$ | — | $4 \cdot 10^{-24}$ | $p_{N_2}=260$–1600, $T=293$ | ⩾30 % RD | [96] |
| (112) | $2N(^4S) \xrightarrow{(M)} N_2(B'^3\Sigma_u^-) \rightarrow N_2(X'\Sigma_g^+)+h\nu$ | $k^2_{CL}$ | $N_2$ | — | $2 \cdot 10^{-21}$ | $p_{N_2}>133$, $T=293$ | 1 | [98] |

**Notes.** (1) Accuracy no better than ×2; (2) Accuracy no better than ×3; RD denotes recommended data.

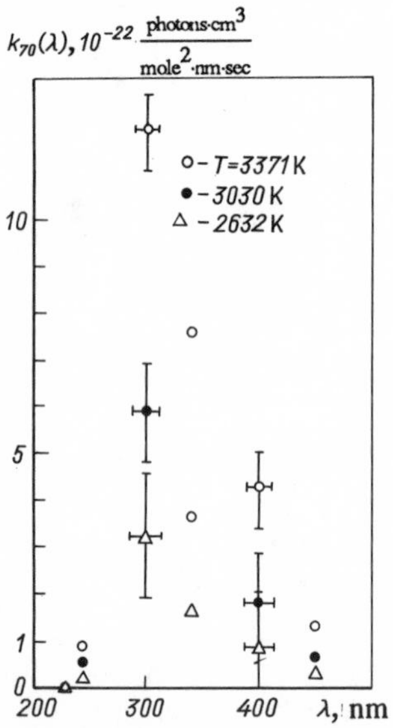


**Fig. 6.** Spectral dependence of the rate constant for chemiluminescence in the reaction $2O(^3P) \xrightarrow{(M?)} O_2(B^3\Sigma_u^-) \rightarrow O_2(X^3\Sigma_g^-) + h\nu$ [58].

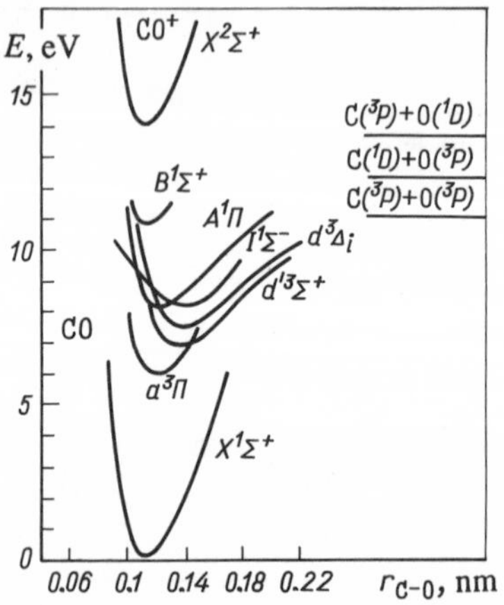


**Fig. 7.** Potential energy curves for the CO molecule [6].

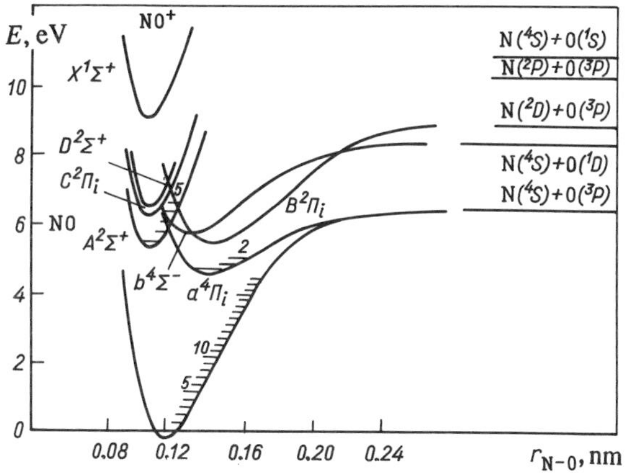


**Fig. 8.** Potential energy curves for the NO molecule [40].

## 3.2. Carbon Monoxide CO

In the spectral region up to the first ionization potential (14.04 eV, $\lambda$ = 88.2 nm), the following photodissociation processes are possible for CO (Fig. 7):

$$\lambda_{thr}\ (\text{nm})$$

$$CO(X^1\Sigma^+) + h\nu \rightarrow C(^3P) + O(^3P) \quad 111.8, \tag{75}$$

$$\rightarrow C(^1D) + O(^3P) \quad 100.1^*, \tag{76}$$

and

$$\rightarrow C(^3P) + O(^1D) \quad 89.8^* \tag{77}$$

where the reactions with an asterisk are forbidden by the Wigner rule.

Of these reactions, only one is allowed by the Wigner rule. Since the threshold for dissociation of CO lies in the far UV, where even the transmission of LiF is low and there are no strong sources of monochromatic radiation, the photodissociation of CO has not been studied. Information on the luminescence of CO in reactions involving electronically excited CO molecules can be found in [1]. Chemiluminescence owing to recombination of fragments of CO has not been investigated.

### 3.3. Nitric Oxide NO [24, 63, 65, 75, 77–79, 81]

The potential energy curves of the ground $X^2\Pi$ state and lower excited states of NO, into which optical transitions are allowed, $A^2\Sigma^+$, $B^2\Pi$, $C^2\Pi$, $D^2\Sigma^+$, etc., have a minimum at similar values of $r_{N-O}$ (Fig. 8); hence, at temperatures close to 300 K the absorption spectrum of NO is caused by transitions into bound states, evidently with a small contribution from transitions into repulsive states. The fluorescence of NO from these states is well studied, their spectroscopic constants and Einstein coefficients are known, and Rydberg–Klein–Rees (RKR) curves have been constructed for them [1, 4, 36–38, 40, 41].

All the rotational levels of the $A^2\Sigma^+{}_{v\,\geq\,4}$, $B^2\Pi_{v\,\geq\,7}$, and $C^2\Pi_{v'\,\geq\,1}$ states predissociate, as do some rotational levels of the $C^2\Pi_{v'\,=\,0}$ state [1, 4]. Since the rate at which they predissociate is not very high (see below), however, under laboratory conditions the photodissociation of NO is usually determined by secondary process involving excited states of NO, rather than by predissociation of these states [1, 4]. In any case, the yield of the photodissociation products of NO has not been measured in any of the papers devoted to the photolysis of NO. We note in conclusion that for photons with energies below the first ionization potential of NO (9.266 eV, $\lambda$ = 133.8 nm), the following photodissociation processes are possible for NO:

$$\lambda_{thr}\ (\mathrm{nm})$$

$$NO\,(X^2\Pi) + h\nu \rightarrow N\,(^4S) + O\,(^3P)\quad 191.5, \tag{78}$$

$$\rightarrow N\,(^4S) + O\,(^1D)\quad 146.4, \tag{79}$$

and

$$\rightarrow N\,(^2D) + O\,(^3P)\quad 139.6. \tag{80}$$

*Chemiluminescence during recombination of fragments of NO* [24, 64, 65, 75, 77–79, 81]. The following chemiluminescence processes involving fragments of NO in the ground states of $N(^4S)$ and $O(^3P)$ have been observed and studied to some extent:

$$N\,(^4S) + O\,(^3P) \rightarrow NO\,(C^2\Pi)_{v'=0} \rightarrow NO\,(X^2\Pi)_{v''} + h\nu, \tag{81}$$

in the $\delta$-bands of NO with $\lambda$ 191–250 nm;

$$N\,(^4S) + O\,(^3P) \rightarrow NO\,(C^2\Pi)_{v'=0} \rightarrow NO\,(A^2\Sigma^+)_{v''=0} + h\nu, \tag{82}$$

with $\lambda \simeq 1224$ nm;

$$N(^4S) + O(^3P) \rightarrow NO(C^2\Pi)_{v'=0} \xrightarrow{h\nu} NO(A^2\Sigma^+)_{v''=0} \rightarrow NO(X^2\Pi) + h\nu, \quad (83)$$

$$N(^4S) + O(^3P) \rightarrow NO(C^2\Pi) \xrightarrow{N_2} \cdots \rightarrow NO(A^2\Sigma^+)_{v'} \rightarrow NO(X^2\Pi) + h\nu, \quad (84)$$

in the γ-bands of NO with $\lambda$ = 200–300 nm;

$$N(^4S) + O(^3P) \xrightarrow{(M)} NO(B^2\Pi) \rightarrow NO(X^2\Pi) + h\nu, \quad (85)$$

in the β-bands of NO with $\lambda$ = 200–430 nm; and

$$N(^4S) + O(^3P) \xrightarrow{(M?)} NO(b^4\Sigma^-) \rightarrow NO(a^4\Pi) + h\nu, \quad (86)$$

in the Ogawa bands with $\lambda$ = 800–1000 nm.

Since bound states converge to $N(^4S) + O(^1D)$ and $N(^2D) + O(^3P)$ (Fig. 8), chemiluminescence should also occur when atoms in these states recombine (see the analysis at the beginning of Sec. 3.) These processes have not yet been observed.

The mechanism for chemiluminescence during recombination of $N(^4S) + O(^3P)$ atoms through the $C^2\Pi$ state is the simplest and has been fairly well studied. Since, in terms of energy, that is the highest state which can be populated during recombination of $N(^4S)$ and $O(^3P)$ ($T \simeq$ 300 K), while cascade transitions can occur during recombination of $N(^4S)$ and $O(^3P)$, it is logical to examine chemiluminescence processes in the system $N(^4S) + O(^3P) + M$ by reviewing published data on reactions (81) and (82).

*The δ-bands of NO, the band at* 1224 nm [24, 64, 65]. The $NO(C^2\Pi)$ state, which does not correlate with the ground states of the initial products (Fig. 8), is populated through the $NO(a^4\pi)$ state. The rotational levels of the $v' = 0$ vibrational level are then populated. The higher vibrational levels are populated with an activation energy $E_{act} > 0$. The short-wavelength limit for chemiluminescence in the δ-bands of NO, therefore, corresponds to the dissociation energy of $NO(X^2\Pi)$ (6.497 eV, $\lambda$ = 191.5 nm). The long-wavelength limit extends all the way to 250 nm [24, 64, 65]. The rate of predissociation of $NO(C^2\Pi)_{v'=0}$ is fairly high ($7 \cdot 10^8$ sec$^{-1}$ [66], $3 \cdot 10^8$ sec$^{-1}$ [67, 68]; also, see [1]). Thus, quenching of the chemiluminescence by third-body molecules is observed only at high pressures (2.6–5.0 kPa) for $N_2$, $CO_2$, CO, and NO [69, 70]:

$$NO(C^2\Pi) + M \rightarrow NO(X^2\Pi) + M, \quad (87)$$

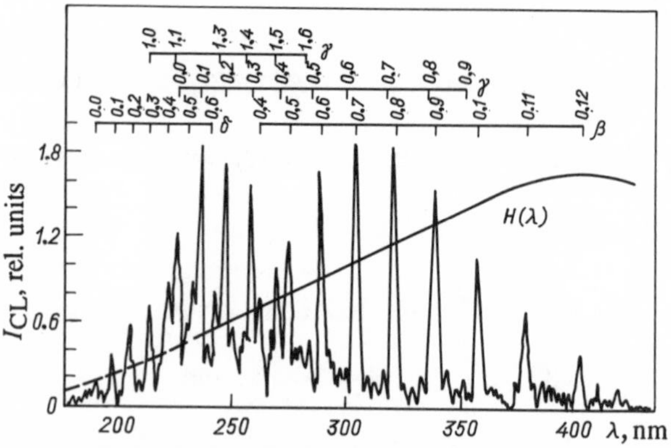


**Fig. 9.** An afterglow spectrum of NO: $p$ = 133 Pa; [O] = [N] = $2 \cdot 10^{14}$ cm$^{-3}$; $T$ = 293 K; $H(\lambda)$ is the relative transmittance of the instrument [65]. The band attributions are shown in the upper part of the figure.

with $k_{87} = (3\text{–}10) \cdot 10^{-10}$ cm$^3$/sec (M = $N_2$, $CO_2$, CO, NO), $k_{87} \ll 10^{-10}$ cm$^3$/sec (M = He, Ar) [71, 72]. When $p_M$ = 2.6–5.0 kPa, the rate of collisionally induced transitions

$$\mathrm{NO}\,(a^4\Pi) \underset{M}{\overset{M}{\rightleftarrows}} \mathrm{NO}\,(C^2\Pi)_{v'=0} \tag{88}$$

is lower than the rate of spontaneous transitions,

$$\mathrm{NO}\,(a^4\Pi) \rightleftarrows \mathrm{NO}\,(C^2\Pi)_{v'=0} \tag{89}$$

[66–68, 71], while vibrational relaxation within the $a^4\Pi$ state makes it impossible to populate $\mathrm{NO}(C^2\Pi)_{v'=0}$ because of energy considerations.

Therefore, radiative recombination in reactions (81) and (82) for $p_M <$ 26 kPa is a typical case of inverse predissociation, reaction (89), and

$$\mathrm{N}\,(^4S) + \mathrm{O}\,(^3P) \rightleftarrows (a^4\Pi), \tag{90}$$

$$\mathrm{N}\,(C^2\Pi)_{v'=0} \rightarrow \mathrm{NO}\,(A^2\Sigma^+)_{v'=0} + h\nu, \tag{91}$$

$$\rightarrow \mathrm{NO}\,(X^2\Pi)_{v''} + h\nu, \tag{92}$$

with $k_{-89} = 7 \cdot 10^8$ sec$^{-1}$, $k_{91} = (3.6 \pm 0.5) \cdot 10^7$ sec$^{-1}$ [72], $k_{92} = 5.1 \cdot 10^7$ sec$^{-1}$ [72], $k_{91} + k_{92} = (7.2 \pm 2.0) \cdot 10^7$ sec$^{-1}$ [66].

None of the references cited above gives a spectral distribution of the rate constant for chemiluminescence; however, an afterglow spectrum of NO, together with a transmission curve for the apparatus, has been published (Fig. 9) [24, 65]. A chemiluminescence spectrum for reaction (82) has been published [73] without correction for the instrument transmission. Data on the rate of $k_{81}$ are given in Table 2. The term $k_{82}$ has not been measured, although presumably it is roughly 1.5 times smaller than $k_{81}$ (see $k_{91}$ and $k_{92}$ in [72]). The activation energy of reactions (81) and (82) is negative, as is to be expected from the potential energy curves of NO given in Fig. 8 (see Table 2).

*The γ-bands of NO* [64, 65, 75]. Studies of the reaction

$$N(^4S) + O(^3P) \xrightarrow{(M?)} NO(A^2\Sigma^+) \rightarrow NO(X^2\Pi) + h\nu \tag{93}$$

are far less complete. The possibility has not been excluded [64, 65, 75] that the main processes describing chemiluminescence in the γ-bands of NO in the presence of M = $N_2$ are reactions (89)–(91) and (94)–(96), i.e.,

$$NO(C^2\Pi)_{v'=0} + N_2 \rightarrow NO(X^2\Pi) + N_2(A^3\Sigma_u^+), \tag{94}$$

$$N_2(A^3\Sigma_u^+) + NO(X^2\Pi) \rightarrow N_2(X^1\Sigma_g^+) + NO(A^2\Sigma_u^+)_{v=0\div 2}, \tag{95}$$

and

$$NO(A^2\Sigma_u^+)_{v'=0\div 2} \rightarrow NO(X^2\Pi)_{v''=0\div 8} + h\nu, \tag{96}$$

in the γ-bands of NO with λ = 200–320 nm.

In the opinion of the authors of [64, 65, 75], the rate constant for chemiluminescence in the γ-bands of NO is given by

$$k_{93} = \frac{J_{93}}{[O]\,[N]} = a + b\,[N_2], \tag{97}$$

where $a = 1.18\cdot 10^{-17}(T/300)^{-0.35}$ photon·cm³/(mole²·sec); $b = 2.12\cdot 10^{-34}$ photon·cm⁶/(mole³·sec) [65], $b/a = 1.8\cdot 10^{-17}$ cm³ [65], $1.02\cdot 10^{-17}$ cm³ [63], and $1.16\cdot 10^{-17}$ cm³ [75].

It is evident that $b$ = const in reactions (89)–(91) and (94)–(96) if $NO(C^2\Pi)$ is not deactivated by collisions with molecules other than $N_2$ and all the $N_2(A^3\Sigma_u^+)$ molecules transfer their excitation energy to $NO(X^2\Pi)$ molecules. Note that the concentration of the latter may be very low. In particular, under steady-state conditions with $[N(^4S)] = [O(^3P)] = 10^{13}$ cm$^{-3}$, $[N_2] = 10^{17}$ cm$^{-3}$ (standard conditions for a homogeneous reactor [25, 30, 76]), and a rate of removal of NO from the reaction zone on the order of $u = 0.1$ sec$^{-1}$, the concentration $[NO] = (5 \cdot 10^{-33}[N][O] \times [N_2])/u \simeq 5 \cdot 10^{11}$ cm$^{-3}$ « $[N(^4S)]$ [61] [here $5 \cdot 10^{-33}$ cm$^6$/sec is the approximate value of the rate constant for formation of NO in the reaction $N(^4S) + O(^3P) + M$]. Since the rate of deactivation of $N_2(A^3\Sigma_u^+)$ in the reaction

$$N_2(A^3\Sigma_u^+) + N(^4S) \rightarrow \text{products} \tag{98}$$

is large ($k_{98} \simeq 5 \cdot 10^{-11}$ cm$^3$/sec [33]), it is clear that under these conditions $b[N_2] \ll a$ for reactions (89)–(91) and (94)–(96), even if $k_{95} = 3 \cdot 10^{-10}$ cm$^3$/sec.

Since the $v' = 0$ vibrational level of the $NO(A^2\Sigma_u^+)$ state is mainly populated in the radiative transition (91) and the $v' = 0$–2 levels, in reactions (94) and (96), we may assume that the chemiluminescence spectrum in the system $N(^4S) + O(^3P) + M$ should depend on the composition of the mixture. This effect, however, has not been investigated. As noted above, only unanalyzed spectra, obtained for $p_{N_2} = 530$ and 33 Pa with different spectral resolutions, are available [24]. Introducing $M = CO_2$, $O_2$, $N_2O$, and $H_2O$ into the composition of the reacting mixture leads to quenching of reaction (93), as the rates of these processes are high [75].

It is obvious that reaction (93) has not been studied in adequate detail: it is unclear whether the $A^2\Sigma^+$ state is populated by inverse predissociation, including that induced by collisions ["replacement" by reactions (94) and (95)], or what the mechanisms and rate constants for these processes are under different experimental conditions (M, $p_M$, $T$).

*The β-bands of NO* [65, 75, 77–79]. Practically all the data on reaction (85), except for the less complete information given in [24, 42, 64, 65], have been obtained by Campbell et al. [75, 77–80]. Young and coworkers [24, 64] found, and it was subsequently confirmed [65], that in the system $N(^4S) + O(^3P) + N_2$ the intensity of chemiluminescence in reaction (85) obeys $J_{85} \propto [O]\,[N]\,[N_2]$. Even then it was noticed that of the four vibrational levels of $NO(B^2\Pi)_{v'=0-3}$ observed in reaction (85), the $v' = 3$ level behaves differently from the others. Neglecting this effect, they obtained

$$k_{85} = \frac{J_{85}}{[O[\;[N]\;[N_2]} = 3.09 \cdot 10^{-34}\;\;(T/300)^{-1.40}$$

photon·cm$^6$/(mole$^3$·sec) [65]. The mechanism for chemiluminescence in the β-bands of NO is the following [24, 64, 65, 80]:

$$N(^4S) + O(^3P) \underset{M}{\overset{M}{\rightleftarrows}} NO^*, \tag{99}$$

$$NO^* + M \rightarrow NO(B^2\Pi) + M, \tag{100}$$

$$NO^* \rightarrow Q \rightarrow \text{product}, \tag{101}$$

$$NO(B^2\Pi) \rightarrow NO(X^2\Pi) + h\nu \tag{102}$$

(here NO* denotes the $b^4\Sigma^-$ and/or $a^4\Pi$ states, and Q is a deactivator particle), with

$$J_{85} = k_{85}[O][N][M] = \frac{k_{99}k_{100}[O][N][M]}{k_{-99} + k_{100} + k_{101}\dfrac{[Q]}{[M]}}. \tag{103}$$

When M = Q = $N_2$ and $k_{100} + k_{101} \gg k_{-99}$, we have

$$J_{85} = \frac{k_{99}k_{100}}{k_{100} + k_{101}}[N][O][M]. \tag{104}$$

This simple mechanism, however, is not completely consistent with later experimental data [75, 77–79], according to which:

a) although the intensity of chemiluminescence from the $v' = 3$ level increases as $p_{N_2}$ is raised, there are, nevertheless, signs that it is being quenched [75, 79] (because of vibrational relaxation?);

b) $J_{85}/[O][N][N_2] \propto T^{-n_v}$, where $n_v = 1.6 \pm 0.2$, $0.18 \pm 0.05$, $0.15 \pm 0.05$, and $1.5 \pm 0.2$ for $v' = 0, 1, 2, 3$, respectively [78, 79]; and

c) $N(^4S)$ and $O(^3P)$ atoms quench chemiluminescence from the $v' = 0$ level. Increasing $[O(^3P)]$ leads to a rise in chemiluminescence from the $v' = 1, 2$ levels, while the intensity of chemiluminescence from $v' = 3$ is independent of $[O(^3P)]$ and $[N(^4S)]$ [77–79].

Most of these peculiarities of the kinetics of reaction (85) have been explained [79] in terms of stepwise filling of $v' = 0$–3 levels of the $NO(B^2\Pi)$ state through the $a^4\Pi$ and $b^4\Sigma$ states (Fig. 8). The specific behavior of $O(^3P)$ and $N(^4S)$ atoms has been explained [77, 78] by the high efficiency of these atoms in reactions with NO* or in stimulating forbidden transitions. Both the experimental data and the mechanism for reaction (85) require additional study.

*Other transitions* [81]. Optical transitions from the $NO(b^4\Sigma^-)$ state have also been observed during recombination of $O(^3P)$ and $N(^4S)$ [81]. This state appears, in particular, in the Ogawa bands [reaction (86)]. The mechanism for this reaction has been studied in most detail by Campbell et

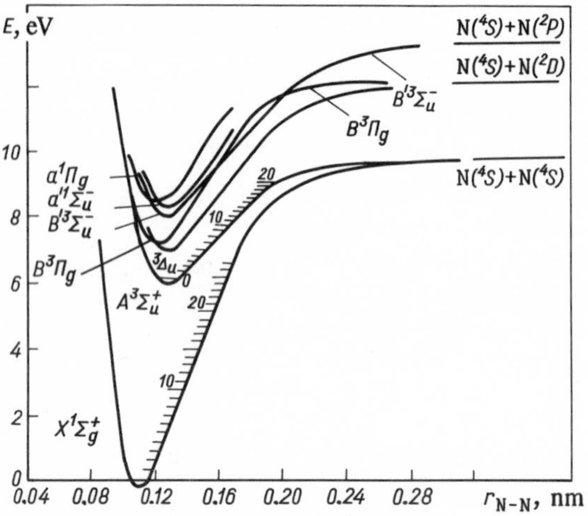

Fig. 10. Potential energy curves for the $N_2$ molecule [40].

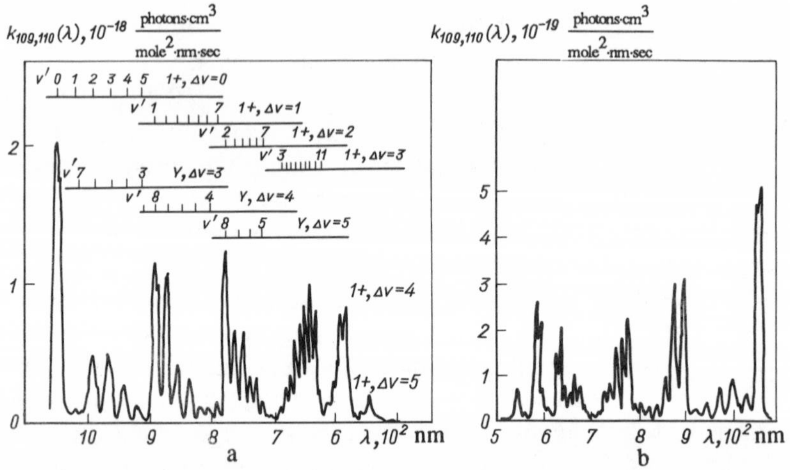

Fig. 11. The spectral distribution of the rate constant for chemiluminescence in reactions (109) and (110). The spectral resolution is 2 nm. The band attributions are shown in the upper part of the figure [$k_{109,110}(\lambda)$ must be multiplied by 2]: a) $p_{N_2}$ = 14.6 Pa, $p_{He}$ = 532 Pa; b) $p_{N_2}$ = 332 Pa, $p_{He}$ = 890 Pa; $T$ = 293 K [30].

al. [81] They found that $k_{86}$ is independent of $p_M$ and equals $2.3 \cdot 10^{-17}$ photon·cm$^3$/(mole$^2$·sec) ($T$ = 300 K). The kinetics and mechanism of this process, as well as of all the others, require further investigation.

*Summary*. The main shortcomings of all the experiments on chemiluminescence in the reaction $N(^4S) + O(^3P)$ are the poor accuracy of the measurements of the atomic densities and the narrow range over which they have been measured. The sources of these shortcomings are the method of obtaining $O(^3P)$ by titration of $N(^4S)$ with nitrous oxide and the lack of a method for measuring $O(^3P)$ in a mixture with $N(^4S)$. Reliable kinetic data on reactions (81)–(86) could be obtained if methods were developed for independently producing $O(^3P)$ and $N(^4S)$ and for measuring their concentrations.

## 3.4. Molecular Nitrogen [8, 9, 24, 30–33, 85, 86, 92, 96–98]

For photon energies below the first ionization potential (15.58 eV, λ = 79.596 nm [13]), the following channels for photodissociation of $N_2$ are energetically possible (Fig. 10):

$$\lambda_{thr}\ (\text{nm})$$

$$N_2(X^1\Sigma_g^+) + h\nu \to N(^4S) + N(^4S)\quad 127.0, \tag{105}$$

$$\to N(^2D) + N(^4S)\quad 102.0, \tag{106}$$

$$\to N(^2P) + N(^4S)\quad 96.7, \tag{107}$$

and

$$\to N(^2D) + N(^2D)\quad 85.7. \tag{108}$$

Since the photodissociation products of $N_2$ – $N(^4S)$, $N(^2D)$, and $N(^2P)$ – are mostly in the ground or metastable states, they are rather difficult to detect during photolysis of $N_2$ in this region of the VUV spectrum. At longer wavelengths, the chemiluminescence methods [82], for example, would be convenient. There are practically no intense monochromatic sources for λ < 120 nm. Thus, the primary photolysis processes for $N_2$ have not been studied.

*Chemiluminescence during recombination of atomic nitrogen* [8, 9, 24, 30–33, 85, 86, 92, 96–98]. Experimental data are available only for chemiluminescence reactions that take place during recombination of nitrogen atoms in the ground state:

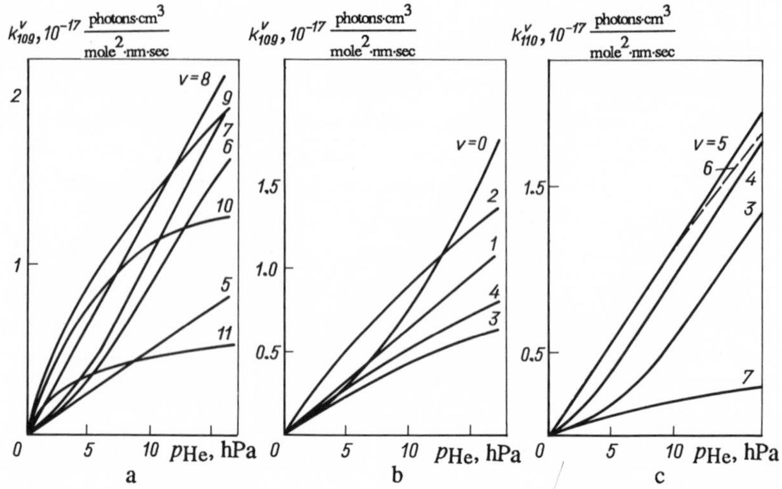


**Fig. 12.** The variation with the helium pressure of the rate constants for chemiluminescence from individual vibrational levels of the $N_2(B^3\Pi_g)_v$ and $N_2(B'^3\Sigma_u^-)_v$ states to all vibrational levels of lower states $k_{109}{}^v$ (a, b) and $K_{110}{}^v$ (c); $p_N = 0$ and $T = 293$ K [$k_{109,110}(\lambda)$ must be multiplied by 2]: a) $v = 5$–$11$; b) $v = 0$–$4$; c) $v = 3$–$7$ [31].

$$2\mathrm{N}\,(^4S) \xrightarrow{(M)} \mathrm{N_2}\,(B^3\Pi_g)_{v'=0-12} \rightarrow \mathrm{N_2}\,(A^3\Sigma_u^+)_{v''=0-8} + h\nu, \tag{109}$$

with $\lambda = 500$–$1050$ nm, the first positive system of $N_2$;

$$2\mathrm{N}\,(^4S) \xrightarrow{(M)} \mathrm{N_2}\,(B'^3\Sigma_u^-)_{v'=3-7} \rightarrow \mathrm{N_2}\,(B^3\Pi_g) + h\nu, \tag{110}$$

with $\lambda = 730$–$1010$ nm, the $Y$ bands of $N_2$;

$$2\mathrm{N}\,(^4S) \xrightarrow{(M)} \mathrm{N_2}\,(A^3\Sigma_u^+)_{v'=0,1} \rightarrow \mathrm{N_2}\,(X^1\Sigma_g^+)_{v''=3-10} + h\nu, \tag{111}$$

with $\lambda = 230$–$350$ nm, the Vegard–Kaplan bands; and

$$2\mathrm{N}\,(^4S) \xrightarrow{(M)} (B'^3\Sigma_u^-) \rightarrow \mathrm{N_2}\,(X^1\Sigma_g^+) + h\nu, \tag{112}$$

with $\lambda = 126$–$142$ nm.

The reactions

$$2\mathrm{N}\,(^4S) \xrightarrow{(M?)} \mathrm{N_2}\,(a^1\Pi_g) \rightarrow \mathrm{N_2}\,(X^1\Sigma_g^+) + h\nu \quad [83,\ 84] \tag{113}$$

with $\lambda = 130$–$180$ nm, the Lyman–Birge–Hopfield bands, and

$$2\mathrm{N}\,(^4S) \xrightarrow{(M?)} \mathrm{N_2}\,(a'^1\Sigma_u^-) \rightarrow \mathrm{N_2}\,(X^1\Sigma_g^+) + h\nu \quad [83,\ 84] \tag{114}$$

with $\lambda$ = 130–180 nm, have also been observed. Quantitative data on these reactions, however, are lacking.

*Chemiluminescence in the visible and near IR* [8, 9, 24, 30–32, 85, 86, 92]. The mechanism for chemiluminescent processes in the system $N(^4S) + N(^4S) + M$ accompanied by emission in the visible and IR [reactions (109) and (110); Fig. 11] is exceedingly complex and is still not known completely. It has been reliably established that the intensity of chemiluminescence from the vibrational levels of the $B^3\Pi_g$ and $B'^3\Sigma_u^-$ states has the form $I_{109,110}^{v}(\lambda) \propto [N(^4S)]^2$ up to $[N(^4S)] = 10^{13}$ cm$^{-3}$ [30]. For $[N(^4S)] > 10^{14}$ cm$^{-3}$ and $p_{N_2} \geq 0.13$ hPa, quenching of the chemiluminescence by nitrogen atoms for a number of the $N_2(B^3\Pi_g)_v$ levels has been observed and attributed to deactivation of intermediate metastable levels of $N_2$ [85, 86]. The dependence of $I_{109,110}^{v}$ on the type of particles serving as a third body and their pressure is very complicated and is different for all the levels. This happens because, immediately after the high vibrational levels of $N_2(A^3\Sigma_u^+)$ are populated by recombination of $N(^4S)$ atoms in the reaction

$$2N\,(^4S) \rightleftarrows N_2\,(A^3\Sigma_u^+)_{v^*}\,, \tag{115}$$

the following processes occur: vibrational relaxation over the $N_2(A, B, B', W(^3\Delta_u))_v$ states,

$$N_2\,(A,\ B,\ B',\ W)_v + M \rightarrow N_2\,(A,\ B,\ B',\ W)_{v-\Delta v} + M \tag{116}$$

(transitions among these states induced by collisions with M with small energy losses,

$$N_2\,(A,\ B,\ B',\ W)_{v_1} + M \xrightarrow{-\Delta E} N_2\,(A,\ B,\ B',\ W)_{v_2} + M, \tag{117}$$

are equivalent to vibrational relaxation [87–89]), the radiative transitions

$$N_2\,(B^3\Pi_g)_{v'} \rightarrow N_2\,(A^3\Sigma_u^+)_{v''} + h\nu \tag{118}$$

and

$$N_2\,(B'^3\Sigma_u^-)_{v'} \rightarrow N_2\,(B^3\Pi_g)_{v''} + h\nu \tag{119}$$

and, finally, deactivation of the $N_2(A, B, B', W)$ states, i.e., collisions with large energy losses,

$$N_2\,(A,\ B,\ B',\ W) + M \rightarrow \text{products}\,. \tag{120}$$

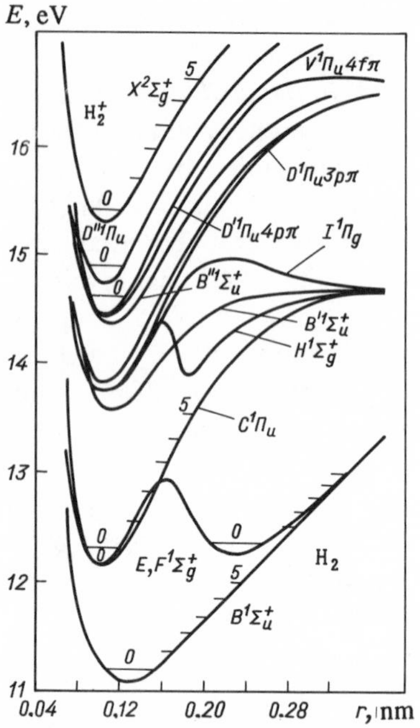


**Fig. 13.** Potential energy curves for the singlet states of the $H_2$ molecule [63].

In the presence of light inert gases, i.e., He [31, 32], Ar [90], and, presumably, Ne, the pressure dependences of the rate constants for chemiluminescence in reactions (109) and (110) are similar (see Fig. 12 and the literature cited in [31]). It may be stated [31, 32] that the upper levels of $N_2(B^3\Pi_g)_{v=6-12}$ and $N_2(B'^3\Sigma_u^-)_{v=2-8}$ are populated by reactions (115)–(117), while the lower levels of $N_2(B^3\Pi_g)_{v=0-4}$ are populated by the radiative transitions (118) and (119), with reaction (118) followed by processes of the type

$$N_2(A^3\Sigma_u^+)_{v=7-11} + M \rightarrow N_2(B^3\Pi_g)_{v=0-3} + M. \quad (121)$$

Collisional deactivation of $N_2(A, B, B', W)$ does not occur, at least in collisions with He [31, 32].

The spectral distribution of the rate constants for chemiluminescence in reactions (109) and (11) can be obtained by using data from [31] (see Fig. 12 and the values of the Einstein coefficients for the transitions (118) and (119) given in [91]). Values of $k_{109}$ and $k_{110}$ are given in Table 2.

Adding the gases M = $N_2$ [30, 31, 92], $O_2$, CO, and $CH_4$ [93], which deactivate $N_2(A, B, B',W)_v$, to an $N(^4S)$ + X (X = He, Ne, Ar) mixture at moderate pressures $p_M$ leads to more complicated chemiluminescence kinetics as new deactivation channels appear. The kinetics of these processes has been examined in most detail for M = $N_2$ [30, 31, 92]. For $p_{He}$ = 0.4–1.6 kPa and $p_N$ = 0–40 Pa, quenching of the chemiluminescence from individual vibrational levels obeys the Stern–Volmer law

$$\frac{{}^0k^v_{109,\,110}}{k^v_{109,\,110}} - 1 = \alpha(v,\, p_{He})\ [N_2],$$

where ${}^0k^v_{109,110}$ is the value of $k^v_{109,110}$ for $p_{N_2}$ = 0. (Values of $\alpha(v, p_{He})$ are given in [92]). When $p_{N_2}$ > 130 Pa, the values of $k_{109}(\lambda)$ and $k_{110}(\lambda)$ are independent of $p_{N_2}$ (see Fig. 11b) because the rates of the processes (116), (117), and (120) induced by collisions with M are much greater than the rates of the spontaneous transitions [in particular, the radiative transitions (118) and (119)]. Deactivation of $N_2(B^3\Pi_g)$ in collisions with $N_2(X^1\Sigma_g^+)_{v=0}$ occurs as a result of the transfer of electronic–vibrational–rotational energy to the initially unexcited molecule:

$$N_2(B^3\Pi_g)_v + N_2(X^1\Sigma_g^+)_{v=0} \rightarrow N_2(X^1\Sigma_g^+)_{v=1-3} + N_2(A,\, B,\, B',\, W)_{v-\Delta v}. \quad (122)$$

*Other data*. Values of $k_{109}$ and $k_{110}$ have been obtained [24] with poor accuracy (Table 2), but they match those given in [30] to within the measurement errors. The values of $k_{109}$ and $k_{110}$ obtained in [85] for an Ar + $N_2$ mixture can be used only for $p_{Ar}$ = 730 Pa and $p_{N_2}$ = 9.3 Pa. We assume that these values, as are those of $k_{55}$ and $k_{56}$ for M = $N_2$, are too low by roughly a factor of 3.

*Chemiluminescence in the UV* [33, 96, 97]. The kinetics of recombination with radiation in the UV is considerably simpler. When M = He (and, apparently, Ne and Ar), the lower vibrational levels $N_2(A^3\Sigma_u^+)$ are populated through radiative transitions from the vibrational levels of $N_2(B^3\Pi_g)$ [33]. Reaction (111) is monomolecular with respect to [$N(^4S)$] because of the large radiative lifetime of $N_2(A^3\Sigma_u^+)_v$ ($\tau_{v'=0,1} \simeq$ 2 sec) [94] and rate constant for deactivation of $N_2(A^3\Sigma_u^+)_v$ through collisions with $N(^4S)$ which is given by $k_{98} \simeq 5\cdot10^{-11}$ sec ($v$ = 0) and $6.5\cdot10^{-11}$ cm$^3$/sec ($v$ = 1) [33, 95]. Light inert gas atoms and $N_2(X^1\Sigma_g^+)$ are poor for deactivating $N_2(A^3\Sigma_u^+)_{v=0-3}$, and the order of reaction (111) with respect to [He] and [$N_2$] is completely determined by the kinetics of the emission from the $N_2(B^3\Pi_g)_v$ level to the $N_2(A^3\Sigma_u^+)_{v=0}$ (M = He) level or of the population, relaxation, and deactivation of $N_2(A, B, B', W)_v$ (M = $N_2$). Chemiluminescence from the vibrational levels $N_2(A^3\Sigma_u^+)_{v>1}$ in the Veg-

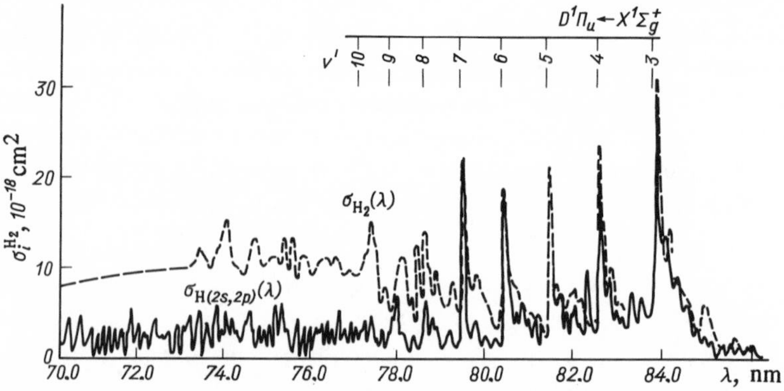


**Fig. 14.** The spectral dependence of the total and partial absorption cross sections of $H_2$, corresponding to photodissociation into H(1*s*) and H(2*s*, 2*p*) ($T = 293$ K) [99].

ard–Kaplan bands has not been observed, apparently because $k_{98}$ ($v > 1$) is large. The report of its occurrence with M = $N_2$ [96] is probably erroneous (see [33, 97]).

*Chemiluminescence in the VUV* [83, 84, 98]. The mechanism for processes leading to filling and emptying of the vibrational levels of the $N_2(a^1\Pi_g, a'^1\Sigma_u^-)$ states is unknown. The rate constant for chemiluminescence in reaction (112) has been estimated (Table 2). In principle, $k_{112}$ could be calculated exactly using data for reaction (110) if the Einstein coefficients for the transitions $N_2(B'^3\Sigma_u^-)_v \rightarrow X(^1\Sigma_g^+)_{v''}$ were known.

## 3.5. Molecular Hydrogen [1, 99–101]

For photon energies below the first ionization potential (15.43 eV, λ = 80.37 nm [13]), only one channel for photodissociation of $H_2$ is possible (Fig. 13):

$$\lambda_{thr}\ (\text{nm})\ (T = 0\ \text{K})$$

$$H_2(X^1\Sigma_g^+) + h\nu \rightarrow H(1s) + H(2s,\ 2p)\quad 84.47. \tag{123}$$

According to the data at our disposal, reaction (123) has been studied in greatest detail by Mentall and Gentien [99]. They found [99] that in the wavelength range λ = 84.47–80.37 nm, the main photoprocess is reaction (123) with $\Phi_H^{H_2}(\lambda) \simeq 1$ (Fig. 14), although fluorescence with a yield of up to $\Phi_{fl}^{H_2}(\lambda) \leq 0.5$ is observed at a few absorption lines of $H_2$ corresponding to transitions into the states $H_2(4p\pi\,^1\Pi_u)$, $5p\sigma\,^1\Sigma_u^+$, $7p\sigma\,^1\Sigma_u^+$, and

$8p\sigma^1\Sigma_u^+$ [100, 101]. H(2*s*, 2*p*) hydrogen atoms are formed with a probability of ~80% in the H(2*s*) metastable state [99], which has a radiative lifetime of $\tau_{rad} \simeq 7$ sec [102]. Collisions with $H_2(^1\Sigma_g^+)$ lead to the transitions H(2*s* → 2*p*) (with an energy difference of $\Delta E \simeq 0.03$ eV between the two states [62]) or H(2*s* → 1*s*):

$$H(2s) + H_2(X'\Sigma_g^+) \rightarrow H(2p) + H_2(X'\Sigma_g^+) \tag{124}$$

and

$$\rightarrow H(1s) + H_2(X'\Sigma_g^+), \tag{125}$$

with $k_{124}/k_{125} = 2$ [99] or 1 [103] ($T = 293$ K) and $k_{124} = 1.4 \cdot 10^{-9}$ $cm^3$/sec [99, 104]. The radiative lifetime of H(2*p*) is given by $\tau_{rad} = 6$ nsec [1].

The only electronically excited state of $H_2$ that converges to H(1*s*) + H(1*s*) is the repulsive $H_2(a^3\Sigma_u^+)$ state. As far as we know, chemiluminescence during collisions of hydrogen atoms in other states has not been studied.

## 4. CONCLUSION

Research on the photodissociation of molecules and on the recombination of their fragments accompanied by radiation is of great significance from a purely scientific standpoint, in addition to its practical interest for aeronomy, the physics and chemistry of plasmas, etc. This research should be quantitative in character and be comprehensive, in the sense of providing a complete description of the properties of the excited states of molecules. This approach makes it possible to obtain a fairly large amount of information about these properties which is needed for creating models of molecular photodissociation and recombination of the fragments and for calculating the potential energy surfaces of these molecules. Since the major products of molecular photodissociation are atoms and radicals in the ground and metastable electronic states, we suggest that in the near future, primary attention should be devoted to developing methods for studying the molecular photodissociation processes involving the production of these fragments and their recombination.

## REFERENCES

1. H. Okabe, *The Photochemistry of Small Molecules*, Wiley, New York (1978).

2. G. Herzberg, *The Spectra and Structure of Diatomic Molecules* [Russian translation], IL, Moscow (1949).
3. G. Herzberg, *The Electronic Spectra and Structure of Polyatomic Molecules* [Russian translation], Mir, Moscow (1969).
4. J. Calvert and J. Pitts, *Photochemistry*, Wiley, New York (1966).
5. M. McEwan and L. Phillips, *The Chemistry of the Atmosphere*, Halsted Press, New York (1975).
6. J. R. McNesby and H. Okabe, *Advances in Photochemistry*, Vol. 3, Wiley, New York (1964), pp. 157–240.
7. J. W. Rabelais, J. M. McDonald, W. Scherr, et al., *Chem. Rev.* **71**, 73–108 (1971).
8. M. Clyne, in: *The Physical Chemistry of Fast Reactions*, B. P. Levitt (ed.), Plenum Press, New York (1973).
9. T. Carrington and D. Garwin, *Excited Particles in Chemical Kinetics* [Russian translation], Mir, Moscow (1973), pp. 123–131.
10. M. F. Golde and B. A. Thrush, *Rep. Prog. Phys.* **36**, 1285–1364 (1973).
11. M. F. Golde and B. A. Thrush, *Adv. At. Mol. Phys.* **11**, 361–409 (1975).
12. *Photochemistry. Specialist Periodical Reports*, Vols. 1–15, Chemistry Society, London (1970–1984).
13. *Binding Energies of Chemical Bonds. Ionization Potentials and Electron Affinity* [in Russian], Nauka, Moscow (1974).
14. J. Berkowitz, *Photoabsorption, Photoionization, and Photoelectron Spectroscopy*, Academic Press, New York (1979).
15. A. M. Pravilov, *Khim. Vys. Energ.* **15**, 343–346 (1981).
16. L. E. Brus and V. E. Bondybey, *J. Chem. Phys.* **65**, 71–76 (1976).
17. W. A. Noyes and P. A. Leighton, *The Photochemistry of Gases*, Dover, New York (1966).
18. A. M. Pravilov and S. E. Ryabov, *Khim. Fiz.* **3**, 939–949 (1984).
19. A. M. Pravilov, I. I. Sidorov, and V. A. Skorokhodov, *Khim. Fiz.* **3**, 537–545 (1984).
20. T. G. Slanger and G. Black, *J. Chem. Phys.* **60**, 468–477 (1973).
21. D. L. Baulch, R. A. Cox, R. F. Hampson, Jr., et al., *J. Phys. Chem. Ref. Data* **9**, 295–471 (1980).
22. I. Deperasinska, J. A. Beswick, and A. Tramer, *J. Chem. Phys.* **71**, 2477–2487 (1979).
23. A. M. Pravilov, S. V. Protopopov, I. I. Sidorov, and V. A. Skorokhodov, *Khim. Fiz.* **3**, 1279–1286 (1984).
24. R. A. Young and R. L. Sharpless, *J. Chem. Phys.* **39**, 1071–1102 (1963).
25. A. M. Pravilov, L. G. Smirnova, and I. O. Sumbaev, *Zh. Fiz. Khim.* **52**, 1863–1866 (1978).
26. A. M. Pravilov and L. G. Smirnova, *Kinet. Katal.* **19**, 1115–1122 (1978).
27. A. M. Pravilov and L. G. Smirnova, *Kinet. Katal.* **22**, 107–115 (1981).
28. A. M. Pravilov and L. G. Smirnova, *Kinet. Katal.* **22**, 559–563 (1981).
29. A. M. Pravilov and L. G. Smirnova, *Kinet. Katal.* **22**, 832–836 (1981).
30. A. F. Vilesov, A. M. Pravilov, I. I. Sidorov, and L. G. Smirnova, *Khim. Fiz.*, No. 10, 1376–1384 (1983).
31. D. S. Bystrov, A. F. Vilesov, A. M. Pravilov, and L. G. Smirnova, *Khim. Fiz.* **6**, No. 9 (1987).
32. A. F. Vilesov, A. M. Pravilov, and L. G. Smirnova, *Khim. Fiz.* **6**, No. 9 (1987).

33. A. M. Pravilov, L. G. Smirnova, and A. F. Vilesov, *Chem. Phys. Lett.* **109**, 343–346 (1984).
34. M. Menzinger, *Adv. Chem. Phys.* **42**, 1–62 (1980).
35. K. Watanabe, in: *Studies of the Upper Atmosphere by Rocket and Satellite* [Russian translation], IL, Moscow (1961), pp. 280–302.
36. R.V. Khokhlov (ed.), *Probabilities of Optical Transitions in Diatomic Molecules* [in Russian], Nauka, Moscow (1980).
37. K. P. Huber and G. Herzberg, *Constants of Diatomic Molecules* [Russian translation], Mir, Moscow (1984).
38. D. K. Hsu and W. M. H. Smith, *Spectrosc. Lett.* **10**, No. 4, 181–262 (1977).
39. S. N. Suchard and J. E. Melzer (eds.), *Spectroscopy Data. Vol. 2. Homonuclear Diatomic Molecules*, IFI/Plenum, New York (1976).
40. F. S. Gilmore, *J. Quant. Spectrosc. Radiat. Transfer* **5**, 369–390 (1965).
41. H. H. Michels, in: *Adv. Chem. Phys.* **45**, Part II, 225–284 (1981).
42. N. K. Bibinov, I. P. Vinogradov, and A. M. Privilov, *Opt. Spektrosk.* **53**, 831–836 (1982).
43. N. H. F. Beeb, E. W. Thulstrup, and A. Andersen, *J. Chem. Phys.* **64**, 2080–2093 (1976).
44. R. J. Buenker and S. D. Peyerimhoff, *Chem. Phys.* **8**, 324–337 (1975).
45. R. J. Buenker and S. D. Peyerimhoff, *Chem. Phys. Lett.* **34**, 225–232 (1975).
46. D. C. Cartwright, W. J. Hunt, W. Williams, at al., *Phys. Rev. A* **8**, 2436–2448 (1973).
47. R. P. Saxon and B. Liu, *J. Chem. Phys.* **73**, 870–875 (1980).
48. R. P. Saxon and B. Liu, *J. Chem. Phys.* **73**, 876–880 (1980).
49. R. P. Saxon and B. Liu, *J. Chem. Phys.* **67**, 5432–5441 (1977).
50. M. Ogawa, *J. Chem. Phys.* **54**, 2550–2556 (1971).
51. N. Washida, Y. Mori, and Y. Tanaka, *J. Chem. Phys.* **54**, 1119–1122 (1971).
52. L. Lee, T. G. Slanger, G. Black, et al., *J. Chem. Phys.* **67**, 5602–5606 (1977).
53. A. M. Pravilov, S. E. Ryabov, and I. O. Shul'pyakov, *Khim. Fiz.* **3**, 680–683 (1984).
54. A. M. Pravilov and I. O. Shul'pyakov, *Khim. Vys. Energ.* **19**, 511–516 (1985).
55. G. M. Lawrence and M. J. McEwan, *J. Geophys. Res.* **78**, 8314–8319 (1973).
56. R. A. Young and G. Black, *J. Chem. Phys.* **44**, 3741–3751 (1966).
57. J. McNeal and S. C. Durana, *J. Chem. Phys.* **51**, 2955–2960 (1969).
58. B. F. Myers and E. R. Bartle, *J. Chem. Phys.* **48**, 3935–3944 (1968).
59. T. G. Slanger, *J. Chem. Phys.* **69**, 4779–4791 (1978).
60. R. D. Kenner, E. A. Ogryslo, and S. Turley, *J. Photochem.* **10**, 199–204 (1979).
61. V. N. Kondrat'ev, *Rate Constants for Gas-Phase Reactions. A Handbook* [in Russian], Nauka, Moscow (1970).
62. B. M. Smirnov and G. V. Shlyapnikov, in: *Plasma Chemistry*, Vol. 3, B.M. Smirnov (ed.) [in Russian], Atomizdat, Moscow (1976), pp. 130–188.
63. A. A. Radtsig and B. M. Smirnov, *Handbook of Atomic and Molecular Physics* [in Russian], Atomizdat, Moscow (1980).
64. R. A. Young and R. L. Sharpless, *Discuss. Faraday Soc.*, No. 33, 228–256 (1962).
65. R. W. F. Gross and N. Cohen, *J. Chem. Phys.* **48**, 2582–2588 (1968).
66. S. Yagi, T. Hikida, and I. Mory, *Chem. Phys. Lett.* **56**, 113–116 (1978).
67. H. J. Smith and F. H. Read, *J. Phys. B* **11**, 3263–3272 (1978).
68. O. Benoist d'Azy, R. Lopez–Delgado, and A. Tramer, *Chem. Phys.* **9**, 327–338 (1975).
69. A. B. Callear and I. W. M. Smith, *Discuss. Faraday Soc.* **37**, 96–111 (1964).

70. A. B. Callear and I. W. M. Smith, *Trans. Faraday Soc.* **61**, 2383–2394 (1965).
71. A. B. Callear and M. J. Pilling, *Trans. Faraday Soc.* **66**, 1886–1906 (1970).
72. A. B. Callear and M. J. Pilling, *Trans. Faraday Soc.* **66**, 1618–1634 (1970).
73. T. W. Dingle, P. A. Freedman, B. Gelernt, at al., *Chem. Phys.* **8**, 171–177 (1975).
74. C. A. Barth, W. J. Schado, and J. Kaplan, *J. Chem. Phys.* **30**, 347–348 (1959).
75. I. M. Campbell and R. Mason, *J. Photochem.* **8**, 321–336 (1978).
76. A. M. Pravilov and L. G. Smirnova, in: *Physicochemical Processes in Low–Temperature Plasmas* [in Russian], Izd. INKhS AN SSSR, Moscow (1985), pp. 86–114.
77. I. M. Campbell and S. B. Neal, *Faraday Discuss. Chem. Soc.*, No. 53, 72–81 (1972).
78. I. M. Campbell and R. S. Mason, *J. Photochem.* **8**, 375–383 (1978).
79. I. M. Campbell, S. B. Neal, M. F. Golde, at al., *Chem. Phys. Lett.* **8**, 612–614 (1971).
80. I. M. Campbell and B. A. Thrush, *Proc. R. Soc. London* **A296**, 222–232 (1967).
81. I. M. Campbell and R. S. Mason, *J. Photochem.* **5**, 383–400 (1976).
82. R. A. Young, G. Black, and T. G. Slanger, *J. Chem. Phys.* **49**, 4769–4796 (1968).
83. M. F. Golde and B. A. Thrush, *Proc. R. Soc. London* **A330**, 79–85 (1972).
84. I. M. Campbell and B. A. Thrush, *Trans. Faraday Soc.* **65**, 32–40 (1969).
85. E. M. Gartner and B. A. Thrush, *Proc. R. Soc. London* **A346**, 121–136 (1975).
86. T. Yamashita, *Sci. Light* **22**, 112–126 (1973).
87. A. Rotem, I. Nadler, and S. Rosenwaks, *J. Chem. Phys.* **76**, 2109–2111 (1982).
88. N. Sadeghi and D. W. Setser, *Chem. Phys. Lett.* **77**, 304–308 (1981).
89. N. Sadeghi and D. W. Setser, *J. Chem. Phys.* **79**, 2710–2726 (1983).
90. R. L. Brown, *J. Chem. Phys.* **52**, 4604–4617 (1970).
91. H. J. Werner, J. Kalcher, and E. A. Reinsch, *J. Chem. Phys.* **81**, 2420–2431 (1984).
92. A. F. Vilesov, A. M. Pravilov, and L. G. Smirnova, *Khim. Fiz.* **6**, No. 10 (1987).
93. K. D. Bayes and G. B. Kistiakowsky, *J. Chem. Phys.* **32**, 992–1000 (1960).
94. D. E. Shemansky, *J. Chem. Phys.* **51**, 689–700 (1969).
95. J. A. Meyer, D. W. Setser, and D. H. Steelman, *J. Phys. Chem.* **74**, 2238–2240 (1970).
96. W. Brennen, R. V. Gutowski and E. C. Shane, *Chem. Phys. Lett.* **27**, 138–140 (1974).
97. J. F. Noxon, *J. Chem. Phys.* **36**, 926–940 (1962).
98. M. F. Golde and B. A. Thrush, *Proc. R. Soc. London* **A330**, 121–130 (1972).
99. J. E. Mentall and E. P. Gentien, *J. Chem. Phys.* **52**, 5641–5645 (1972).
100. J. Breton, P. M. Guyon, and M. Glass–Maujean, *Phys. Rev. A* **21**, 1909–1913 (1980).
101. P. M. Guyon, J. Breton, and M. Glass–Maujean, *Chem. Phys. Lett.* **68**, 314–319 (1979).
102. J. Shapiro and G. Breit, *Phys. Rev.* **113**, 179–186 (1969).
103. F. J. Comes and U. Wenning, *Z. Naturforsch.* **24**, 587–596 (1969).
104. W. L. Fite, R. F. Brackman, D. G. Hummer, et al., *Phys. Rev.* **116**, 363–367 (1959).

# MECHANISMS FOR ELECTRONIC RELAXATION IN ATOM–MOLECULE MEDIA

**A. N. Dvoryankin, L. B. Ibragimov, Yu. A. Kulagin, and L. A. Shelepin**

## 1. INTRODUCTION

Recently a large class of problems has arisen in gas dynamics, plasma physics and chemistry, laser chemistry, photochemistry, etc., which cannot be solved without drawing on information about the kinetics of electronically excited states of atoms and molecules. A great detail of experimental material has been accumulated on electronic processes involving electronically excited particles and the volume of this data increases from year to year. In order to organize access to this material by a large number of users and include these data in the AVOGADRO data base, they must be systematized on the basis of modern ideas about the deactivation of electronically excited atoms and molecules.

The basis of the modern theory of atom–molecule collisions is the adiabatic approximation, according to which it is possible to separate the electronic and nuclear motions because of the large difference in the characteristic times. This, in turn, makes it possible to separate the translational (*T*), rotational (*R*), vibrational (*V*), and electronic (*E*) degrees of freedom. This approach has been extremely successful in the analysis of rotational and vibrational kinetics. The adiabatic approximation, however, clearly fails for processes involving changes in the electronic state of the reagents. In electronically nonadiabatic processes (ENP), mixing of the electronic and nuclear degrees of freedom takes place, so that many channels are typically available and the distribution of reaction products is complicated. The foundations of the theoretical analysis and interpretation of experimental data on ENP have been reviewed elsewhere [1–6]. It is important to emphasize that a rigorous description of ENP involving molecules is possible only when information is available on the structure of the electronic terms, i.e., on the potential energy surfaces (PES) of the collision complex. As opposed to atom–atom collisions, however, the PES correlating with electronically excited states of the reagent and products cannot be calculated yet for polyatomic collision complexes in most cases of practical interest. Thus, in order to describe many ENP it is necessary, in principle, to draw upon experimental information.

Thus, additional demands are placed on experimental measurements of the rates of processes involving ENP, such as the need to monitor the states of the reagents and the final reaction products. Experimental data on the deactivation of electronically excited states contain only the cross sections or rate constants, without any indication of the possible reaction channels or the distribution of the deactivation products over these channels. The effect of nonequilibrium populations in the vibrational and rotational states of the molecules on the integrated cross sections for ENP is also not usually evaluated. Thus, the questions of a well-founded classification of the processes, adequate experimental setup, and obtaining the maximum amount of information from the available set of experimental data are of fundamental importance for the theory of ENP.

The purpose of this paper is to formulate the basic principles for classifying and analyzing ENP in atom–molecule systems while taking the above-mentioned circumstances into account. Specifically, we examine ENP involving quenching and transfer of the energy of electronic excitation in gaseous media containing atoms and diatomic molecules. In the proposed classification scheme the main criterion is $\Delta E_c$, the overall change in the total internal energy (electronic–vibrational–rotational) of the colliding particles. Further classification is based, in accordance with the change in the internal energy of the electronically excited states, on the type of interaction which leads to the ENP, and on other factors. The characteristic features of the processes which appear as a consequence of the ENP mechanism are indicated for the different classification groups.

In many cases, the available experimental data are not sufficient to uniquely establish the mechanism for ENP. As a specific example, we examine experimental values of the rate constants of ENP for singlet oxygen. Some questions about the kinetics of processes which occur in the presence of ENP are also discussed briefly.

## 2. BASIC PRINCIPLES FOR THE CLASSIFICATION OF ENP IN HEAVY PARTICLE COLLISIONS

In general, an ENP during a particle collision has the form

$$A_{n_1} + B_{n_2} \rightarrow A_{n_1'} + B_{n_2'} + \Delta E_c, \tag{1}$$

where $A$ and $B$ are the collision partners, $n_i$ and $n_i'$ are the sets of quantum numbers characterizing the internal states of the particles before and after the collision, and $\Delta E_c = E_{n1} + E_{n2} - (E_{n1}' + E_{n2}')$ is the overall change in the internal electronic–vibrational–rotational energy of the colliding particles. For molecular particles the quantum numbers break up into the groups $n = e, v, J$, which characterize the electronic, vibrational, and rotational states of the molecule.

The basic criteria according to which ENP can be classified may be divided into two groups: phenomenological and dynamical.

*Phenomenological* criteria characterize the type of particles (molecule, atom, ion), the type of excitation of the partners (electronic, electronic–vibrational, etc.), the redistribution of the energy of the particles involved in a collision [the change $\Delta E_c$ in the overall internal energy in the reaction (1), the change $\Delta E_{\text{int}} = E_{n1} - E_{n1}'$ in the internal energy of a particle, the sign of the energy change].

*Dynamical* criteria reflect the specific features of the electronic–vibrational transitions, the type and characteristics of the interparticle interaction leading to an ENP, and the type of nuclear motions that interact effectively with the electronic degrees of freedom. The dynamic criteria are far more complicated to determine experimentally and their interpretation may depend on the theoretical model used to describe ENP.

The dynamical criteria include the following:

1) selection rules: involving conservation of or changes in the quantum numbers that characterize the state of the particles before and after a collision;

2) the temperature dependence of the cross sections for ENP;

3) the dependence of the cross sections on the parameters which determine the various interactions between colliding particles: charges, polarizability, multipole angular momenta, spin angular momenta, atomic number, etc.;

4) the distribution of the products of an ENP over the energetically permitted channels, the width of the energy spectrum of the products;

5) the dependence of the ENP cross sections on the mixing coefficients for different electronic states in a molecule; and

6) the dependence of the differential cross sections for ENP on the kinetic energies of the colliding particles and on the scattering angles.

As our understanding of the mechanisms for ENP develops, the list of dynamical criteria may be refined and supplemented.

In order to systematize the data on deactivation of electronically excited states of diatomic molecules in thermal collisions with neutral particles, here we introduce a classification of ENP that is based on both phenomenological and dynamical criteria. As a basis we have taken the overall change in the internal energy, $\Delta E_c$, and the type of interaction (long- or short-range) leading to the ENP between the collision partners. In terms of the total change in the overall internal energy, $\Delta E_c$, ENP can be divided into resonance ($\Delta E_c < 0.01$ eV), quasiresonance ($\Delta E_c < 0.1$ eV), and nonresonance ($\Delta E_c \geq 1$ eV). Nonresonance processes take place mainly as a result of a short-range exchange-type interaction between the collision partners. In many cases, resonance and quasiresonance processes ($\Delta E_c < 0.1$ eV) occur because of long-range interactions and, as a rule, are characterized by large cross sections ($\sigma \sim 0.5$–$10$ nm$^2$). Further subdivision within this group has been based on the change in the internal energy of an electronically excited diatomic molecule, $\Delta E_{\text{int}} = E_{n1} - E_{n1}'$, during the collision process:

1) $\Delta E_{\text{int}} \leq 0.01$ eV. This amount of energy change is typical for radiationless collisional intramolecular electronic transitions (CIET) between close electronic states of a diatomic molecule. Depending on the type of perturbation leading to mixing of the nearby electronic states between which the collision induces a transition, these processes can be divided into CIET under the influence of internal nonadiabatic perturbations intrinsic to the molecule itself and CIET under the influence of external perturbations caused by the interaction with a collision partner; and

2) $\Delta E_{\text{int}} \geq 1$ eV. Transitions of this type occur during exchange of excitation energy between a diatomic molecule and its collision partner. Depending on the distribution of the initial energy of electronic excitation among the degrees of freedom of the reaction products, we can distinguish *E–E*, *E–EV*, and *E–V* exchange. Dynamical criteria can be used for further classification within these groups. The ultimate purpose of this classification is to choose a set of parameters (determined experimentally or theoretically) which will allow us to find the driving mechanism for an ENP, to describe its characteristic features, and to predict the dependence of the cross section for the ENP on the experimental conditions: tempera-

ture, pressure, degree of excitation and properties of the collision partners. At present, this program cannot be carried out because of incomplete experimental data and an inadequate theoretical understanding of nonadiabatic transitions in polyatomic collision complexes. Thus, in the following we shall be concerned primarily with a limited class of processes for which fairly detailed experimental data are available and an adequate theoretical description exists.

The complexity of analyzing data on the deactivation of electronically excited molecular states, assigning these processes to certain groups, and isolating specific mechanisms by which the electronic energy of molecules relaxes will be demonstrated for the example of reactions involving singlet oxygen.

## 3. RESONANCE AND QUASIRESONANCE ENP

### 3.1. Role of Long-Range Interactions in ENP

The interactions between atomic and molecular particles can be divided roughly into two qualitatively different classes according to their properties: long-range interactions, which are electromagnetic in nature, and short-range interactions of an exchange character which arise when there is significant overlap of the electron shells of the colliding atom–molecule particles. Long-range electrostatic interactions (multipole, polarization, etc.) generally correspond to attraction between the colliding particles. Short-range interactions correspond to repulsion if both collision partners have closed electron shells. In electronically excited atoms and molecules, however, the electron shells are never closed. Thus, if one or both of the collision partners are in electronically excited states, then an electron belonging to one of the particles can completely or partially move to a lower-lying vacant orbital of the other particle when there is significant overlap of the electron shells. This produces strong attractive forces between the particles. In order to find the interaction between colliding particles at small distances, one must calculate the electronic terms of a polyatomic (in general) collision complex, a complicated problem which is at the limit of capability of modern quantum chemistry, even in the case of three atoms (a collision of an atom with a diatomic molecule). At intermediate distances, exchange interactions are determined by the overlap of the electron shells of the colliding particles and have an exponential-power law asymptotic behavior of the type $U_{exch}(R) \sim R^m \exp(-\gamma R)$. Efficient asymptotic methods have been developed for calculating them [4].

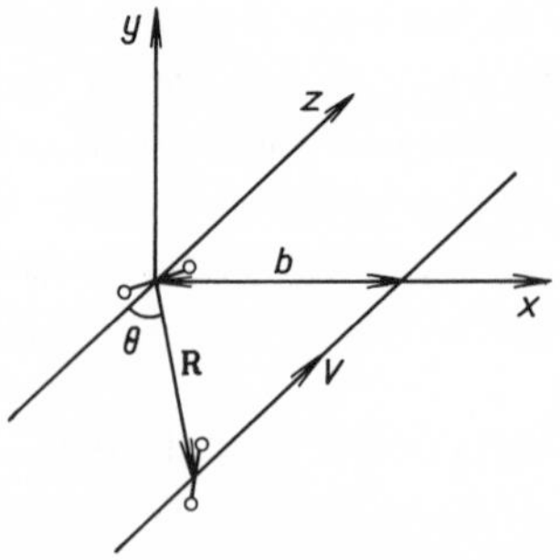


**Fig. 1.** The parameters of the trajectory of the relative motion of colliding particles.

At large distances $R > 1/\gamma$, exchange interactions can be neglected. The interaction energy is determined by the long-range forces and has an inverse power law dependence on the distance, $U_{LR}(R) \sim R^{-n}$.

Determining the type of interaction leading to an ENP plays an important role in the interpretation of experimental data and in constructing an adequate model for the ENP.

As will be shown below, the cross sections for many ENP have certain characteristic features: they depend resonantly on the change in the total internal energy, obey the selection rules for multipole transitions, and correlate with the electrostatic characteristics of the collision partners (polarizability, multipole moments). The combination of these traits makes it possible to attribute them unambiguously to ENP originating in long-range interactions between the colliding particles.

At large distances, the operator for the electrostatic interaction between atom–molecule particles is conveniently expanded in a multipole series in reciprocal powers of the distance between them [7]:

$$U_{AB}(R) = \sum_{l_1, l_2} U_{l_1 l_2} / R^{l_1 + l_2 + 1}. \tag{2}$$

Each term in the sum (2) describes the interaction of the $2l_1$-pole moment of particle $A$ with the $2l_2$-pole moment of particle $B$, and the corresponding coefficients are given by [7]

$$U_{l_1 l_2} = C_{l_1 l_2} \sum_{m, m_1, m_2} \langle l_1 m_1 l_2 m_2 \mid l_1 l_2 l m \rangle \, {}^{A}Q_{l_1 m_1} \, {}^{B}Q_{l_2 m_2} Y^{*}_{lm}(\Omega), \tag{3}$$

where $l = l_1 + l_2$; $C_{l_1 l_2}$ is a numerical coefficient; $\langle l_1 m_1 l_2 m_2 \mid l_1 l_2 l m \rangle$ are the Clebsch–Gordan coefficients; $Q_{lm}$ are the components of the operators of the multipole moments; $Y_{lm}^*(\Omega)$ are the spherical harmonics; and $\Omega \equiv (\varphi, \theta)$ are the angles specifying the direction of the radius vector **R** that characterizes the relative position of the colliding particles.

The main contribution to the cross section for ENP resulting from multipole interactions is from large impact parameters $b$, for which the trajectory of the relative motion of the particles can be regarded as rectilinear: $R = (b^2 + V^2t^2)^{1/2}$, $\varphi = 0$, and $\cos\theta = -Vt/\sqrt{b^2 + V^2t^2}$, where $V$ is the relative velocity of the particles (Fig. 1). The transition probability in this case can be calculated using perturbation theory. This sort of approximation is widely used for calculating the cross sections and rate constants for rotational transitions that take place without changes in the electronic state of the colliding molecules [7–9]. Calculations similar to those of Rabitz and Gordon [9] show that for collisions involving changes in the electronic states of the partners, the probability of an electronic transition is given by first order time-dependent perturbation theory as

$$P_{n,n'} = A_{l_1l_2} \frac{\langle J_1\Lambda_1 l_1\Delta\Lambda_1 \mid J_1'\Lambda_1'\rangle^2 \langle J_2\Lambda_2 l_2\Delta\Lambda_2 \mid J_2'\Lambda_2'\rangle^2 R_l\left(\frac{\omega b}{V}\right)}{V^2b^{2l}}, \qquad (4)$$

where $|n\rangle = |J_1\Lambda_1e_1v_1\rangle\,|J_2\Lambda_2e_2v_2\rangle$ and $|n'\rangle = |J_1'\Lambda_1'e_1'v_1'\rangle \times |J_2'\Lambda_2'e_2'v_2'\rangle$ are the initial and final states of the system of colliding particles. The quantity $A_{l_1l_2}$ determines the efficiency of the transition and is given by

$$A_{l_1l_2} = \frac{2^{2l+1}\,[(l-1)!]^2}{(2l_1+1)!\,(2l_2+1)!\,\hbar^2} \langle e_1\Lambda_1v_1 \mid Q^{l_1}_{\Delta\Lambda_1} \mid e_1'\Lambda_1'v_1'\rangle^2 \langle e_2\Lambda_2v_2 \mid Q^{l_2}_{\Delta\Lambda_2} \mid e_2'\Lambda_2'v_2'\rangle^2, \qquad (5)$$

where $J$, $\Lambda$, $e$, and $v$ are the angular momentum, projection of the electronic angular momentum on the axis of the molecule, and the electronic and vibrational quantum numbers that specify its state, and $\langle e\Lambda v \mid Q_{\Delta\Lambda}{}^l \mid e'\Lambda'v''\rangle$ are the matrix elements of the components of the transition $2l$-pole moment of the molecule.

The dependence of the transition probability on the resonance defect $\Delta E_c$ and the parameters of the trajectory, $V$ and $b$, is given by the resonance function $R_l(\xi)$ which depends on the Massey parameter $\xi$, defined as the product of the transition rate $\omega = \Delta E_c/\hbar$ and the characteristic collision time $\tau = b/V$, or $\xi = \omega\tau = \Delta E_c b/\hbar V$. Typical shapes of the resonance functions for dipole–dipole $R_2(\xi)$ and quadrupole–quadrupole $R_4(\xi)$ interactions are shown in Fig. 2. They are normalized to unity at $\xi = 0$, i.e., $R(0) = 1$. For larger values of $\xi$, they fall off exponentially as $R_2(\xi) \propto$

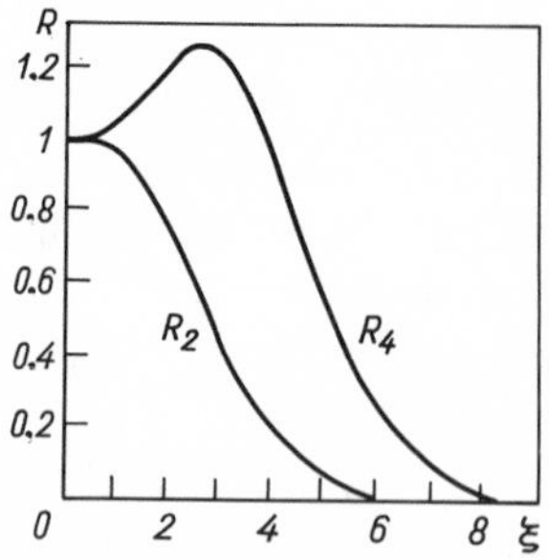


**Fig. 2.** The resonance functions for dipole-dipole ($R_2$) and quadrupole-quadrupole ($R_4$) interactions between collision partners.

exp (–2ξ). The rapid drop in the resonance functions as $\Delta E_c$ increases also explains the resonant character of the transitions owing to long-range interactions.

## 3.2. Resonant and Quasiresonant *E–E* and *E–V* Exchange Processes

This group includes radiationless deactivation of electronically excited states of atoms and molecules in collisions

$$A^*_{n_1} + B_{n_2} \rightarrow A_{n'_1} + B^*_{n'_2}$$

where the change in the internal energy of each of the collision partners is large ($|\Delta E_{n_1 n_1'}|$, $|\Delta E_{n_2 n_2'}| \geq 1$ eV), while the overall change in the internal energy is small ($\Delta E_c \ll 1$ eV) and the initial energy of electronic excitation goes mainly into electronic (*E–E* processes) or vibrational (*E–V* processes) excitation of the collision partner.

The processes in which the main features of long-range interactions show up most distinctly are ENP owing to dipole–dipole interactions between the collision partners. In this case, the main contribution to the cross section is from large values of the impact parameter *b*, at which the short-range interaction and higher-order multipole interactions can be neglected and only the first order of perturbation theory need be retained in the calculations. The characteristic features of these processes can be observed in experiments and are easily interpreted.

In the case of higher multipole interactions, the description is more complicated. Higher orders of perturbation theory make a significant contribution [9] and the interpretation of experimental data becomes more complicated.

A typical example of ENP owing to quadrupole–quadrupole and quadrupole–dipole interactions is the quenching of an electronically excited state of the $O_2$ molecule ($b^1\Sigma_g^+$) by diatomic molecules considered in Section 3.

Some examples of *E–E* exchange in dipole–dipole interactions that have been studied in a fair amount of detail are ENP during collisions of electronically excited molecules with alkali metal atoms in a gaseous medium at $T = 700$–$1000$ K. The following reactions have been studied experimentally in [10–12], respectively:

$$\mathrm{Li}_2(A^1\Sigma_u^+, v, J) + \mathrm{Li}(^2S_{1/2}) \rightarrow \mathrm{Li}_2(X^1\Sigma_g^+, v', J') + \mathrm{Li}(^2P), \tag{6}$$

$$\mathrm{LiH}(A^1\Sigma^+, v, J) + \mathrm{Li}(^2S_{1/2}) \rightarrow \mathrm{LiH}(X^1\Sigma^+, v', J') + \mathrm{Li}(^2P), \tag{7}$$

and

$$\mathrm{Na}_2(A^1\Sigma_u^+, v, J) + K(^2S_{1/2}) \rightarrow \mathrm{Na}_2(X^1\Sigma_g^+, v', J') \rightarrow K(^2P). \tag{8}$$

These processes were investigated by selectively populating various vibrational levels of the electronically excited state of the molecule.

The rate constant for reaction (6) was determined from the intensity of emission by excited atoms, and for reactions (7) and (8), from the rate at which the excited state of the molecule was quenched.

According to Eqs. (4) and (5), the probability of the transition $|n\rangle = |AvJ\rangle\,|S\rangle \rightarrow |n'\rangle = |Xv'J'\rangle\,|P\rangle$ is given by

$$P_{nn'} = \frac{8}{9\hbar^2 V^2 b^4}\,|\langle Av\,|\,d_1|Xv'\rangle|^2\,|\langle P\,|\,d_2\,|\,S\rangle|^2\,|\langle J\,0\,1\,0\,|\,J'\,0\rangle|^2 \times R_2(\omega_{nn'}\,b/V), \tag{9}$$

where $\langle Av\,|\,d_1\,|Xv'\rangle$ and $\langle P\,|\,d_2\,|\,S\rangle$ are the matrix elements of the transition dipole moments of the molecule and atom.

The probability of quenching the electronically excited state of the molecule $|AvJ\rangle$ (and, therefore, the probability of appearance of the electronically excited state of the atom) is obtained from Eq. (9) by summing over the possible final states of the molecule, $J' = J \pm 1$, and has the characteristic form

$$P_{\mathrm{qu}} = P_{S\rightarrow P} \sim \left[R_2\left(\frac{\omega_J^+ b}{V}\right) + R_2\left(\frac{\omega_J^- b}{V}\right)\right], \tag{10}$$

where $\omega_J^{\pm} = |E_{AvJ} + E_S - E_{Xv'J\pm1} - E_P|\hbar^{-1}$ is the rate of the corresponding branch of the transition. Thus, the dependences of the cross sections

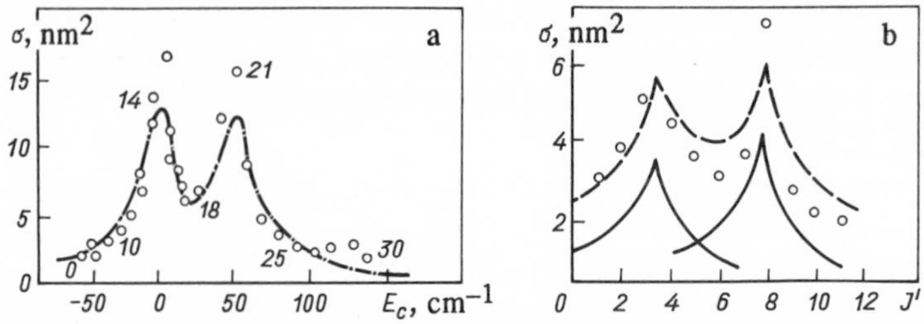


**Fig. 3.** The dependence of the cross sections for quenching of $Li_2(A^+\Sigma^+)$ on the resonance defect $\Delta E_c$ (a) and of $LiH(A^+\Sigma^+)$ on the angular momentum $J'$ (b): the points denote experimental data and the dot-dashed curve, the calculations of [10]; the smooth curve denotes a calculation assuming only a dipole-dipole interaction of the reagents and the dashed curve, a calculation shifted by 1.6 $nm^2$ [11]

for reactions (6) and (7) on the initial rotational quantum number $J$ (for fixed $v$ and $v'$) have two peaks corresponding to the resonances in the $P$- and $R$-branches. These characteristic maxima can be seen clearly in Fig. 3.

The cross section for the process $Li_2(A, v = 17, J) + Li(^2S) \rightarrow Li_2(X, v' = 9, J' = J \pm 1) + Li(^2P)$ is plotted in Fig. 3a as a function of the resonance defect $\Delta E_c$ for the $P$-branch of the $J' = J - 1$ transition. The dashed curve in Fig. 3a corresponds to a calculation using the probability (10) and the points are experimental data for $T \sim 10^3$ K. The numbers next to the points are the angular momenta $J$ of the electronically excited states of the molecules. The peak at $J' = 14$ corresponds to the resonance in the $P$-branch and that at $J' = 21$, to the resonance in the $R$-branch. The cross sections for the resonance and nonresonance transitions differ substantially, with $\sigma_{res}(J = 21) = (11 \pm 1.5)$ $nm^2$ and $\sigma_{nonr}(J = 27) = (1.5 \pm 0.5)$ $nm^2$. The characteristic width of the peaks is ~10 $cm^{-1}$. An analogous dependence of the cross sections on the resonance defect has been observed for the reaction $LiH(A, v = 5, J) + Li(^2S) \rightarrow LiH(X, v' = 11, J' = J \pm 1) + Li(^2P)$. The smooth curve of Fig. 3b corresponds to the values of $\sigma_{res}$ obtained from Eq. (9). It is clear that the locations and shapes of the maxima in the calculated curve correspond to those in the experiment. The experimental data are fitted well by the formula $\sigma_{exp} = \sigma_{res} + 1.6$ $nm^2$. This means that there is an additional mechanism for quenching of $LiH(A)$, for which the cross section is $\sigma_{nonr} \approx 1.6$ $nm^2$ and depends nonresonantly on the change $\Delta E_c$ in the total internal energy.

Reaction (8) has been studied in less detail, but the same features have been observed in it: the selection rules characteristic of dipole interactions

($\Delta\Lambda = \pm 1.0$; $\Delta J = \pm 1$), a large cross section ($\sigma \geq 1$ nm$^2$) which depends on the transition dipole moment, and a resonance dependence of the cross section on the net change $\Delta E_c$ in internal energy.

## 3.3. Collisional Intramolecular Electronic Transitions (CIET)

Along with $E$–$E$ and $E$–$V$ exchange processes, the subclass of quasiresonant ENP caused by the long-range interaction between the collision partners includes collisional intramolecular electronic transitions (CIET). This group of processes includes radiationless collisional electronic transitions between electronic–vibrational–rotational states of a molecule with similar energies of the type

$$\left.\begin{aligned} A_{n_1}(e, v, J) + B_{n_2} \to A_{n_1'}(e', v', J') + B_{n_2'}, \\ \Delta E_c = |E_n(e, v, J) - E_{n'}(e', v', J')| \lesssim 0.01 \text{ eV.} \end{aligned}\right\} \quad (11)$$

Further classification within this group is naturally related to the character of the interaction which mixes the electronic states of the molecule being studied. In terms of this criterion we can distinguish two types of process:

1) CIET (external), where the electron wave functions of the molecule $|e_n\rangle$ and $|e_{n'}\rangle$ undergo mixing as a result of a $V$ interaction with the collision partner $\langle e_n|V|e_{n'}\rangle$. Here the matrix element is nonzero. The basic features of this type of ENP are analogous to those of the $E$–$E$ and $E$–$V$ exchange processes discussed above. The $V$ interaction leading to mixing the molecular states can be either an electrostatic interaction (if the mixed states have the same multiplicity) or a long-range magnetic interaction (spin–orbital, in the case of transitions between states with different multiplicities).

2) CIET (internal), a group which includes transitions between vibrational–rotational states belonging to different electronic terms which are mixed as a result of internal nonadiabatic perturbations intrinsic to the molecule itself and do not involve collisions. The perturbed states are effective channels for electron relaxation. The ENP which proceed through mixed states differ substantially from CIET (external) and are examined in more detail below.

Combination CIET can also be distinguished. This group includes transitions between electronic states which have been mixed either through a collisional interaction with a collision partner or through internal pertur-

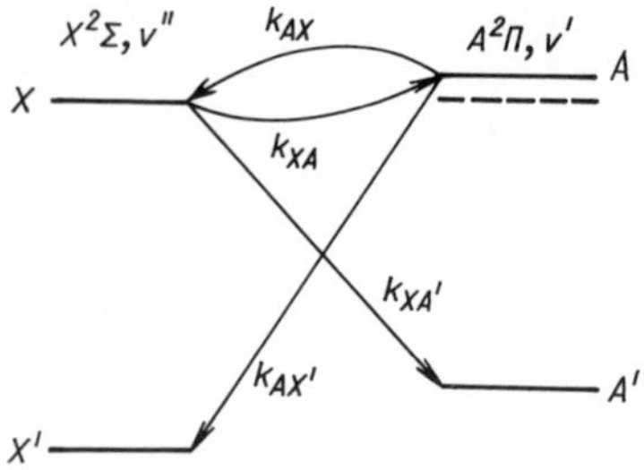


**Fig. 4.** A scheme for populating the vibrational level $A'$ from level $A$ through a collisional transition into the vibrational-rotational manifold $X$ and back (cascade transition).

**TABLE 1.** Rate Constants for Cascade Transitions in CN

| $v$, $A^2\Pi$ | 3 | 4 | 5 | 6 | 7 | 8 | 9 |
|---|---|---|---|---|---|---|---|
| $k_{AX'}$, $10^{-11}$ cm$^3$sec$^{-1}$ | 3 | 3.4 | 7.0 | 11 | 14 | 20 | 20 |
| $k_{XA'}$, $10^{-13}$ cm$^3$sec$^{-1}$ | 2 | 6 | 6 | 7 | 3 | 11 | 14 |

bations. As a rule, it is difficult to isolate these two factors when interpreting experimental data.

In accordance with Eq. (9), the transition probability for CIET (external) is proportional to the squares of the matrix elements:

$$P_{nn'} \sim |\langle n_1 | Q_{l_1} | n_1' \rangle|^2 \, |\langle n_2 | Q_{l_2} | n_2' \rangle|^2. \tag{12}$$

In many experiments, the rotational state $A$ and the state $B$ of the collision partner must be averaged over the initial $J_1$, $n_2$ and summed over the final $J_1'$, $n_2'$ states. This leads to a smearing out (over an interval ~$kT$) of the resonant character of the transition. Making the usual approximation

$$|\langle e_1 v_1 | Q_{l_1} | e_1' v_1' \rangle|^2 = \bar{Q}^2_{l_1} q_{v_1 v_1'}, \tag{13}$$

where $q_{v_1 v_1'}$ is the Franck–Condon factor and $\bar{Q}_{l_1}$ is the effective value of the matrix element for the electronic $2l_1$-pole transition between the electronic states $e_1$ and $e_1'$, we obtain the following estimate for the transition probability:

$$P_{n_1 \to n_1'} \sim C_B \bar{Q}^2_{l_1} q_{v_1 v_1'} \exp\left(-\frac{\omega_{n_1 n_1'}}{kT}\right). \tag{14}$$

This formula is often used for interpreting experimental data. The rate in Eq. (14) is determined solely by the energy difference between the initial $|e_1 v_1\rangle$ and final $|e_1' v_1'\rangle$ electronic–vibrational states of molecule $A$ and the factor $C_B$ takes the properties of the collision partner $B$ into account.

The features of this type of mechanism have been observed in a study of the dependence on the Ar buffer gas pressure of the rate at which the fluorescence of CN($A^2\Pi$) in the gaseous phase is quenched [13]. The vibrational levels $v = 3$–$9$ were excited by a tunable laser. The time dependence of the emission from these levels could not be explained by vibrational relaxation within the same electronic state because its rate is so low at $T = 300$ K. The experimental results were interpreted [13] in terms of a kinetic scheme that includes fast intramolecular cascade transitions through the vibrational levels of the ground state $X^2\Sigma^+$ during collisions.

Figure 4 is a schematic illustration of these transitions, through which the vibrational level $A'$ belonging to the electronic term $A^2\Pi$ is not populated directly from the higher level $A$, but through transitions from $A$ to the vibrational–rotational level $X$ of the $X^2\Sigma^+$ term, which has a similar energy, followed by a transition back to the level $A'$. A cascade transition of the type

$$A^2\Pi\,(v) \to X^2\Sigma^+\,(v'') \to A^2\Pi\,(v-1)$$

is equivalent to vibrational relaxation in the excited electronic state $A^2\Pi$. This approach made it possible [13] to determine the cross sections for the individual stages of such transitions among these vibrational levels of the $A^2\Pi$ state and the nearest levels of the $X^2\Sigma^+$ state (Table 1). Here $k_{AX}$ is the rate constant for the collisional transition between $v$, $A^2\Pi$ and the nearest (in terms of energy) vibrational state $v''$, $X^2\Sigma^+$; $k_{XA'}$ is the rate constant for the transition from the latter level, $v''$, $X^2\Sigma^+$, to $(v-1)$, $A^2\Pi$. The large difference between $k_{AX}$ and $k_{XA'}$ is explained by the fact that in this case $\Delta E_{AX} \ll \Delta E_{XA'}$.

The conclusions reached by Katayama et al. [13] about the nature of this process lie within the framework of the concept of the CIET (external) mechanism discussed in this section. The cross sections obtained in the experiment can be approximated by Eq. (14). The rate constants given in

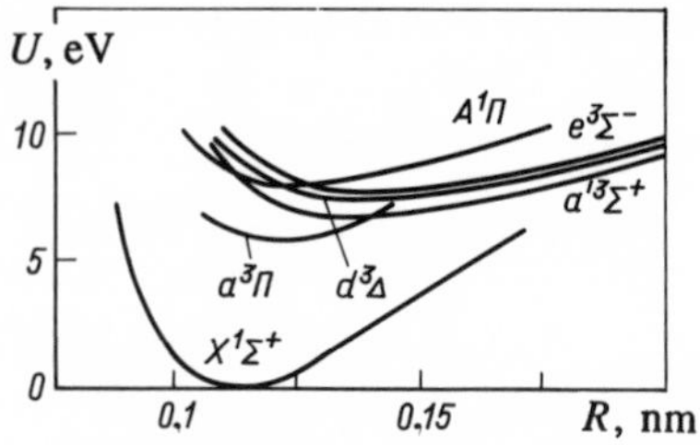


**Fig. 5.** Potential curves for the CO molecule

Table 1 correlate with the Franck–Condon factors and the energy gap $\Delta E_B$. Of all the vibrational levels $v$ of the $A^2\Pi$ state that were studied, the smallest energy gaps $\Delta E_B$ between these levels and the nearest (in terms of energy) vibrational levels of the $X^2\Sigma^+$ state were observed for $v$ equal to 8 and 9. Consequently, the values of $k_{AX}$ and $k_{XA'}$ for these levels are the largest.

In CIET (internal) processes, mixed states that belong simultaneously to two or several electronic states play a primary role. Such states are observed if the terms of the different electronic states come close to one another or intersect, as well as when the vibrational–rotational levels belonging to different electronic states have the same energy (random resonance). The wave function of a mixed state is a linear combination of the adiabatic electronic–vibrational–rotational wave functions [14]:

$$| n_1\rangle = \alpha \mid evJ\rangle + \beta \mid e'v'J'\rangle. \tag{15}$$

As a rule, one of the coefficients is of order unity ($\alpha \sim 1$) and the other is small ($\beta \ll 1$), so that the state of a molecule described by the wave function $| n_1 \rangle$ can be characterized by the quantum numbers $n_1 = (e, v, J)$ for the adiabatic approximation.

Some of the most typical and well-studied processes of the CIET (internal) type involve transitions between electronically excited singlet states of the $CO(A^1\Pi)$ molecule and the triplets $d^3\Delta$, $e^3\Sigma^-$, and $a'^3\Sigma^+$ whose electronic terms intersect (Fig. 5).

In the many experiments on fluorescence quenching in CO, internal perturbations between electronic states have been studied thoroughly and the existence of reversible CIET has been established and used to explain the complicated time variation of the radiation from the tested electronic–vibrational–rotational levels when the pressure of the partner gas is varied.

Cross sections σ for these processes have been obtained for various mixed electronic states. Table 2 shows the rate constants $k$ for CIET in CO. Listed there are the collision partner $B$, the vibrational level of the electronic state that is populated in the experiment, and the perturbing state into which the transition takes place. The data were obtained for a temperature of 300 K. A review of the literature on CIET processes in CO [15–23] shows that $k$ is large for them: $10^{-12}$–$10^{-9}$ cm$^3$sec$^{-1}$.

A number of characteristic features have been observed, including:

1. the "mixing" of the perturbed levels obeys the selection rules for perturbations (for a given type of coupling in the molecule);

**TABLE 2.** The Rate Constants for CIET (inner) among the Electronically Excited $A^1\Pi$, $e^3\Sigma^-$, and $d^3\Delta$ States of the CO Molecule ($k$ is the rate constant for the direct CIET and $\bar{k}$ that for the inverse CIET)

| Electronic state | Perturbing electronic state | B | $k$, $10^{10}$ cm$^3$sec$^{-1}$ | $\bar{k}$, $10^{10}$ cm$^3$sec$^{-1}$ | References |
|---|---|---|---|---|---|
| $A^1\Pi$, $v=0$ | $e^3\Sigma^-$, $v'=1$ | He<br>Ar<br>Ne<br>Kr | 0.77<br>1.7<br>0.77<br>2.5 | 0.45<br>0.62<br>0.45<br>1.05 | [15] |
| $A^1\Pi$, $v=0$ | $d^3\Delta_1$, $v'=4$ | Ar | 1.58 | 0.57 | [16] |
| $A^1\Pi$, $v=1$ | $d^3\Delta_1$, $v'=5$ | Ar | 1.44 | 0.12 | [16] |
| $A^1\Pi$, $v=2$, $J=27$ | $e^3\Sigma^-$, $v'=4$, $J'=27$ | Ar<br>CO | —<br>10 | 1<br>40 | [18] |
| $A^1\Pi$, $v=3$ | $d^3\Delta$, $v'=8$ | Ar | 0.36* | — | — |
| $A^1\Pi$, $v=6$ | $d^3\Delta$, $v'=12$;<br>$a'^3\Sigma^+$, $v'=17$ | Ar | 0.9* | — | — |
| $d^3\Delta$, $v=7$ | $e^3\Sigma^-$, $v'=4$ | He<br>Ar<br>$H_2$ | $1.6\cdot10^{-3}$<br>$2.6\cdot10^{-2}$<br>$6.2\cdot10^{-2}$ | — | [21] |
| $d^3\Delta$, $v=7$ | $e^3\Sigma^-$, $v'=4$ | $N_2$<br>CO | $2.9\cdot10^{-2}$<br>$3\cdot10^{-2}$ | — | [21] |
| $d^3\Delta_1$, $v=5$ | $d^3\Delta_{2.3}$ | He<br>Ar | 0.42<br>1.4 | — | [20] |

*The values of $k$ have been calculated using the σ obtained in [17].

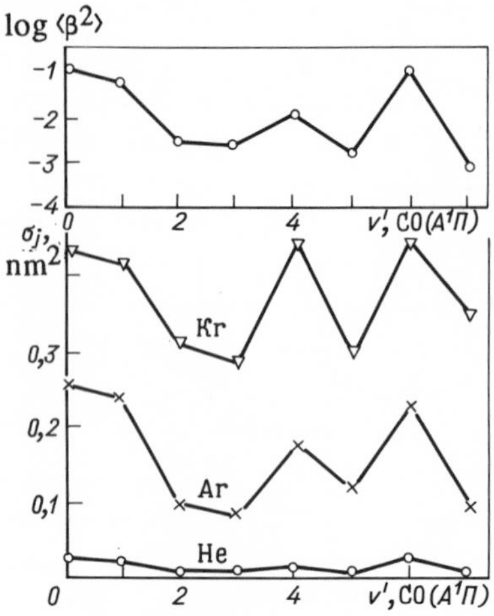


**Fig. 6.** The correlation of the mixing coefficients $\langle\beta^2\rangle$ and quenching cross sections for vibrational levels of the $CO(A^1\Pi)$ state according to [17]. Each of the vibrational levels $v = 0$–7 has mixed states with electronic terms $d^3\Delta$, $e^3\Sigma^-$, or $a'^3\Sigma^+$ (see Table 2).

2. the cross section $\sigma$ correlates with the mixing coefficients $\beta^2$ for the levels [17];

3. for small values of $\beta^2$, $\sigma$ depends on $\beta^2$ linearly, while for large $\beta^2$ ($\beta^2 \geq 0.1$) saturation is observed;

4. the cross sections $\sigma$ are correlated with the polarizability of the collision partner.

Several of these features are illustrated in Fig. 6. The data are those of Grimbert et al. [17], who have studied the effect of the pressure of the inert gases on the quenching of fluorescence in CO from selectively excited $v = 0$–7 levels of the $A^1\Pi$ state. There is a clear correlation between $\sigma$ and the averaged mixing coefficient $\beta^2$ for a given vibrational level. The dependence on the collision partner also shows up here: the cross section $\sigma$ increases with the atomic number. In a study of the CIET $d^3\Delta(v = 7) \rightarrow e^3\Sigma^-$ $(v' = 4)$ [21], this dependence has been related to the polarizability of the partner.

The basic characteristics of these transitions are described by a model proposed by Freed [14]. This model implies that the cross section for CIET (internal) is given approximately by the product of the cross section for rotational relaxation and the square of the absolute value of the mixing coefficient:

$$\sigma \approx \beta^2 \sigma_{rr} . \tag{16}$$

A similar dependence has been observed for the cross sections of CIET in the CO molecule.

Equation (16) also explains the dependence of $\sigma$ on the polarizability of the collision partner, since the cross section for rotational relaxation $\sigma_{rr}$ depends on the polarizability.

CIET (internal) processes have also been observed in other diatomic molecules, including for electronic states with the same multiplicity. The clearest example is the CN molecule. Broida and coworkers [24–28] have carried out a cycle of studies on CIET in $CN(A^2\Pi)(v = 10) \rightarrow B^2\Sigma^+ (v' = 0)$, both between perturbed and between unperturbed rotational levels of these combining electronic states. It was found that the cross section for a transition between excited levels is much greater than between pure levels belonging only to the $^2\Pi$- or $^2\Sigma$-states. Thus, in the first case, for the transition between excited states $A^2\Pi, v = 10, K = 4 \rightarrow B^2\Sigma^+, v' = 0, K' = 4$ the rate constant is $k = 4 \cdot 10^{-10}$ cm$^3$sec$^{-1}$ [25] and, in the second case, $k \sim 10^{-12}$ cm$^3$sec$^{-1}$.

The Freed model for mixed states helps explain the observed intensity anomalies in the spectrum of cyanogen in the afterglow of active nitrogen at pressures below 1 kPa with $T = 300$ K. These anomalies in the violet spectrum are associated with preferential formation of CN in the $A^2\Pi$ state. At low pressures, the rates of the CIET and rotational relaxation are low, hence it is primarily excited levels of the $B^2\Sigma^+$ state that are populated. The CIET (internal) model makes it possible to describe the kinetics of these processes and offers a satisfactory explanation of the experimentally observed line intensity ratios.

Note that if the mixed states in collisional transitions have the same multiplicity, then CIET of the second type (external) are possible, simultaneously with CIET (internal).

The absence of mixing among levels of the triplet and singlet states in another molecule, $C_2$, leads to small cross sections for the transition

$$C_2(a^3\Pi) + M \rightarrow C_2(X^1\Sigma^+) + M \tag{17}$$

in collisions with Ar atoms and $CO_2$ and $N_2$ molecules. Despite the small energy difference between the $a^3\Pi$ state and the ground $X^1\Sigma^+$ state (only

**TABLE 3.** Rate Constants for Quenching of $C_2$ ($a^3\Pi$) by Different Partners

| $M$ | $N_2$ | $CO_2$ | $CF_4$, Ar | Kr | Xe | $O_2$ |
|---|---|---|---|---|---|---|
| $k$, $cm^3 sec^{-1}$ | $3 \cdot 10^{-14}$ | $3 \cdot 10^{-14}$ | $3 \cdot 10^{-14}$ | $2 \cdot 10^{-13}$ | $4.5 \cdot 10^{-12}$ | $2.7 \cdot 10^{-11}$ |

**TABLE 4.** Rate Constants for Vibrational Relaxation in Electronically Excited States of Diatomic Molecules ($T$ = 300 K)

| Molecule, state | $M$ | $k_{vib}$, $cm^3 sec^{-1}$ | Notes |
|---|---|---|---|
| $N_2$ ($C^3\Pi$, $v=1$)<br>$N_2$ ($C^3\Pi$, $v=2$)<br>$N_2$ ($B^3\Pi$, $v=1$) | $N_2$<br>$N_2$<br>Ar | $2 \cdot 10^{-11}$<br>$0.66 \cdot 10^{-11}$<br>$9 \cdot 10^{-12}$ | Interaction with other electronic states |
| $N_2$ ($A^3\Sigma_u^+$, $v=1$) | Ar<br>$N_2$ | $1.8 \cdot 10^{-18}$<br>$5.3 \cdot 10^{-17}$ | A metastable state. For small $v$ there is no interaction with other states |
| CO ($A^1\Pi$, $v=1$) | He<br>Ne<br>Ar | $6.5 \cdot 10^{-12}$<br>$6.5 \cdot 10^{-12}$<br>$9.9 \cdot 10^{-11}$ | Interaction with the $d^3\Delta$, $v'$ = 4 and $e^3\Sigma^-$, $v'$ = 1 states |
| CO ($A^1\Pi$, $v=2$) | He<br>Ne<br>Ar | $2.4 \cdot 10^{-12}$<br>$2 \cdot 10^{-12}$<br>$3.5 \cdot 10^{-11}$ | Interaction with the $d^3\Delta$, $v'$ = 5 and $a'^3\Sigma^+$, $v'$ = 10 states |
| CO ($a^3\Pi$, $v=1$) | He | $4.4 \cdot 10^{-14}$–<br>$4.4 \cdot 10^{-13}$ | No interactions |
| NO ($A^2\Sigma^+$, $v=1$) | He<br>Ar | $8.5 \cdot 10^{-15}$<br>$1.1 \cdot 10^{-13}$ | The $A^2\Sigma^+$, $v$ = 0.1 state is not perturbed by other electronic states |
| CN ($A^2\Pi$, $v=3 \div 9$) | Ar | $9.1 \cdot 10^{-14}$–<br>$2.1 \cdot 10^{-13}$ | Intersystem cascade transitions occur |
| $C_2$ ($d^3\Pi_g$, $v=3,4$) | He | $<5 \cdot 10^{-13}$ | No interactions |
| $S_2$ ($B^3\Sigma_u^-$, $v=4$) | Ar | $1 \cdot 10^{-11}$ | Interaction with the $B''^3\Pi_u$ state |

0.075 eV), an experimental value of the rate constant $k$ for reaction (17) of less than $10^{-14}$ $cm^3sec^{-1}$ has been obtained [29].

Replacing light partners by heavy (Kr, Xe) and singlet partners ($N_2$, $CO_2$) by oxygen $O_2$, which has a triplet electronic ground state, leads to an increase in the rate constant for quenching of $C_2(a^3\Pi)$ by several orders of magnitude [29, 30] (Table 3). This quenching behavior in $C_2(a^3\Pi)$ molecules indicates that CIET (external) can also take place as a result of a spin–orbit interaction between the collision partners.

The difference between CIET (internal) and CIET (external) processes shows up in the role played by the collision partner. In the case of CIET (internal), this role reduces basically to collisional induction of the transition. The quantities which determine the cross section are the internal characteristics of the molecule under study. In the case of CIET (external), the properties of the collision partner on which the interaction between the particles depends (for example, the spin or a multipole moment) will determine the cross section to a significant degree.

## 3.4. Vibrational Relaxation of Electronically Excited States of Diatomic Molecules and CIET

An analysis of published experimental data showed that all these types of perturbations, which mix nearby electronic states, not only manifest themselves through quenching, but also have an effect on the vibrational relaxation of electronically excited states of the molecules. The rate constants for vibrational relaxation, $k_{vib}$, of individual electronic states depend strongly on the presence of random resonances or interactions with neighboring states. Table 4 shows the rates of vibrational relaxation for excited electronic states of several diatomic molecules obtained using the estimates for the vibrational transition probabilities, $P_{v,v-1}$, given by Marcous et al. [32].

In the paper by Katayama et al. [13] examined above, the quenching of the fluorescence of $CN(A^2\Pi, v = 3–9)$ in argon is explained in terms of intramolecular cascade transitions through the ground $X^2\Sigma^+$ state during collisions. Essentially, the rate constant for the cascade process also determines $k_{vib}$ for CN($A$), especially at low temperatures and in an inert gas medium, $k_{vib} \sim k_{XA'} \sim 10^{-12}$–$10^{-13}$ $cm^2sec^{-1}$, and depends on the energy gap $\Delta E$, which is equal to the vibrational quantum.

It is apparent from an analysis of the values of $k_{vib}$ for different electronic states of molecules that the largest $k_{vib}$ occur in states that have been

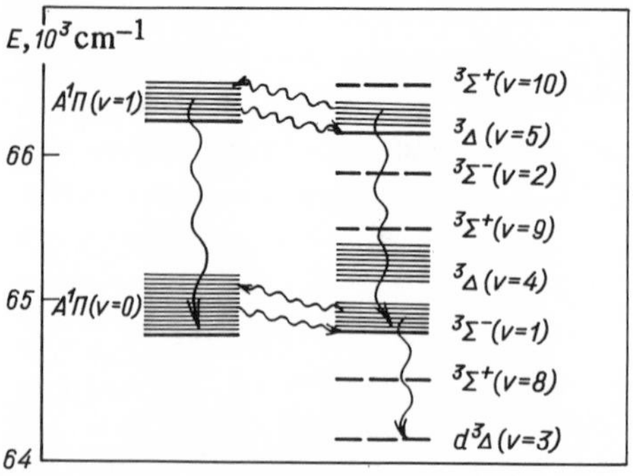


**Fig. 7.** A diagram of electronically excited levels of CO and collisionally induced transitions

perturbed by a neighboring state: $N_2(C^3\Pi)$, $N_2(B^3\Pi)$, $CO(A^1\Pi)$, $S_2(B^3\Sigma^-)$. For them, $k_{vib} \sim 10^{-12}$–$10^{-11}$ cm$^3$sec$^{-1}$.

Lavollee and Tramer [16] have studied the dependence of the fluorescence spectrum of selectively populated levels of $CO(A^1\Pi, v = 1)$ on the pressure of an argon buffer gas. Time-resolved radiation was observed in the (1–0), (0–1), and (1–6) bands of the $A^1\Pi \rightarrow X^1\Sigma^+$ transition. The experimental results were explained in terms of indirect vibrational relaxation in a three-step process: CIET (internal) from the $v' = 1$ level of the $A^1\Pi$ singlet to one of the triplets $d^3\Delta$, $e^3\Sigma^-$, or $a'^3\Sigma^+$; vibrational relaxation in the triplet or triplets having smaller vibrational quanta than the $A^1\Pi$ state; and inverse CIET (internal) from the triplet to the $v = 0$, $A^1\Pi$ vibrational level (Fig. 7). The necessary condition for this sort of process is that the probabilities of both the forward and reverse CIET processes be high enough. For $CO(A^1\Pi)$ and its neighboring levels $d^3\Delta$ and $e^3\Sigma^-$, this condition is satisfied.

The values of $k_{vib}$ given in Table 4 for electronically excited states of molecules that do not have internal perturbations are considerably lower. These are the $CO(a^3\Pi)$, $NO(A^2\Sigma^+)$, and $N_2(A^3\Sigma_u^+, v = 1)$ states. The metastable $A^3\Sigma_u^+$ state of nitrogen, whose lowest vibrational levels do not intersect other electronic states, including the ground state, are especially characteristic in this regard. The value $k_{vib} \sim 10^{-18}$ cm$^3$sec$^{-1}$ for that state is comparable to the rate of vibrational relaxation of (homonuclear) diatomic molecules in the ground state. Vibrational relaxation of $NO(A^2\Sigma^+)$ may proceed via cascade transitions through high vibrational levels of the ground state, but this has not been confirmed experimentally.

# 4. ELECTRONIC KINETICS OF ATOM–MOLECULE SYSTEMS

## 4.1. Analysis of Data on the Rate Constants of ENP for the Example of Quenching of Singlet Oxygen

The analysis of elementary processes is the most important ingredient of kinetics. The probabilities of these processes are the parameters in the kinetic equations for specific systems. One peculiarity of the electronic kinetics is that the data on the kinetics appear, in turn, to be important for analyzing the elementary processes. This is a consequence of the ambiguous interpretation of much experimental data, especially those on the quenching of fluorescence or vibrational relaxation of electronically excited states. The states of the products often are not identified, while the reaction channels and the dependence of their cross sections on the parameters that determine the mixing of the electronic states and the interaction with the collision partner are unknown. Understanding the character of these processes requires careful experimental measurements in which the products and their states are identified. If this is not done, then only the net rate of a process is determined and it becomes extremely difficult to estimate the relative contributions of the different channels.

As a typical example of the situation that arises in real cases, let us consider the quenching of the low-lying electronically excited $O_2(a^1\Delta_g)$ and $O_2(b^1\Sigma_g^+)$ states of the oxygen molecule. Methods for producing and detecting singlet oxygen and analyzing experimental data on the rates of processes in oxygen–iodine mixtures have been reviewed by Didyukov et al. [33]. Here we present a brief analysis of the features of processes that lead to deactivation of these states.

First, we examine the low-lying metastable state $O_2(a^1\Delta_g)$ which has an energy of 0.98 eV and can serve as an efficient reservoir of electronic excitation energy.

In most cases, the basic type of quenching mechanism for processes in which the final reaction products are known is transfer of energy to the collision partner through *E–E* or *E–V* exchange. For instance, in studies of the quenching of NO, CO, and HF molecules based on observation of IR radiation, it has been shown that the primary mechanism for quenching is *E–V* exchange. In this process the $v'' = 2$–4 vibrational levels of the NO molecule [40, 41], the $v'' = 2$ level of CO [41], and the $v'' = 2$ and 3 levels of HF [42] are excited.

The processes of practical importance for $O_2(a^1\Delta_g)$ are $E$–$E$ exchange reactions, of which the fastest is the quasiresonant reaction

$$O_2(a^1\Delta_g) + I(^2P_{3/2}) \underset{k_{-1}}{\overset{k_1}{\rightleftarrows}} O_2(X^3\Sigma_g^-) + I(^2P_{1/2}) + \Delta E_1, \tag{18}$$

for which the rate constant $k_1 = 7.6 \cdot 10^{-11}$ cm$^3$sec$^{-1}$ at $T = 300$ K, with $k_1 = 7.6 \cdot 10^{-14}$ cm$^3$sec$^{-1}$, $k_{-1}/k_1 = (4/3)\exp(-400/T)$ [33, 43] and $\Delta E_1 =$ 400 K. The reactions

$$O_2(a^1\Delta_g) + O_2(a^1\Delta_g) \xrightarrow{k_2} O_2(b^1\Sigma_g^+) + O_2(X^3\Sigma_g^-) + \Delta E_2 \tag{19}$$

and

$$O_2(a^1\Delta_g) + I(^2P_{1/2}) \xrightarrow{k_3} O_2(b^1\Sigma_g^+) + I(^2P_{3/2}) + \Delta E_3 \tag{20}$$

have also been well studied. The rate constants for these processes are $k_2 = 2 \cdot 10^{-17}$ cm$^3$sec$^{-1}$ [44] and $k_3 = 1.1 \cdot 10^{-13}$ cm$^3$sec$^{-1}$ [44], with $E_2 =$ 3800 K and $E_3 = 3400$ K.

The rates of these $E$–$E$ exchange processes differ by several orders of magnitude. This difference can be attributed to the difference in the energy defects of the reactions and to the interaction of the collision partners. In fact, in the fast reaction (18), the resonance defect is $\Delta E_1 = 400$ K, while in the slower reactions (19) and (20), $\Delta E_{2,3} > 3000$ K. Reaction (18) apparently occurs because of long-range spin–orbit interactions, while reactions (19) and (20), which have large resonance defects, occur because of short-range interactions between the collision partners.

The $O_2(a^1\Delta_g)$ molecule is itself a good quenching agent. One example is the efficient quenching of electronically excited states of $O(^1S)$ atoms. This process has been studied by several authors [45–47]. The measured overall rate constant was $2 \cdot 10^{-10}$ cm$^3$sec$^{-1}$. The reaction channels for this process have been found to be the following [47]:

$$O(^1S) + O_2(a^1\Delta_g) \rightarrow O(^1D) + O_2(b^1\Sigma_g^+), \tag{21}$$

$$\rightarrow O(^3P) + O(^3P) + O(^3P), \tag{22}$$

$$\rightarrow O(^3P) + O_2(A^3\Sigma_u^+,\ A'^3\Delta_u), \tag{23}$$

with the contribution of the first channel to the overall reaction being 18% and that of the second, 17%, with the remainder from the third channel. Clearly we are dealing in this case with an ENP caused by short-range interactions. Evidence for this type of reaction is provided by its nonresonant character and the existence of many channels.

**TABLE 5.** Rate Constants for Quenching of $O_2(a^1\Delta_g)$ and $O_2(b^1\Sigma_g{}^+)$

| $M$ | $k_a$, $cm^3sec^{-1}$ | $k_b$, $cm^3sec^{-1}$ |
|---|---|---|
| Ar | $8.3\cdot10^{-21}$ | $5.8\cdot10^{-18}$ |
| He | $8\cdot10^{-21}$ | $10^{-16}$ |
| $N_2$ | $3\cdot10^{-21}$ | $2.3\cdot10^{-15}$*; $5.9\cdot10^{-20}\,T^{1.6}\exp(428.4/T)$ |
| $CO_2$ | $(3.8\pm0.2)\cdot10^{-19}$ | $(4.5\pm0.3)\cdot10^{-13}$ |
| $O_2$ | $(3.0\pm0.4)\cdot10^{-18}$ | $1.69\cdot10^{-15}$*; $4.3\cdot10^{-22}\,T^{2.4}\exp(-241/T)$ |
| $H_2O$ | $(5.6\pm0.4)\cdot10^{-18}$ | $(6.7\pm0.5)\cdot10^{-12}$ |
| HCl | $(13.1\pm0.6)\cdot10^{-18}$ | $7.3\cdot10^{-13}$*; $6.6\cdot10^{-15}\exp(T/103)$ |
| $H_2$ | $(4.5\pm0.2)\cdot10^{-18}$ | $7.9\cdot10^{-13}$*; $5.1\cdot10^{-12}\exp(-550/T)$ |
| $O_3$ | $4\cdot10^{-15}$ | $(1.8\pm0.2)\cdot10^{-11}$ |
| $I_2\,(A^3\Pi)$ | $3\cdot10^{-11}$ | $(2.0\pm0.3)\cdot10^{-11}$ |
| $I\,(^2P_{1/2})$ | $1.3\cdot10^{-13}$ | — |
| $I\,(^2P_{3/2})$ | $7.7\cdot10^{-11}$ | — |
| O· | $7\cdot10^{-16}$ | $(8.0\pm2)\cdot10^{-14}$ |
| $O\,(^1S)$ | $3\cdot10^{-10}$ | — |

*The value of $k_b$ for $T = 300$ K.

These examples show that the mechanisms for quenching a given state vary for different collision partners.

The $O_2(b^1\Sigma_g{}^+)$ state has an energy of 1.6 eV. Intense emission in the (0–0) band of the $b \to X$ transition is often observed in afterglow systems. Clearly, the main path for formation of $O_2(b^1\Sigma_g{}^+)$ is the recombination of O atoms with energy transfer of the type [48]

$$O(^1D) + O_2 \to O(^3P) + O_2(b^1\Sigma_g^+,\ v' = 0,\ 1). \tag{24}$$

The state $O_2(b^1\Sigma_g{}^+)$, as well as $O_2(a^1\Delta_g)$, is inefficiently quenched if the partner is an inert gas atom or a small molecule (Table 5). Quenching by ozone, on the other hand, is very fast [23]. The most reliable values of the rate constants for quenching by different components have been obtained by Kohse-Höinghaus and Stuhl [49, 33], who took special steps to reduce the influence of impurity quenching agents.

Little information is available on the reaction channels for quenching of $O_2(b^1\Sigma_g{}^+)$. It has been shown in several papers that the predominant mechanism in the interaction with molecular quenching agents is E–V exchange:

$$O_2(b^1\Sigma_g^+) + M(v = 0) \to O_2(a^1\Delta_g) + M(v). \tag{25}$$

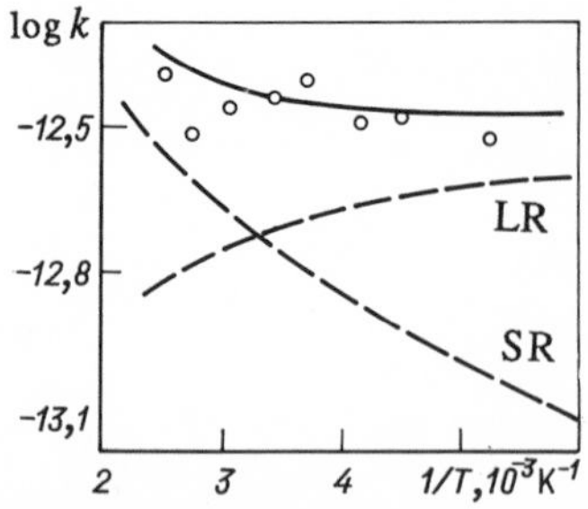


**Fig. 8.** Experimental and theoretical values of the rate constant $k$ ($cm^3 sec^{-1}$) for process (26) as functions of the temperature $T$: points (experiment), curve LR (calculations [54] with long-range interactions); curve SR (with short-range interactions); smooth curve (their sum).

Significant vibrational excitation of each of the partners $M$ = HCl, DCl, $CO_2$, $N_2O$, $H_2O$, $D_2O$, and $NH_4$ has been observed [41, 50]. The fraction of the electronic excitation energy converted to vibrations of the partner and of the $O_2$ was 75–80%. The $O_2(b^1\Sigma_g{}^+)$ was not quenched to the electronic ground state in any of the cases that were studied [51].

Reaction (25) has been studied theoretically [52–54] under the assumption that the transition is caused by long-range electrostatic interactions, so that the earlier calculations [53], which included only short-range exchange interactions, yielded values that were one or two orders of magnitude lower than the experiments.

In these calculations [53, 54] it was assumed that the electronic transition in the $O_2$ molecule is caused by the interaction of the intermediate quadrupole $O_2(b^1\Sigma_g{}^+ \rightarrow a^1\Delta_g)$ with a multipole moment (intermediate dipole or quadrupole) of the collision partner. The calculations showed that second-order perturbation theory makes a large contribution to the transition probability. The rate constants calculated in this way are in good agreement with the experimental values for a number of different collision partners ($M$ = $H_2$, $D_2$, HD, NO, CO, HBr).

A comparison of calculated rate constants with only the long-range or only the short-range interactions included shows that for the above collision partners, the long-range electrostatic interactions make a dominant contribution to the cross section for reaction (25) at room temperature (Table 6). The calculations show that about 95% of the products of reaction (25) are formed in a definite vibrational state and the quenching rate correlates well with the vibrational state $O_2(a, v' = 0, 1, 2)$ into which the

**TABLE 6.** Experimental and Computed Values of the Rate Constants for the Process $O_2(b^1\Sigma_g^+, v = 0) + M \rightarrow O_2(a^1\Delta_g, v) + M$ [53, 54]

| $M$ | Rate constant for reaction (25), $cm^3sec^{-1}$ | | | State of the reaction products | |
|---|---|---|---|---|---|
| | Experiment | SR* | LR* | $O_2$ $(a^1\Delta_g, v)$ | $M$ $(v)$* |
| $H_2$ | $8.3\cdot10^{-13}$ | $1\cdot10^{-13}$ | $7.6\cdot10^{-13}$ | 0 | 1 |
| HD | $2.3\cdot10^{-13}$ | $5\cdot10^{-15}$ | $1.7\cdot10^{-13}$ | 1 | 1 |
| $D_2$ | $2\cdot10^{-14}$ | $1.5\cdot10^{-15}$ | $2\cdot10^{-14}$ | 1 | 1 |
| NO | $5\cdot10^{-15}$ | — | $1.6\cdot10^{-14}$ | 1 | 2 |
| CO | $3.7\cdot10^{-15}$ | $5.2\cdot10^{-17}$ | $4.2\cdot10^{-15}$ | 2 | 1 |
| HBr | $3.8\cdot10^{-13}$ | $4\cdot10^{-14}$ | $2.7\cdot10^{-13}$ | 0 | 2 |

*The calculated vibrational states of the reaction products are listed in the last column. SR and LR denote calculations in which only the short-range and long-range interactions, respectively, are taken into account.

**Table 7.** Classification of Resonant ($\Delta E_c$ < 0.01 eV) ENP Owing to Long-Range Interactions

| Groups of processes (proposed classification) | Change in internal energy of electronically excited particle, $\Delta E_c$, eV | Typical values of the rate constants and cross sections | | Parameters determining the cross section |
|---|---|---|---|---|
| | | $k$, $cm^3sec^{-1}$ | $\sigma$, $nm^2$ | |
| Energy exchange: *E–E*-, *E–EV*-, and *E–V*-processes | $\gtrsim 1$ | $10^{-14}$—$10^{-8}$ | $2\cdot10^{-5}$—20 | Resonance defect $\Delta E_c$ and matrix elements of electrostatic interaction, which mix the states between which transitions occur. These matrix elements, in turn, depend on electronic multipole moments for the transition, Franck–Condon factors, the polarizability of the collision partners, etc. |
| CIET (external) | <0,01 | $10^{-15}$—$10^{-12}$ | $2\cdot10^{-6}$—$2\cdot10^{-3}$ | |
| CIET (internal) | <0,01 | $10^{-13}$—$10^{-9}$ | $2\cdot10^{-4}$—2 | Cross sections determined by the mixing coefficients and the cross section for rotational relaxation. Mixing coefficients depend on nonadiabatic perturbations intrinsic to the molecules. |

oxygen molecule goes. Then about 90% of the difference in the electronic excitation energy, $\Delta E_e = E(O_2(b)) - E(O_2(a))$, goes into vibrational degrees of freedom of the reaction products.

The temperature dependences of the experimental and calculated rate constants for the reaction

$$O_2(b^1\Sigma_g^+) + HBr \rightarrow O_2(a^1\Delta_g) + HBr(v) \quad (26)$$

are compared in Fig. 8 [54]. It is clear that the roles of the long- and short-range interactions in the quenching of $O_2(b^1\Sigma_g^+)$ are different at different temperatures and that the electrostatic interactions predominate at room temperature.

We recommend the data on quenching of the singlet $a^1\Delta_g$ and $b^1\Sigma_g^+$ states given in Table 5 for use in kinetic calculations. The justification for choosing these values [33] is an analysis of experimental data on the rate of reactions involving $O_2(a^1\Delta_g, b^1\Sigma_g^+)$ molecules and a comparison of calculated values of the gain coefficient in an oxygen–iodine medium with experiment. The rate constants are given for $T$ = 300 K. Recommended temperature dependences are given for the partners $N_2$, $O_2$, HCl, and $H_2$.

## 4.2. Electronic Kinetics of Simple Atom–Molecule Media

As they have high rates, resonance and quasiresonance processes play a dominant role in the evolution of the distribution function for the populations of the electronic states of atoms and molecules in nonequilibrium media. A determination of the fast and slow processes, along with the establishment of a hierarchy of characteristic times, makes it possible, in principle, to develop a qualitative picture of the kinetics [6]. The feasibility of such a distinction for electronic relaxation has been examined [55–60] for the example of simple atomic and molecular media containing molecules and atoms of a single element ($N_2$–N, $O_2$–O, $Cl_2$–Cl). There it was shown that, because of fast quasiresonant *E–E* exchange processes, a nonequilibrium quasistationary distribution function may develop in the lower metastable electronically excited states of the atoms and molecules, which form a unified atom–molecule system.

Metastable electronically excited states of atoms and diatomic molecules play an important and, in many cases, a controlling role in the kinetics of atom–molecule media. Their multiplicity, as a rule, differs from that of lower-lying states and the nondiagonal matrix elements of the operator for the electrostatic interactions between wave functions of states with different multiplicity are equal to zero. Thus, the metastable states have immense radiative lifetimes ($\tau \sim 0.1$–$10^3$ sec). The same matrix elements, however, determine the rate of quenching of the metastable states through long-range interactions with collision partners. The rate constants and cross sections for quenching of metastables in collisions with particles having a closed electron shell are small, with $k_{qu} \sim 10^{-18}$–$10^{-16}$ cm$^3$sec$^{-1}$ ($\sigma_{qu} = 10^{-5}$–$10^{-3}$ nm$^2$), and deactivation of metastables proceeds mainly through collisions among themselves or with radicals [58]. Due to their large radiative lifetimes and low quenching rates, the low-lying metastable states of homonuclear diatomic molecules ($N_2$, $O_2$) often become a bottleneck for electronic relaxation. Thus, it is possible to create large densities ($n \sim 10^{12}$–$10^{16}$ cm$^{-3}$) of metastables with substantial energies ($E \sim 1$–6 eV), which can be used as a reservoir of electronic excitation energy.

## 5. CONCLUSION

In this article we have discussed electronically nonadiabatic processes (ENP) for the radiationless quenching of electronically excited states of diatomic molecules through collisions with neutral particles. For resonant ENPs caused by the long-range interaction between the collision partners, a classification scheme based on phenomenological and dynamic criteria was proposed. The major groups of processes and the cross sections and rate constants characteristic of them are shown in Table 7. Within these groups, further classification is possible in terms of dynamic features, i.e., in terms of selection rules and of peculiarities in the dependences of the cross sections on the temperature, pressure, and characteristics of the collision partner. One should strive to set up experiments which make it possible to determine the dynamic features, as they contain important information on the specifics of the interaction leading to an ENP. The significance of the proposed classification is not exhausted by the need for ordering in the experimental data to be entered in the data base of the AVOGADRO system. Identification of elementary processes aids in understanding their mechanisms and provides an opportunity for establishing the quantitative dependences of the characteristic cross sections on the parameters of the medium and for separating the fast and slow processes leading to relaxation of electronically excited states of atoms and molecules.

Just as the widely used quasistationary nonequilibrium population distributions of vibrational kinetics were obtained from a semiquantitative model for the probabilities of *V–V*-, *V–V'*-, and *V–T*-processes that allowed these probabilities to be compared, constructing a hierarchy of characteristic times in electron kinetics offers the hope of obtaining a qualitative description of the population distribution of electronically excited states of atoms and molecules in a relaxing gas.

## REFERENCES


1. E. E. Nikitin and S. Ya. Umanskii, *Nonadiabatic Transitions in Slow Atomic Collisions* [in Russian], Nauka, Moscow (1978).
2. V. M. Galitskii, E. E. Nikitin, and B. M. Smirnov, *The Theory of Atomic Collisions* [in Russian], Nauka, Moscow (1981).
3. E. E. Nikitin, The Dynamics of Molecular Collisions [in Russian], *Progress in Science and Technology. Series on Kinetics and Catalysis,* Vol. 11, Izd. VINITI AN SSSR, Moscow (1983).
4. B. M. Smirnov, *Asymptotic Methods in the Theory of Atomic Collisions* [in Russian], Atomizdat, Moscow (1973).
5. M. S. Child, "Electronic excitation: nonadiabatic transitions," *in*: R. B. Bernstein (ed.), *Atom–Molecule Collision Theory*, Plenum Press, New York–London (1979), pp. 427–466.
6. B. F. Gordiets, A. I. Osipov, and L. A. Shelepin, *Kinetic Processes in Gases and Molecular Lasers* [in Russian], Nauka, Moscow (1980).
7. C. G. Gray, *Can. J. Phys.* **46**, 136–139 (1968).
8. C. G. Gray and J. von Kranendonk, *Can. J. Phys.* **44**, 2411–2430 (1966).
9. H. A. Rabitz and R. G. Gordon, *J. Chem. Phys.* **53**, 1815–1850 (1970).
10. J. Derouard and N. Sadeghi, *Chem. Phys. Lett.* **111**, 353–359 (1984).
11. K. G. Ibbs, P. H. Wine, K. J. Chung, and L. A. Melton, *J. Chem. Phys.* **74**, 6212–6218 (1981).
12. V. B. Grushevskii and M. L. Yanson, in: *Energy Transfer Processes in Metal Vapors* [in Russian], Izd. Latv. Gos. Univ. im. P. Stuchki, Riga (1985), pp. 13–19.
13. D. H. Katayama, T. A. Miller, and V. E. Bondybey, *J. Chem. Phys.* **71**, 1662–1669 (1979).
14. K. F. Freed, "Collisional effects on electronic relaxation processes. Potential energy surfaces," *Adv. Chem. Phys.* **62**, 207–270 (1980).
15. M. Lavollee and A. Tramer, *Chem. Lett.* **47**, 523–526 (1977).
16. M. Lavollee and A. Tramer, *Chem. Phys.* **45**, 45–51 (1979).
17. D. Grimbert, M. Lavollee, A. Nitzan, and A. Tramer, *Chem. Phys. Lett.* **57**, 45–49 (1978).
18. T. G. Slanger and G. Black, *J. Chem. Phys.* **63**, 969–974 (1975).
19. T. A. Carlson, N. Duric, P. Erman, and M. Larsson, *Z. Phys.* **A287**, 123–136 (1978).
20. T. G. Slanger and G. Black, *J. Chem. Phys.* **58**, 194–202 (1973).
21. T. G. Slanger and G. Black, *J. Chem. Phys.* **58**, 3121–3128 (1973).
22. T. G. Slanger and G. Black, *J. Chem. Phys.* **59**, 4367–4379 (1973).

23. T. G. Slanger, "Reactions of electronically excited diatomic molecules," *in: Reactions of Small Transient Species. Kinetics and Energetics*, Academic Press, London–New York (1983).
24. H. P. Broida and R. L. Golden, *Can. J. Chem.* **38**, 1666–1678 (1960).
25. K. M. Evenson, J. I. Dunn, and H. P. Broida, *Phys. Rev.* **136**, 1566–1575 (1964).
26. K. M. Evenson and H. P. Broida, *J. Chem. Phys.* **44**, 1637–1641 (1966).
27. H. F. Radford and H. P. Broida, *J. Chem. Phys.* **38**, 644–657 (1963).
28. D. W. Pratt and H. P. Broida, *J. Chem. Phys.* **50**, 2181–2185 (1969).
29. H. Reisler, M. S. Maniger, and C. Wittig, *J. Chem. Phys.* **73**, 2280–2286 (1980).
30. M. S. Maniger, H. Reisler, and C. Wittig, *J. Chem. Phys.* **73**, 829–835 (1980).
31. M. R. Taherian and T. G. Slanger, *J. Chem. Phys.* **84**, 3814–3819 (1984).
32. P. I. Marcous, M. van Swaay, and D. W. Setser, *J. Chem. Phys.* **83**, 3168–3173 (1979).
33. A. I. Didyukov, Yu. A. Kulagin, L. A. Shelepin, and V. N. Yarygina, "A review of reaction rates and kinetics of processes in oxygen–iodine gas mixtures," Preprint No. 9, P. N. Lebedev Physics Institute, (1986).
34. J. G. Parker, *J. Chem. Phys.* **67**, 5352–5361 (1977).
35. I. D. Clark and R. P. Wayne, *Chem. Phys. Lett.* **3**, 93–95 (1969).
36. C. Schmidt and H. I. Schiff, *Chem. Phys. Lett.* **23**, 339–342 (1973).
37. R. L. Brown, *J. Geophys. Res.* **75**, 3935–3936 (1970).
38. F. D. Findlay and D. R. Snelling, *J. Chem. Phys.* **54**, 2750–2755 (1971).
39. K. H. Becker, W. Groth, and U. Schurath, *Chem. Phys. Lett.* **14**, 489–492 (1972).
40. E. A. Ogryzlo and B. A. Thrush, *Chem. Phys. Lett.* **23**, 34–36 (1973).
41. C. G. O. Thomas and B. A. Thrush, *Proc. R. Soc. London Ser. A* **356**, 295–306 (1977).
42. S. Madronich, J. R. Wiesenfeld, and G. J. Wolga, *Chem. Phys. Lett.* **46**, 267–270 (1977).
43. R. G. Dervent and B. A. Thrush, *Discuss. Faraday Soc.* **53**, 162–167 (1972).
44. R. E. Heidner, III, C. E. Gardner, T. M. El-Sayed, G. I. Segal, and I. V. V. Kasper, *J. Chem. Phys.* **74**, 5618–5626 (1981).
45. R. D. Kenner, E. A. Ogryzlo, and P. T. Wassel, *Nature (London)* **291**, 398–399 (1981).
46. R. D. Kenner and E. A. Ogryzlo, *J. Photochem.* **18**, 379–382 (1982).
47. T. G. Slanger and G. Black, *Geophys. Res. Lett.* **8**, 535–538 (1981).
48. L. C. Lee and T. G. Slanger, *J. Chem. Phys.* **69**, 4053–4060 (1978).
49. K. Kohse–Höinghaus and F. Stuhl, *J. Chem. Phys.* **72**, 3720–3726 (1980).
50. R. G. O. Thomas and B. A. Thrush, *Proc. R. Soc. London Ser. A* **356**, 287–294 (1977).
51. E. A. Ogryzlo and B. A. Thrush, *Chem. Phys. Lett.* **24**, 314–316 (1974).
52. M. Braitwaite and E. A. Ogryzlo, *Chem. Phys. Lett.* **42**, 158–161 (1976).
53. M. Braitwaite, J. A. Davidson, and E. A. Ogryzlo, *J. Chem. Phys.* **65**, 771–778 (1976).
54. M. Braitwaite, J. A. Davidson, and E. A. Ogryzlo, *Ber. Bunsenges. Phys. Chem.* **81**, 179–181 (1977).
55. Yu. A. Kulagin and L. A. Shelepin, *Pis'ma v Zh. Tekh. Fiz.* **7**, 1073–1077 (1981).
56. Yu. A. Kulagin and L. A. Shelepin, *Zh. Prikl. Spektrosk.* **39**, 827–831 (1983).

57. A. I. Didyukov, Yu. A. Kulagin, S. A. Reshetnyak, and L. A. Shelepin, "Metastable electronic states and the kinetics of active media," in: *Trudy FIAN*, Vol. 144, Moscow (1984), pp. 67–106.
58. A. N. Dvoryankin, Yu. A. Kulagin, and L. A. Shelepin, "Mechanisms for electronic relaxation in atom–molecule media", Preprint No. 107, P. N. Lebedev Physics Institute (1986).
59. A. N. Dvoryankin, L. B. Ibragimova, and L. A. Shelepin, *Kinetic and Gas Dynamic Processes in Nonequilibrium Media* [in Russian], Izd. MGU, Moscow (1984), pp. 17–18.
60. A. N. Dvoryankin, Yu. A. Kulagin, and L. A. Shelepin, *Kinetic and Gas Dynamic Processes in Nonequilibrium Media* [in Russian], Izd. MGU, Moscow (1984), pp. 18–19.

# INTRAMULTIPLET TRANSITIONS IN ALKALI AND ALKALINE-EARTH METAL ATOMS INDUCED BY COLLISIONS WITH INERT GAS ATOMS

E. I. Dashevskaya, A. Z. Devdariani, and A. L. Zagrebin

## 1. INTRODUCTION

Studies of the vapors of the alkali (AM) and alkaline-earth metals (AEM) occupy a special place in optics and the physics of atomic collisions. This is explained by a number of factors, of which the foremost are the comparative ease of excitation, the reliability with which various characteristics of the ground and excited states can be measured, and the practical importance of various devices which use the properties of AM and AEM (quantum frequency standards, quantum magnetometers, metal vapor lasers).

As long ago as the end of the 1920s, Wood observed and for the first time quantitatively characterized the sensitized fluorescence of AM vapors, with and without admixtures of inert gases, caused by collisional transfer of excitation among fine structure levels. Over the next thirty years these processes were regarded as typical examples of collisions involving a small change in the internal energy of atoms (see, for example, the discussion by Massey and Burhop [1]).

Interest in this circle of problems was renewed in the 1960s and 1970s, when new experimental techniques made it possible to determine the excitation transfer cross sections with great precision. In a number of research centers, excitation transfer processes in AM vapors were studied and the cross sections of the associated processes were measured. Unfortunately, until now the theory of excitation transfer has remained at the level of the earlier work of Stückelberg, which is cited uncritically in many books on the theory of collisions [2].

In the early 1950s Kastler proposed the concept of optical pumping and it was first realized in studies with AM vapors. In the early development of this method, it was already clear that this was a powerful means for studying atomic collision processes leading to the destruction of the polarization of atoms in the ground and excited states produced by light. Research in this area yielded extensive information on relaxation processes which, as a rule, supplemented that obtained from experiments on sensitized fluorescence. The experiments, however, were often interpreted theoretically in terms of simplified relaxation mechanisms which were based more on intuition than on physically justified models.

Recently, experiments have been set up with beams of excited AM or AEM atoms or with an ensemble of gas atoms that have been excited by a narrow laser line. The interpretation of these experiments requires a detailed analysis of the dynamics of pairwise collisions that change both the internal state of an atom and the direction of its motion.

In this situation, it seems appropriate to state and solve the theoretical problem of the possible mechanisms for the inelastic collisions of AM and AEM atoms responsible for relaxation during optical pumping and sensitized fluorescence. Part of this general problem involves theoretical studies of intramultiplet mixing, i.e., transitions between fine structure components of a resonantly excited atom M* during a collision with an inert gas atom X. For an AM, the resonantly excited state is the doublet $^2P_j$ ($j$ = 1/2, 3/2) state, while for an AEM, it is the triplet $^3P_j$ ($j$ = 0, 1, or 2) state. It turns out that the theoretical analysis of intramultiplet mixing in the process

$$M^*\,(^{2S+1}P_j) + X\,(^1S_0) \to M^*\,(^{2S+1}P_{j'}) + X\,(^1S_0) + \Delta\varepsilon_{jj'} \qquad (1)$$

under conditions such that no polarization effects are important (i.e., when the initial state is not polarized and we are interested in the cross sections for transitions into all final states with different magnetic quantum numbers), can be formulated in a general way for atoms M in the first and second groups of the periodic table. This generality manifests itself, first of all, through the fact that the reference equations used to calculate the partial transition probabilities are the same for M($^2P_j$) + X collisions as for M($^3P_j$) + X collisions. The difference between these processes shows up only in the stage of setting up the adiabatic terms in these systems by means of reference equations in the nonadiabatic regions. From this point of view, the study of the mechanisms for transitions among the fine structure components of AM and AEM atoms is of a more general character, since this approach can be transferred to processes involving atoms in the fifth and fourth groups (in which one or two holes in the valence shell replace the one or two valence electrons in the valence shell of an AM or AEM atom) with only some small changes.

This review is organized as follows: the features of M* + X collisions at thermal energies which are used in solving the collision problem are discussed in Section 2. The reference equations used to calculate the partial transition probabilities are also examined briefly there. In Section 3, we discuss the structure of the adiabatic terms of the quasimolecules $M^*(^2P) + X$ and $M^*(^3P) + X$, the study of which makes it possible to isolate localized (in the sense of either geometrical extent or energy interval) regions of nonadiabatic coupling. In Section 4, a general picture is given of the mechanisms for intramultiplet mixing in AM atoms and the results of some cross section calculations are presented. Many details are left out here, since they have been examined repeatedly in the literature [3–8]. Finally, in Section 5, we examine the main mechanism for nonadiabatic transitions and present some calculations of cross sections for intramultiplet mixing in AEM atoms.

## 2. THE DYNAMICS OF INTRAMULTIPLET MIXING IN $M(^2P_j)$ AND $M(^3P_j)$ ATOMS AT LOW ENERGIES

At energies above a few hundredths of an electron volt, pairwise M* + X collisions are quasiclassical in nature. In addition, below energies of a few electron volts, a resonance state of an atom M is adiabatically isolated from the other states in the sense that the cross sections for transitions from this state into other states are negligibly small. These two circumstances are decisive in any attempt to simplify the general scattering problem: it can be studied in the quasiclassical approximation in the adiabatic (quasimolecular) basis of electron wave functions made up of states of the $^2P$ and $^3P$ terms of free M* atoms. The problem, nevertheless, is still difficult to solve. For an AM atom the scattering equations form two sets of three coupled equations, while for AEM atoms, they form sets of four and five coupled equations. Although solving such problems numerically in order to determine the scattering matrix presents no fundamental difficulty, that is not justified for two reasons: first, the adiabatic terms are not known with sufficient accuracy and, second, in calculations of the integral cross sections for transitions averaged over the magnetic quantum numbers, many subtle details of the interaction (information about which is contained in the elements of the $S$-matrix) are lost. Thus, it is best to formulate a simpler (and, of course, more approximate) approach in which one neglects, from the start, those effects which are lost in computing the integral (and, more so, averaged over the velocities) cross sections.

This approach is based on separating the regions of nonadiabatic coupling, calculating the transition probabilities in each of these regions, and neglecting all interference terms in the total transition probability in those cases where they oscillate rapidly as the impact parameter is varied.

It is known that in the electronic adiabatic basis, there are, all together, two types of nonadiabatic coupling which are due, respectively, to radial or rotational motion of an M–X pair [9].

Radial motion induces transitions between terms with the same axial symmetry and this interaction region is always localized, i.e., its width is significantly smaller than the characteristic interatomic distances over which the transition takes place. For a $M^*(^2P) + X$ system, there are only two terms with the same symmetry out of the three doubly degenerate molecular terms (two terms with a total projection of the electron angular momentum on the molecular axis of $\Omega = 1/2$ and one with $\Omega = 3/2$). Thus, in the nonadiabatic interaction region, only two states are coupled. For a $M^*(^3P_j) + X$ system, of the three doubly degenerate terms (one term with $\Omega = 2$ and two with $\Omega = 1$), two have the same symmetry, while of the three nondegenerate terms with $\Omega = 0$, one differs from the other two in terms of its symmetry on reflection in the collision plane. Furthermore, in the interaction region only pairs of terms are coupled. This means that the probabilities of transitions induced by radial motion of the partners can be found by solving a two-state problem.

Rotation of a pair induces transitions between terms which differ in the projection of $\Omega$ by $\pm 1$. The subsequent coupling of the terms leads, generally speaking, to a system of equations of higher dimension. In this system, however, one can neglect the interaction between terms belonging to systems with different quantum numbers $\Lambda$. Then the problem reduces to calculating the probabilities of transitions between fine structure components of the $\Pi$-term, a doublet for an AM atom and a triplet for an AEM atom. These components can be regarded as parallel with a high degree of accuracy, which again makes it possible to reduce this case to a two-level problem.

Thus, in all the cases to be examined below, the rather complicated interaction of several states can be reduced to a set of two-level problems, which have been studied in great detail. The transition probabilities for these standard problems can be used to calculate the total transition probability $P$ and then the integral cross section. The following difficulty may arise in this final stage: if the partial transition probabilities $P_{ik}$ in the expression for $P$ fall off sufficiently rapidly with the impact parameter, then the total cross section can be calculated simply by integrating the probability $P(b)$ over all impact parameters $b$ from zero to infinity [in fact the integral is usually cut off at some distance $R_1$, where $P(R_1)$ goes to zero] and interference effects can be neglected for all $b$ which make a significant contribution to the integral. If, on the other hand, the transition probabilities depend only weakly on the impact parameter, then the upper limit of the integral over $b$ cannot be correctly determined without including interference effects. The upper limit can, however, be found with sufficient accuracy from the Firsov criterion [10], which requires that the largest phase difference should be on the order of unity for the trajectory with this critical value of the impact parameter, $b = R^*$. Then the situation is analogous to that encountered in the theory of nonresonant charge exchange [11]. In this way, when the probability calculated without taking interference into effect is denoted by $P(b, v)$, the corresponding cross section can be written in the form

$$\sigma(v) = 2\pi \int_0^{R_{\min}} P(b, v)\, b\, db, \tag{2}$$

where $R_{\min}$ is the minimum of the two values $R_1$ or $R^*$. Without dwelling on the details, we note only that the choice of $R_{\min}$ is usually uniquely related to the energy dependence of the cross section: in the rising portion of the cross section, $R_{\min} = R_1$, while in the falling portion, $R_{\min} = R^*$. Almost all the processes to be examined below are calculated on the rising portion of the cross section, which corresponds to thermal collisions.

We now briefly examine the reference equations used to calculate the partial transition probabilities. Studies of the structure of the terms of M*–X systems including spin–orbit coupling (see Section 3) show that they can behave in only two ways. Terms with the same axial symmetry begin to diverge rapidly as the interatomic distance is reduced, while terms with different symmetries, which belong to the fine components of the Π-term, remain practically parallel over a large part of the variation in $R$.

For sharply diverging terms with the same symmetry, the appropriate reference equations are those from Nikitin's model [9, 12]. The transition probability $p$ for a single pass of the system through the region where the terms diverge slowly is given in terms of the parameters $\Delta\varepsilon$, $\cos\theta$, and $\alpha$, which appear in the expression for the adiabatic splitting of two terms,

$$\Delta U = [A^2 \exp(-2\alpha R) + 2A\Delta\varepsilon \exp(-\alpha R)\cos\theta + \Delta\varepsilon^2]^{1/2} \tag{3}$$

and the exponential function $A\exp(-\alpha R)$, which provides for the rapid change in the interaction as $R$ varies. In addition, the transition probability depends on the velocity $v$ with which the nonadiabatic region is crossed. The explicit formula for $p$ has the form

$$p = \exp[-\xi(1-\cos\theta)]\,\mathrm{sh}\,[\xi(1+\cos\theta)]\,\mathrm{sh}^{-1}[2\xi], \tag{4}$$

where $\xi$ is the characteristic Massey parameter given by $\xi = \pi\Delta\varepsilon/2\hbar\alpha v$. For the regimes with highly nonadiabatic ($\xi \ll 1$) or almost adiabatic ($\xi \gg 1$) crossing of the interaction region, Eq. (4) takes the forms

$$p = \begin{cases} \exp[-2\xi(1-\cos\theta)], & \xi \gg 1, \\ \cos^2(\theta/2), & \xi \ll 1. \end{cases} \tag{5}$$

Note that the transition probability $p$ is determined by only that small portion of the trajectory corresponding to the extent of the nonadiabatic region, $\Delta R \sim 1/\alpha$.

For parallel terms with different symmetries, there is no equation which would yield an expression for the transition probabilities at all values of the parameters. Nevertheless, simple results can be obtained in the limiting cases.

If the term splitting is so small that during the time $\tau_1$ the system moves in the region where the terms experience a Coriolis interaction the phase difference $\tilde{\Delta}\varepsilon\tau_1$ that develops between them is negligible, then they can be regarded as degenerate. The resulting transition probability is expressed simply in terms of the rotation angle $\Phi_\Pi$ of the molecular axis in that region of the quasimolecule's motion where these terms appear as the fine structure of a molecular Π-term. For example, for the two terms $\Omega = 1/2$ and $\Omega = 3/2$ which form the fine structure of the $A^2\Pi$ term of an $M(^2P)$–X system, the transition probability is given by

$$\mathcal{P} = \sin^2\frac{\Phi_\Pi}{2}, \tag{6}$$

where $\Phi_\Pi$ is the angle of rotation of the molecular axis in the region of a Hund type *a* bond (then, of course, the condition $\tilde{\Delta}\varepsilon\tau_1 \ll 1$ means that a type *a* bond actually is converted to a type *b* bond). It is clear from Eq. (6) that the transition probability is determined by an extended portion of the trajectory and, therefore, depends on the curvature of the trajectory in the Π-term.

If the term splitting is so large that a collision is almost adiabatic relative to the shortest time $\tau_2$ for the motion along the trajectory (this might be the time for bending of the trajectory when the atoms M and X approach one another in a $\Pi$-term), then the transition probability is determined by a small portion of the trajectory near the turning point. Since it is clear in advance that $\mathcal{P}$ is small in this case, the problem can be solved by perturbation theory with a specified model for the interaction near the turning point. A realistic model for this case, in which the repulsion of the atomic cores is described by an exponential function exp $(-\alpha_2 R)$, has been discussed elsewhere [13]. With that dependence, the transition probability is given to within exponential accuracy by

$$\mathcal{P} \sim \exp\left[-\frac{2\pi\widetilde{\Delta\varepsilon}}{\hbar\alpha_2 v}\right], \tag{7}$$

where $\tilde{\Delta}\varepsilon$ is the splitting between the $\Omega$-components of the $^2\Pi$ terms and is given by $\tilde{\Delta}\varepsilon = (2/3)\Delta\varepsilon_{1/2\,3/2}$. Finally, it should be noted that, since usually $\tau_2 \ll \tau_1$, there is an intermediate case in which $\tilde{\Delta}\varepsilon\tau_2 \ll 1$, but $\tilde{\Delta}\varepsilon\tau_1 \geq 1$. An approximate expression for the transition probability in this case has been proposed by Dashevskaya and Reznikov [14].

## 3. ADIABATIC TERMS OF M*–X QUASIMOLECULES

Since the *LS*-coupling approximation is satisfied with good accuracy for the isolated M atoms considered in the following discussion, the adiabatic terms of an M*–X quasimolecule which correlate with the states of the separated atoms, $M^*(^{2S+1}P_j) + X(^1S_0)$, can be constructed in two stages: first the terms correlating with the states of a spinless M*(*P*) atom are calculated and then the spin–orbit coupling is included in a limited basis set, the position dependence of which is determined by two functions without spin that correspond to the $\Sigma$- and $\Pi$-states of the M*(*P*) + X system. Here it should be noted that the problems of calculating the terms arising from a one-electron state *np* or from a two-electron state *nsnp* are extremely similar, since in both cases the interaction at large distances is determined by the asymptotic part of the wave function of the *p*-electron. The existence of an exchange interaction between the *s*- and *p*-electrons in the *nsnp* system has no direct connection with this problem, since it only determines the form of the two-electron wave function of a free M atom in the triplet and singlet states. Nevertheless, in practical calculations the quasi-

**TABLE 1.** Multiplet Splittings of Atoms

| Group I | | Group II | | |
|---|---|---|---|---|
| Element | $\Delta\varepsilon$, cm$^{-1}$ | Element | $\Delta\varepsilon_{01}$, cm$^{-1}$ | $\Delta\varepsilon_{12}$, cm$^{-1}$ |
| Li | 0.35 | Be | 0.64 | 2.35 |
| Na | 17.19 | Mg | 20.06 | 40.71 |
| | | Ca | 52.16 | 105.88 |
| K | 57.72 | Zn | 190.08 | 388.93 |
| Rb | 273.60 | Sr | 186.83 | 394.21 |
| | | Cd | 542.11 | 1170.87 |
| Cs | 544.11 | Ba | 370.59 | 878.12 |
| | | Hg | 1767.22 | 4630.68 |

one-electron system $M^*(^2P) + X$ appears simpler, so that far more work has been done on it than on the quasi-two-electron systems $M^*(nsnp\,^3P_j)$ + X (see [15], for example.)

As an illustration of the overall status of calculations of the terms of quasimolecules MX, we shall consider only the systems Na + He and Ne + Ar, which have been studied in greatest detail. These systems are typical in the sense that the dispersive interaction changes in this series from weak to strong, while for fixed X, the qualitative features of the potentials $U_\Sigma$ and $U_\Pi$ are the same for all M*.

The simplest case corresponds to He and Ne, for which the multiple scattering model [15–21] is applicable and yields an analytic expression for the terms of the one-electron system.

In the case of the strongly polarizing Ar atom, for which the scattering length is negative, neither the method of multiple scattering nor a model potential method employing information on electron scattering are applicable. Only the method based on a semiempirical model potential proposed by Baylis [19] yields qualitatively correct results and is confirmed upon comparison with experimental data on the Π-states; nevertheless, the resulting Σ-term is too strongly repulsive [22]. More complicated calculations [23] also appear to overestimate the repulsion in the Σ-state.

Based on the above remarks, we may reach the following general conclusions about calculating $U_\Sigma$ and $U_\Pi$ by the different methods. For weakly polarizing atoms of the inert gases, the multiple scattering method can be used for rapid estimates of $U_\Sigma$ and $U_\Pi$ and it yields quantitatively correct results for He and qualitatively correct results for Ne. The results of the local model potential method [19] are much poorer in these cases. Nevertheless, it is the only method used to calculate M*–X interactions for heavy X. It yields qualitatively correct results, but overestimates the repulsion in the Σ-state. Finally, in the roughest approximation, which can be regarded as a semiempirical approximation, the adiabatic terms $U_\Lambda (\Lambda = \Sigma, \Pi)$ are represented as a superposition of the exchange interaction $U_\Lambda^{exch}(R)$, calculated using the formulas of the asymptotic theory [10, 11, 15], and the asymptotic dispersive attraction $U_\Lambda^{disp}(R)$:

$$U_\Lambda(R) = U_\Lambda^{exch}(R) + U_\Lambda^{disp}(R). \tag{8}$$

Here

$$\left.\begin{aligned} U_\Sigma^{exch}(R) &= gR^{4/\alpha-2}\exp(-\alpha R), \\ U_\Pi^{exch}(R) &= 0, \end{aligned}\right\} \tag{9}$$

where $\alpha$ is given in terms of the ionization potential $I^*$ of the M* atom as $\alpha = 2(2I^*)^{1/2}$, and $g$ is a function which varies slowly with $R$. As for the potentials $U_\Lambda^{disp}$, they are obtained from the well-known asymptotic formulas [10].

Let us now suppose that the adiabatic terms and electron wave functions of a spinless M(P) – X quasimolecule have been found in one way or another. Then the spin–orbit interaction can be taken into account by diagonalizing the energy matrix in a basis set whose position-dependent part corresponds to the wave functions of the spinless system MX. Here the central question is the dependence of the operator for the spin–orbit interaction on the interatomic distance. With these systems there are significant justifications for assuming that at the distances of interest for intramultiplet mixing processes, this operator can be regarded as independent of $R$ [16]. In this case, all matrix elements of this operator can be expressed in terms of the spin–orbit interaction in free AM and AEM atoms. The wave functions $\Psi_k$ constructed in this way are characterized by a quantum number $\Omega$ (a good quantum number) and a quantum number $j$ (a bad quantum number) which characterizes the genealogical origin of the term from a certain fine-structure state of the $P_j$ multiplet of the free atom, i.e., $\Psi_k = |\Omega, (j)\rangle$. If we assume (by no means necessarily) that the molecular wave functions have been constructed from the atomic wave functions of this multiplet, which we shall denote by $|\Omega, j\rangle$, then the functions $|\Omega, (j)\rangle$ can be written as a linear combination of the $|\Omega, j\rangle$. Under these assumptions, we obtain the following expressions for the adiabatic molecular term of the $M^*(^2P)$ + X system including the spin–orbit interaction:

$$\left.\begin{aligned} U_{1,2} &= \frac{1}{2}\left[U_\Sigma + U_\Pi + \Delta\varepsilon \pm \left(\Delta\varepsilon^2 + \frac{2}{3}\Delta\varepsilon\Delta U + \Delta U^2\right)^{1/2}\right], \\ U_3 &= U_\Pi + \Delta\varepsilon, \\ \Delta U &= U_\Sigma - U_\Pi, \end{aligned}\right\} \tag{10}$$

where $\Delta\varepsilon$ is the difference between the energies of the $^2P_{3/2}$ and $^2P_{1/2}$ terms of the free M atom (Table 1) and the indices $k = 1, 2, 3$ correspond to functions of the form [12, 13]

$$\left.\begin{aligned}\Psi_1^{\pm} &= |\pm 1/2,\ (1/2)\rangle = \cos\lambda\ |\pm 1/2,\ 1/2\rangle \pm \sin\lambda\ |\pm 1/2,\ 3/2\rangle,\\ \Psi_2^{\pm} &= |\pm 1/2,\ (3/2)\rangle = \mp\sin\lambda\ |\pm 1/2,\ 1/2\rangle + \cos\lambda\ |\pm 1/2,\ 3/2\rangle,\\ \Psi_3^{\pm} &= |\pm 3/2,\ (3/2)\rangle = |\pm 3/2,\ 3/2\rangle,\end{aligned}\right\} \quad (11)$$

where

$$\tan 2\lambda = 2\sqrt{2}\,\Delta U/(3\Delta\varepsilon + \Delta U). \quad (12)$$

The wave functions (11) undergo a conversion from a Hund type *c* coupling to a type *a* coupling in a small region of width $\Delta R$ centered at $R_1$. The center of this region is found from the condition [15]

$$|\,\Delta U(R_1)\,| = \Delta\varepsilon. \quad (13)$$

In the case of a weak polarization interaction (here and below, this means a situation in which $U_\Sigma^{\text{exch}} \gg |U_\Sigma^{\text{disp}}|$ for distances $R \sim R_1$), Eq. (13) has one root.

For a $M^*(^3P) + X$ system, the adiabatic potentials have the form

$$\left.\begin{aligned}U_{1,2} &= \frac{1}{2}\left[U_\Sigma + U_\Pi + \Delta\varepsilon_{02} \pm \left(\Delta\varepsilon_{02}^2 + \frac{2}{3}\Delta\varepsilon_{02}\Delta U + \Delta U^2\right)^{1/2}\right],\\ U_{3,4} &= \frac{1}{2}\left[U_\Sigma + U_\Pi + \Delta\varepsilon_{01} + \Delta\varepsilon_{02} \pm (\Delta\varepsilon_{12}^2 + \Delta U^2)^{1/2}\right],\\ U_5 &= U_\Pi + \Delta\varepsilon_{01},\\ U_6 &= U_\Pi + \Delta\varepsilon_{02},\end{aligned}\right\} \quad (14)$$

where $\Delta\varepsilon_{jj'}$ is the splitting of the terms of the free atoms specified by the quantum numbers $j$ and $j'$ (Table 1). These terms correspond to the wave functions

$$\left.\begin{aligned}\Psi_1 &= |0^-,\ (2)\rangle = b_1^{(1)}\,|0,\ 0\rangle + b_5^{(1)}\,|0,\ 2\rangle,\\ \Psi_2 &= |0^-,\ (0)\rangle = b_1^{(2)}\,|0,\ 0\rangle + b_5^{(2)}\,|0,\ 2\rangle,\\ \Psi_3 &= |1,\ (2)\rangle = b_3^{(1)}\,|1,\ 1\rangle + b_6^{(1)}\,|1,\ 2\rangle,\\ \Psi_4 &= |1,\ (1)\rangle = b_3^{(2)}\,|1,\ 1\rangle + b_6^{(2)}\,|1,\ 2\rangle,\\ \Psi_5 &= |0^+,\ 1\rangle = |0,\ 1\rangle,\\ \Psi_6 &= |2,\ (2)\rangle\ |2,\ 2\rangle.\end{aligned}\right\} \quad (15)$$

The coefficients $b_l^{(k)}$ are given in [24].

For most pairs M($nsnp\,^3P$)–X [except Be($2^3P$)–Ne, Ar, Kr, Xe], the polarization interaction is weak, so that for each pair of states with the same symmetry $\Omega$, given by $0^-(^3P_2)$, $0^-(^3P_0)$, and $1(^3P_2)$, $1(^3P_1)$ there is only one region of strong coupling, located near internuclear separations of $R_{\text{exch}}^{(0)}$ and $R_{\text{exch}}^{(1)}$, at which the exchange interaction predominates ($\Delta U > 0$) and the conditions [24]

$$\Delta U\left(R_{\text{exch}}^{(0)}\right) = \Delta\varepsilon_{02} \quad \text{and} \quad \Delta U\left(R_{\text{exch}}^{(1)}\right) = \Delta\varepsilon_{12} \tag{16}$$

are met. The half width of the strong coupling regions is $\Delta R_{\text{exch}} \approx 1/\alpha$. Near $R_{\text{exch}}$ a $c \to a$ change in the Hund coupling takes place (here we are considering adiabatic terms in the absence of rotation of the molecular axis).

Figure 1 gives a qualitative picture of the terms of an M($nsnp\,^3P$)–X($^1S_0$) quasimolecule with a weak polarization interaction. This picture is consistent with published calculations of the interaction for the pairs Mg*, Ca*–He, Ne [25, 26] and Cd–Ar, Kr, Xe [27] using the model potential method and with some nonempirical calculations of the terms of Mg*–He [28, 29]. In the undetailed region of Fig. 1 where the dispersive interaction predominates ($R \gg R_{\text{exch}}$), the correlation of the adiabatic terms is the same as the correlation of the terms with a strong polarization interaction (Fig. 2) in the region $R > R_{\text{disp}}$.

The fine splitting $\Delta\varepsilon_{jj'}$ in the Be($2^3P$) atom is small, and for the pairs Be–Ar, Kr, Xe, values of $|\Delta U(R)| > \Delta\varepsilon$ can be attained in the region where the dispersive interaction predominates ($\Delta U < 0$). This leads to the appearance of additional regions where states with the same symmetry are strongly coupled near $R_{\text{disp}}^{(0)}$ and $R_{\text{disp}}^{(1)}$ given by the equations

$$-\Delta U\left(R_{\text{disp}}^{(0)}\right) = \Delta\varepsilon_{02} \quad \text{and} \quad -\Delta U\left(R_{\text{disp}}^{(1)}\right) = \Delta\varepsilon_{12}. \tag{17}$$

These equations have two roots each. The smaller values of $R_{\text{disp}}^{(0,1)}$ are close to $R_{\text{exch}}^{(0,1)}$ and hereinafter are equated to $R_{\text{exch}}^{(0,1)}$. We retain the notation $R_{\text{disp}}^{(0,1)}$ for the larger values. If the strong coupling regions near $R_{\text{exch}}$ and $R_{\text{disp}}$ are separated, then a $c \to a$ change in the type of coupling takes place in a neighborhood of $R_{\text{disp}}$† with a half width of $\Delta R_{\text{disp}} = R_{\text{disp}}/6$ and the strong coupling region near $R_{\text{exch}}$ appears to be a

†In the following, the center of the region where the type of coupling changes is denoted by $R_1$ ($R_1 = R_{\text{exch}}$) or $R_{\text{disp}}$.

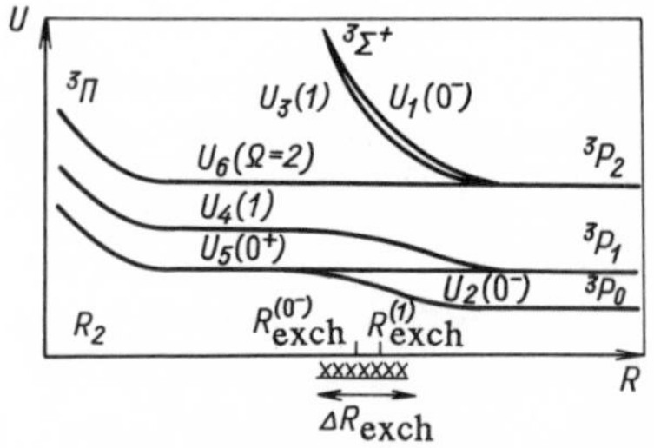


**Fig. 1.** A qualitative picture of the terms of an M($nsnp\,^3P$)–X($^1S_0$) quasimolecule with a weak polarization interaction: in the shaded region $\Delta R_{exch}(R_1)$, a $c \to a$ change in the type of Hund coupling takes place. $R_2$ is the turning point for motion along the Π-term.

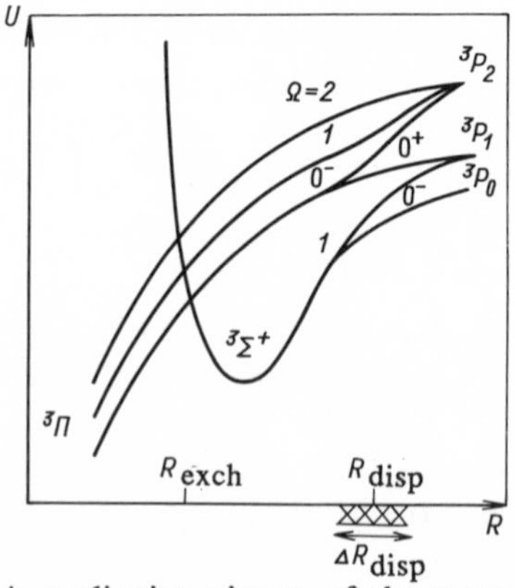


**Fig. 2.** A qualitative picture of the terms of an M($nsnp\,^3P$)–X($^1S_0$) quasimolecule with a strong polarization interaction: in the shaded reaction $\Delta R_{disp}(R_1)$, a $c \to a$ change in the type of Hund coupling takes place. The quasi-intersections of the Ω-components of the $^3\Sigma^+$ and $^3\Pi$ terms ($^3\Sigma_{0^-}{}^+$ and $^3\Pi_{0^-}$; $^3\Sigma_1{}^+$ and $^3\Pi_1$) are replaced by intersections.

quasicrossing of the Ω-components of the $^3\Sigma$- and $^3\Pi$-terms ($^3\Sigma_{0^-}{}^+$ and $^3\Pi_{0^-}$; $^3\Sigma_1{}^+$ and $^3\Pi_1$) with a minimum splitting between the adiabatic terms, $\Delta U_{min} \sim \Delta\varepsilon$. Figure 2 shows a qualitative picture of the terms of an M($nsnp\,^3P$)–X($^1S_0$) quasimolecule with a strong polarization interaction.

For some pairs MX, in particular for Be($2^3P$) + Ne, an intermediate polarization interaction may be realized. Then the main characteristics of the potentials $U_\Sigma$ and $U_\Pi$ in the region near $R_1$ where the type of coupling changes cannot be found by asymptotic methods [the inaccuracies in the asymptotic values of $U_\Sigma$ and $U_\Pi$ (8) are comparable to their difference $\Delta U$] and a more exact calculation of the M($^3P$)–X interaction is necessary.

## 4. MECHANISMS AND CROSS SECTIONS FOR INTRAMULTIPLET MIXING IN M($^2P_j$) + X SYSTEMS

There are two regions of nonadiabatic coupling for the molecular terms of an MX system. The first, centered at $R_1$ with a scale of $\Delta R_1 \sim 1/\alpha$, is caused by the radial motion. Here transitions between terms with $\Omega = 1/2$ are possible. The second is caused by the rotational motion; it extends, generally speaking, over the interval from $R_1$ to the turning point $R_2$ of the quasimolecule as it moves over the Π-term. Transitions between the $\Omega = 1/2$ and $\Omega = 3/2$ components of the $A^2\Pi$ term are possible in this region.

Let $p$ denote the transition probability in the region of $R \sim R_1$ when it is crossed once and $\mathcal{P}$ denote the probability of a transition in the region $R_2 < R < R_1$ as the atoms approach and move apart. Then, when interference is neglected, the following expression for the total probability of a $1/2 \rightarrow 3/2$ transition is obtained [7]:

$$P^{1/2\rightarrow 3/2} = 2p(1-p) + (1-p)^2\mathcal{P}. \tag{18}$$

Next it is appropriate to examine the two limiting cases which differ in the Massey parameter for the mixing process. Here and in the next section the probability $p$ will be calculated in terms of the Nikitin model [Eq. (4)] with two additional corrections (if needed): the first is related to the polarization interaction and involves replacing $\Delta\varepsilon$ by $\Delta\varepsilon_{\rm eff}$ [7]. The second is related to symmetrizing the change in the collision velocity as a result of an inelastic process and involves replacing the velocity $v$ by its average (relative to the initial and final energies) value. Note that only when the second correction is taken into account will the probabilities of upward ($1/2 \rightarrow 3/2$) and downward ($3/2 \rightarrow 1/2$) transitions be different, while the rate constants for these processes satisfy the principle of detailed balancing.

The probability $\mathcal{P}$ is calculated as shown in Section 2 with a correction (if needed) for symmetrizing the velocity.

We shall first consider the almost adiabatic case, where $\Delta\varepsilon/\hbar v \gg 1$. This case is typical of Rb and Cs atoms in collisions with atoms of all the inert gases at temperatures of 300–700 K. When $p \ll 1$ and $\mathcal{P} \ll 1$, Eq. (18) simplifies to

$$P^{1/2\to 3/2} = 2p_{1/2\to 3/2} + \mathcal{P}_{1/2\to 3/2}, \tag{19}$$

where $p$ is calculated in the limit of exponential smallness using Eq. (5) and $\mathcal{P}$ is calculated using Eq. (7) including the corresponding preexponential factor. It follows from Eq. (19) that the mixing cross section is also expressed as the sum of cross sections corresponding to the radial and rotational interactions:

$$\sigma^{1/2\to 3/2}(v) = \sigma_{\text{rad}}^{1/2\to 3/2} + \sigma_{\text{rot}}^{1/2\to 3/2}. \tag{20}$$

Taking the average over a thermal velocity distribution yields the following formulas for the average cross sections:

$$\left.\begin{aligned} \langle\sigma_{\text{rad}}^{1/2\to 3/2}(T)\rangle &= A(T)\exp\left[-(6\pi^2\Delta\varepsilon^2\mu^2/\alpha kT)^{1/3}\right], \\ \langle\sigma_{\text{rot}}^{1/2\to 3/2}(T)\rangle &= a(T)\exp\left[-(6\pi^2\Delta\varepsilon^2\mu^2/\alpha_2 kT)^{1/3}\right]. \end{aligned}\right\} \tag{21}$$

Here $\alpha$ is the parameter in the exponential approximation (3) for the adiabatic terms with $\Omega = 1/2$; $\alpha_2$ is the parameter in the exponential approximation for the interaction potential between M* and X near the turning point on the $\Pi$-term (see Section 2); and, $\mu$ is the reduced mass of the partners M and X. Explicit expressions for the preexponential factors $A(T)$ and $a$(T), which depend much more weakly on the temperature than the explicitly separated exponential factors, are given in [7, 13], for example. For us at this point, it is only important that $A \gg a$ and that $\alpha$ is considerably smaller than $\alpha_2$ ($\alpha_2$ characterizes the interaction of X with the valence electron of M* and $\alpha_2$ the interaction of X with the core $M^+$).

The most interesting feature of the mechanism for intramultiplet mixing under adiabatic conditions is the competition of the radial and rotational interactions, which is possible because of the great difference between the exponents in Eq. (21) and in the values of the preexponential factors. This competition is illustrated in Fig. 3, which summarizes the theoretical and experimental results for almost all pairs except Li + X. Note that in the almost adiabatic case, the average mixing cross sections have a strong positive temperature dependence.

The opposite case of highly nonadiabatic collisions occurs for Li atoms (with all the inert gases) and approximately for Na atoms in collisions with He and Ne atoms. In this limiting case, $p = 1/3$ and $\mathscr{P}$ is given by Eq. (6). Then, Eq. (18) gives

$$p^{1/2 \to 3/2} = \frac{2}{3}\left(1 - \frac{1}{3}\cos\Phi_R\right), \tag{22}$$

which is just the probability for relaxation (to within a factor of 2/3) of the orientation of a spinless atom in the $P$-state, as should be true according to the general theory [16]. The probability (22) depends on the shape of the trajectory in the $\Pi$-term, and only thereby does it come to depend (indeed, very weakly) on the velocity. The cross section depends more noticeably on the velocity, since an additional dependence enters through the upper limit of integration, $R_{\min}$ [see Eq. (2)]. Nevertheless, the overall dependence of the cross section on the velocity or of the average cross section on the temperature is extremely weak and is negative.

We now consider processes of the intermediate type. At temperatures of 300–400 K these include collisions of Na with atoms of the heavy inert gases (Ar, Kr, Xe) and collisions of K with all the inert gas atoms (He, Ne, Ar, Kr, Xe). First, we consider collisions of Na. As the atomic number of X increases, the polarization interaction increases rapidly. For example, in the series He, Ne, Ar, Kr, Xe the depth of the potential well in the $\Pi$-term increases as 0.67, 0.77, 34.2, 59.4, and 89.3 (in units of $10^{-3}$ eV). This means that for these energies, beginning with Ar, the trajectory may be bent into a spiral and may come close to the repulsive core. The exact energy at which bending sets in is known to be 0.8 times the depth of the potential well for a Lennard-Jones potential. Since an increase in the mass of X for a Na–X system leads to a large increase in the reduced mass $\mu$, it is natural to assume that $\mathscr{P} = 1/2$ for a trajectory bent into a spiral. In this case, for the heavy atoms Ar, Kr, and Xe the cross sections should be determined only by the probabilities $p_1$ and radii $R_{\min}$. The parameters $\xi$, calculated for the systems Na–Ar, Kr, Xe at $T = 400$ K, neglecting polarization, are equal to 1.2, 1.3, and 1.3. However, replacing $\Delta\varepsilon$ by $\Delta\varepsilon_{\text{eff}}$, in accordance with the formulas of [7, 12], yields 0.684, 0.584, and 0.4 for $\xi_{\text{eff}}$. If we note that the average value is $\xi = \xi_{\text{eff}} = 0.55$ for the system Na–He, then it is clear that roughly the same conditions $\mathscr{P} = 1/2$ are satisfied for Na–Ar, Kr, Xe collisions as for the Na–He system. This implies that the cross sections for the $1/2 \to 3/2$ transition in these systems are given approximately by

$$\sigma^{1/2 \to 3/2} = \frac{2}{3}\pi R_1^2 \tag{23}$$

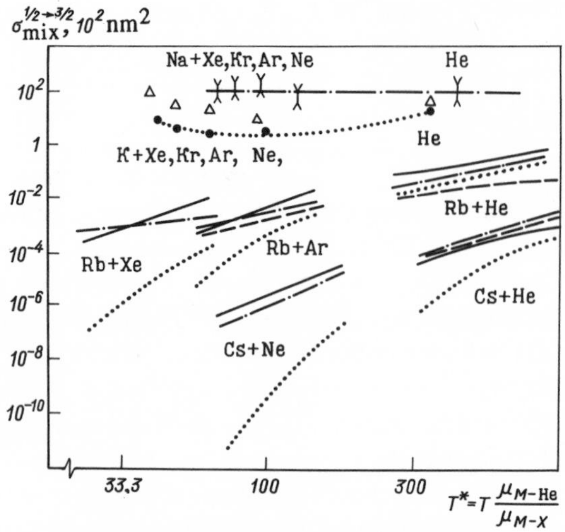


**Fig. 3.** A comparison of theoretical and experimental average cross sections $\langle\sigma^{1/2\to3/2}(T)\rangle$ for intramultiplet mixing with different pairs $M(^2P)$ + X. Experimental cross sections: for Rb, Cs + X [35], smooth curves; for K + X [31], triangles; for Na + X, vertical segments. Theoretical cross sections: $\langle\sigma_{rad}{}^{1/2\to3/2}(T)\rangle$, dot-dashed curves; $\langle\sigma_{rot}{}^{1/2\to3/2}(T)\rangle$, dashed curves; $\langle\sigma_{rad}{}^{1/2\to3/2}\rangle$ + $\langle\sigma_{rot}{}^{1/2\to3/2}\rangle$, dotted curves.

and the entire variation in the cross sections when X is changed depends on the change in $R_1$. Since $R_1$ changes relatively little, these cross sections should be close to the cross section for this transition in Na–He collisions (Fig. 3).

We now turn to K + X collisions. The parameter $\tilde{\Delta}\varepsilon\tau_1$ is large enough for the K + He system that the contribution of the Coriolis interaction to the cross section for the 1/2 → 3/2 transition can be neglected. Then the transition probability can be calculated using the formula

$$P^{1/2\to3/2} = 2p(1-p). \tag{24}$$

For $R_1$ = 14 a.u., the cross section is equal to $30\cdot10^{-2}$ nm$^2$ ($T$ = 365 K) [30], which is roughly a factor of two smaller than the experimental value [31]. If we neglect the Coriolis interaction for the other K + X pairs as well, then as the atomic number of X increases, the discrepancy between the theoretical and experimental cross sections increases, reaching an order of magnitude for K + Xe. In the latter case, the cross section for mixing in K is close to that in Na [8].

One could attempt to eliminate these discrepancies between theory and experiment in an explicit way by including the Coriolis interaction between the components of the $A^2\Pi$ doublet for $R < R_1$. However, as opposed to the highly nonadiabatic collisions of Na + X, where transitions among the fine structure components take place during the movement along the $A^2\Pi$-term over the entire interval $R_2 < R < R_1$, and as opposed to the almost adiabatic collisions of Rb + X and Cs + X, where the $A^2\Pi_{1/2} \rightarrow A^2\Pi_{3/2}$ transitions occur in the region where the partners come closest together (and therefore the model of a repulsive exponential potential is applicable), a description of nonadiabatic transitions for the K + X quasimolecule requires a complete dynamic study of the motion along the $A^2\Pi$-term. This is so because the Massey parameter for heavy atoms X is on the order of unity when the acceleration of the atoms in the bending region is taken into account. Since the adiabatic terms of the K + X system are known only very imprecisely at present, this sort of study has not been carried out.

An overall comparison of theoretical and experimental results is shown in Fig. 3, where the abscissa represents the effective temperature $T^* = T\mu_{M-He}/\mu_{M-X}$. The experimental cross sections for Na [32–34] correspond to $T = 400$ K. The maximum difference in the measured cross sections from various authors is as much as 100%. On the scale of the figure, their variation with changing X is small and can be explained qualitatively by the increasing contribution from bending as the atomic number of X increases. The horizontal dot–dashed curve represents a theoretical estimate in which the radial and rotational contributions to mixing have been set approximately equal. The results for K + X, which correspond to $T = 365$ K, show the discrepancy between theory and experiment mentioned above. The dotted curve drawn through the calculated points means that only the contribution of the radial nonadiabatic coupling has been taken into account. The experimental data for Rb and Cs are taken from [35].

## 5. MECHANISM AND CROSS SECTIONS FOR INTRAMULTIPLET MIXING IN M($^3P_j$) + X SYSTEMS

The possibility of nonadiabatic transitions between terms of an M($^3P$) + X quasimolecule is noticeably greater than for M($^2P$) + X. However, as with the AM, in the case of the AEM two mechanisms should be considered: a) transitions between states of the same symmetry $0^-(^3P_2) \rightarrow 0^-(^3P_0)$ and $1(^3P_2) \rightarrow 1(^3P_1)$ induced by the relative radial motion of the atoms in the region where the type of coupling undergoes a change near $R_1$ (their probabilities are $p_{0-0-}$ and $p_{11}$); and, b) Coriolis transitions $\Omega \rightarrow \Omega'$ between the $\Omega$-components of the $^3\Pi$-term (their probability is $\mathcal{P}_{\Omega\Omega'}{}^{\Pi}$) and between the $\Omega$-components of the $^3\Sigma^+$-term (their probability is $\mathcal{P}_{\Omega\Omega'}{}^{\Sigma}$) induced by rotation of the molecular axis in the region $R \leq R_1$.

In terms of the relationship among the polarization and exchange interactions near $R_1$ and the Massey parameter $\xi^k = \tilde{\Delta}\varepsilon R_1/v$ [7],† which characterizes the ratio of the spin–orbit splitting $\tilde{\Delta}\varepsilon_2$ to the Coriolis interaction $V^k \sim \omega = vb/R^2$ between the $\Omega$-components of the $^3\Pi$-term in the region $R < R_1$, three characteristic cases stand out among thermal collisions between different pairs $M(^3P) + X$: 1) almost adiabatic collision conditions ($\xi^k \gg 1$) with a weak polarization interaction; 2) quasiresonant collision conditions ($\xi^k \ll 1$) with a weak polarization interaction; and 3) quasiresonant collision conditions with a strong polarization interaction. A fourth case (almost adiabatic collision conditions with a strong polarization interaction) is possible only for collisions of $Be(2^3P)$ + Ar, Kr, Xe at low temperatures ($T \ll 200$ K).

## 5.1. Intramultiplet Transitions with Almost Adiabatic Collision Conditions and a Weak Polarization Interaction

Almost adiabatic collision conditions with a weak polarization interaction are typical for the pairs Ca, Zn, Sr, Cd, Ba, Hg + X over a wide range of thermal energies and for the pairs Mg + Ne, Ar, Kr, Xe when $T \leq$ 500 K. The Massey parameters for the radial and Coriolis (between the $\Omega$-components of the $^3\Pi$-terms) transitions are $\xi = \pi\Delta\varepsilon/2\alpha v \gg 1$ and $\xi^k = \tilde{\Delta}\varepsilon R_1/v \gg 1$, while the probabilities $p_{if}$ and $\mathcal{P}_{if}{}^{\Pi}$ of the corresponding $U_i \rightarrow U_f$ transitions ($i_2 f = 1$–6) (14) are small. When $\xi^k \gg 1$, the Coriolis transitions $^3\Pi_\Omega \rightarrow {}^3\Pi_{\Omega'}$ are localized near the turning point $R_2$ (see Fig. 1), so that the regions of nonadiabaticity near $R_1$ and $R_2$ do not overlap. Including the Coriolis transitions $^3\Sigma_\Omega \rightarrow {}^3\Sigma_{\Omega'}, {}^3\Pi_{\Omega'}$ is unimportant [24, 36]. The cross sections of the transitions $^3P_j \rightarrow {}^3P_j'$ averaged over a Maxwellian distribution $\langle \sigma^{j\rightarrow j'}(\Pi) \rangle$ can be written as the sum of contributions from the radial $\langle \sigma_{if}{}^{\mathrm{rad}}(T) \rangle$ and Coriolis $\langle \sigma_{if}{}^{\mathrm{rot}}(T) \rangle$ mechanisms [24, 39]:

†In this and the following sections, atomic units are used with $e_{\mathrm{el}} = m_{\mathrm{el}} = \hbar = 1$.

$$\left.\begin{aligned}
&\langle\sigma^{2\to1}\rangle=\langle\sigma^{2\to1}_{\rm rad}\rangle+\langle\sigma^{2\to1}_{\rm rot}\rangle,\\
&\langle\sigma^{2\to1}_{\rm rad}\rangle=\frac{2}{5}\langle\sigma^{\rm rad}_{34}\rangle;\quad \langle\sigma^{2\to1}_{\rm rot}\rangle=\frac{2}{5}\langle\sigma^{\rm rot}_{64}\rangle,\\
&\langle\sigma^{2\to0}\rangle=\langle\sigma^{2\to0}_{\rm rad}\rangle=\frac{1}{5}\langle\sigma^{\rm rad}_{12}\rangle,\\
&\langle\sigma^{1\to0}\rangle=\langle\sigma^{1\to0}_{\rm rot}\rangle=\frac{2}{3}\langle\sigma^{\rm rot}_{42}\rangle.
\end{aligned}\right\}\qquad(25)$$

The second-order contribution of the Coriolis transitions $^3\Pi_2 \to {}^3\Pi_0$ to the cross section $\langle\sigma^{2\to0}(T)\rangle$ is small, since the probabilities $\mathscr{P}_{62}{}^{\Pi} \approx 1/2\mathscr{P}_{64}{}^{\Pi}\mathscr{P}_{42}{}^{\Pi}$ [24]; these transitions can be important only for the pairs Ba, Cd, Hg + X, where $\langle\sigma_1{}^{\rm rad}\rangle$, $\langle\sigma_{64}{}^{\rm rot}\rangle \ll 10^{-20}$ cm$^2$.

For calculating the contributions of the radial transitions, it is natural to use the exponential model of Nikitin (see Section 2), since the exchange interaction, which can be approximated over a limited range of variation in $R$ by an exponent, is predominant in the transition region.

The polarization shift of the terms in the transition region is taken into account as with the alkali metals, i.e., the variation in $U_{\Sigma,\Pi}{}^{\rm disp}(R)$ is assumed to be slow compared to $U_\Sigma{}^{\rm exch}(R)$:

$$\left.\begin{aligned}
&U^{\rm disp}_{\Sigma}(R)-U^{\rm disp}_{\Pi}(R)-U^{\rm disp}_{\Sigma}(R_{\rm exch})-U^{\rm disp}_{\Pi}(R_{\rm exch})=\delta_\varepsilon,\\
&U^{\rm disp}_{\Sigma}(R)+U^{\rm disp}_{\Pi}(R)-U^{\rm disp}_{\Sigma}(R_{\rm exch})+U^{\rm disp}_{\Pi}(R_{\rm exch})=2D,
\end{aligned}\right\}\qquad(26)$$

where $\delta_\varepsilon$, $D < 0$, and $|\delta_\varepsilon|$, $|D| \ll \Delta\varepsilon$. Depending on the transition, $R_{\rm exch} = R_{\rm exch}{}^{(0)}$, $R_{\rm exch}{}^{(1)}$.

The parameters of Nikitin's exponential model, $\Delta\varepsilon$ and $\cos\theta$, are written in terms of $\delta_\varepsilon$, $\Delta\varepsilon_{12}$, and $\Delta\varepsilon_{02}$ as

$$\left.\begin{aligned}
&\Delta\varepsilon_0=\left(\Delta\varepsilon_{02}^2+\frac{2}{3}\,\Delta\varepsilon_{02}\delta_\varepsilon+\delta_\varepsilon^2\right)^{1/2},\\
&\cos\theta_0=-\left(\frac{1}{3}\,\Delta\varepsilon_{02}+\delta_\varepsilon\right)\Big/\,\Delta\varepsilon_0\geqslant-1/3
\end{aligned}\right\}\qquad(27)$$

for the $\Omega = 0^- \to 0^-$ transition and as

$$\left.\begin{aligned}
&\Delta\varepsilon_1=\left(\Delta\varepsilon_{12}^2+\delta_\varepsilon^2\right)^{1/2},\\
&\cos\theta_1=-\,\delta_\varepsilon/\Delta\varepsilon_1\geqslant0
\end{aligned}\right\}\qquad(28)$$

for the $\Omega = 1 \to 1$ transition. In the model, the average term is approximated by

$$\bar{U}(R)=B\exp(-\alpha R)+D,\quad B=2A.\qquad(29)$$

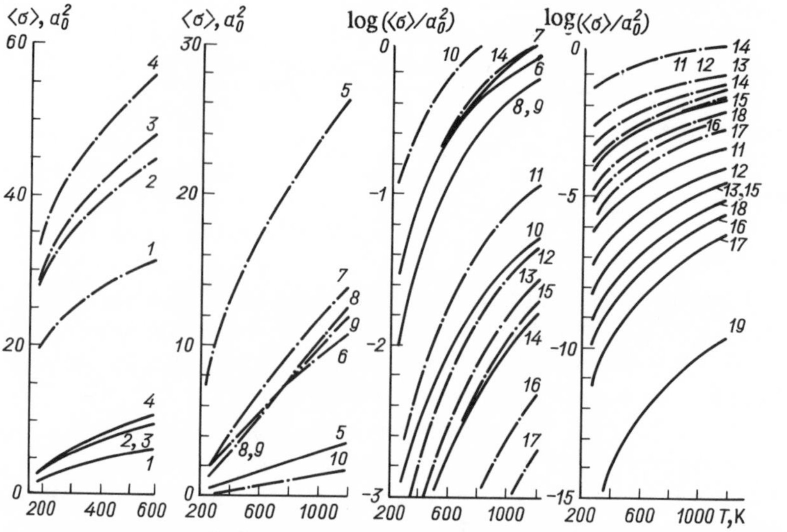


**Fig. 4.** Cross sections $\langle\sigma^{2\to0}(T)\rangle$ averaged over a Maxwellian distribution (smooth curves) and $\langle\sigma_{rad}{}^{2\to1}(T)\rangle$ (dot-dashed curves) for intramolecular transitions: 1–4) Mg + Ne, Ar, Kr, Xe; 5–9) Ca + He, Ne, Ar, Kr, Xe; 10–13) Zn + He, Ar, Kr, Xe; 14–17) Sr + He, Ar, Kr, Xe; 18–19) Ba + He, Ar.

The cross sections averaged over a Maxwellian distribution, $\langle\sigma_{12}{}^{rad}\rangle$ and $\langle\sigma_{34}{}^{rad}\rangle$, are written as follows for $\Gamma_1 \gg 1$ [24, 40]:

$$\left.\begin{aligned}
&\langle\sigma_{12}^{rad}(T)\rangle = 4\pi\left(R_{exch}^{(0)}\right)^2\left(\frac{\pi}{3}\right)^{1/2}\Gamma_1^{1/3}\exp\left[-3\Gamma_1^{2/3}-\frac{E_A+D}{T}+\frac{\Delta\varepsilon_{02}}{2T}\right],\\
&\Gamma_1=\zeta(1-\cos\theta_0)\,T^{-1/2},\qquad E_A=[B\Delta\varepsilon_0(1+\cos\theta_0)]/(2A),\\
&\zeta=(\pi\Delta\varepsilon_0/2\alpha)\,(\mu/2)^{1/2}.
\end{aligned}\right\}\qquad(30)$$

In calculating the temperature dependences (shown in Fig. 4) of the cross sections $\langle\sigma^{2\to0}(T)\rangle$ and $\langle\sigma_{rad}{}^{2\to1}(T)\rangle$ ($200 \le T \le 1000$ K) [24, 37, 39] with the aid of Eq. (30) and the formulas of [41] for Mg, Ca + X, the potentials $U_{\Sigma,\Pi}$ calculated near $R_{exch}$ by an asymptotic method [15, 42, 43] have been used. The ionization potentials of the M($^3P$) and X($^1S_0$) atoms are very different. Thus, the parameters of the exponential model obtained on the basis of an asymptotic calculation are not sufficiently reliable. An analysis [37] shows that the scatter in the cross section $\langle\sigma_{rad}\rangle$ for the pairs Mg, Ca + X associated with the inaccuracy in the interaction parameters is 200–300%, and that for the pairs Zn, Sr + X, the scatter increases to an order of magnitude.

An analysis of the Coriolis transitions $U_6 \to U_4$ and $U_4 \to U_2$ can be limited to the first order of perturbation theory and coupling with a third state can be neglected. A model problem for the Coriolis coupling of two repulsive exponential terms with a splitting constant of $\tilde{\Delta}\varepsilon$ with $\xi^k \gg 1$ has been solved [13] in the framework of perturbation theory. If the upper adiabatic term $U_6$ (or $U_4$ for the $U_4 \to U_2$ transition) near the turning point $R_2$ is approximated by the expression $U_6(R) - U_6(R \to \infty) = C \exp(-\alpha_2 R) - U_0^{(6)}$ [or $U_4(R) - U_4(R \to \infty) = C \exp(-\alpha_2 R) - U_0^{(4)}$], then [13, 24]

$$\left.\begin{aligned}
&\langle \sigma_{64}^{\text{rot}}(T)\rangle = \pi R_2^2 \mid \langle {}^3\Pi_2 \mid \hat{S}_x \mid {}^3\Pi_1 \rangle \mid \frac{8\pi^2(\alpha_2 R_2)^2}{\exp(\alpha_2 R_2)} \left(\frac{\pi}{3}\right)^{1/2} \Gamma_2^{-1/3} \left[\frac{1}{2}\right. \\
&\left. + F(y^*)\right]^2 \exp\left[-3\Gamma_2^{2/3} + \frac{\tilde{\Delta}\varepsilon_{64}}{2T} + \frac{U_0^{(6)}}{T}\right], \\
&\Gamma_2 = (\pi \tilde{\Delta}\varepsilon_{64}/\alpha_2)(\mu/2T)^{1/2}; \\
&y^* = (2\tilde{\Delta}\varepsilon_{64})/(\alpha_2 v^* \exp(\alpha_2 R_2/2)), \\
&v^* = (2\pi \tilde{\Delta}\varepsilon_{64} T/\alpha_2 \mu)^{1/2},
\end{aligned}\right\} \quad (31)$$

where $\langle {}^3\Pi_2 | \hat{S}_x | {}^3\Pi_1 \rangle$ $1/\sqrt{2}$, $\tilde{\Delta}\varepsilon_{64}$ is the splitting between the terms $U_6$ and $U_4$ with $R < R_{\text{exch}} - \Delta R_{\text{exch}}$, and $F(y)$ is given in [7, 13]. The expression for $\langle \sigma_{42}^{\text{rot}} \rangle$ differs from Eq. (31) in that $\langle {}^3\Pi_2 | \hat{S}_x | {}^3\Pi_1 \rangle$ is replaced by $\langle {}^3\Pi_1 | \hat{S}_x | {}^3\Pi_{0-} \rangle = 1/2$, $\tilde{\Delta}\varepsilon_{64}$ by $\tilde{\Delta}\varepsilon_{42}$, and $U_0^{(6)}$ by $U_0^{(4)}$. These transitions are localized in a neighborhood of $R$ with a half width of $\Delta R_2 = 1/\alpha_2$ [13].

Numerical evaluation of the cross sections $\langle \sigma_{64}^{\text{rot}} \rangle$ and $\langle \sigma_{42}^{\text{rot}} \rangle$ is difficult because of a lack of data on the potential $U_\Pi$ in the repulsive region $R \sim 5a_0$, where the asymptotic methods are not applicable. Comparing the cross sections $\langle \sigma_{\text{rad}}^{2\to 1} \rangle$ and the values of $\langle \sigma_{\text{rot}}^{2\to 1} \rangle$ for reasonable interaction parameters $\alpha_2$, $R_2$, and $U_0^{(6)}$ leads to the conclusion that for $T \sim 300$ K, the total cross sections $\langle \sigma_{\text{rad}}^{2\to 1}(T) \rangle$ for the pairs Mg + Ne, Ar, Kr, Xe; Ca + X; and Zn, Sr + He are determined basically by the contribution of the radial mechanism, $\langle \sigma_{\text{rad}}^{2\to 1}(T) \rangle$ (Fig. 4). For the pairs Zn, Sr + Ne, Ar, Kr, Xe and Cd, Ba, Hg + X, the contribution of the Coriolis transitions is important.

## 5.2. Intramultiplet Transitions under Quasiresonant Collision Conditions

Quasiresonant conditions are typical for Be($2^3P$) + X pairs, even at thermal energies $E \geq 0.02$ eV ($T \geq 300$ K), and for Mg($3^3P$) + He pairs at energies $E > 0.15$ eV. For Be, Mg + He, the polarization interaction is weak and for Be + Ar, Kr, Xe, it is strong. It might be expected that for the pairs Be($2^3P$) + Ar, Kr, Xe, as for the pairs Li($^2P$) + Ar, Kr, Xe [44], the regions of strong coupling near $R_{\mathrm{disp}}$ and $R_{\mathrm{exch}}$ should be separated (Fig. 4). The region of the quasicrossing of the terms $^3\Pi_\Omega$ and $^3\Sigma_\Omega$ near $R_{\mathrm{exch}}$ (Fig. 4) in Be($2^3P$) + Ar, Kr, Xe collisions, as is the region of the quasi-crossing of the terms $^2\Pi_{1/2}$ and $^2\Sigma_{1/2}$ in Li($2^2P$) + Ar, Kr, Xe collisions [44], is passed through nonadiabatically, since the parameter in the Landau–Zener model is $\xi_{\mathrm{LZ}} = 2\pi a^2/(\Delta F v)$ and the probability $p = \exp(-\xi_{\mathrm{LZ}}) \approx 1$. Thus, for both strong and weak polarization interactions under quasiresonant collision conditions, the main mechanisms for the transitions in reaction (1) are the same.

Formulas for the total transition probabilities $p^{j \to j'}$ are obtained by summing over all possible sequences of transitions $\Omega \to \Omega'$ [36, 38]:

$$
\left.
\begin{aligned}
P^{2\to} &= \frac{2}{5}\,[\mathcal{P}^{\Pi}_{21} p_{11} + \mathcal{P}^{\Pi}_{20+}] + \frac{2}{5}\,[p_{11}\mathcal{P}^{\Sigma}_{11}(1-p_{11}) \\
&+ (1-p_{11})\,\mathcal{P}^{\Pi}_{11} p_{11} + (1-p_{11})\,\mathcal{P}^{\Pi}_{10+}] + \frac{1}{5}\,[p_{0-0-} - \mathcal{P}^{\Sigma}_{0-1}(1-p_{11}) \\
&+ (1-p_{0-0-})\,\mathcal{P}^{\Pi}_{0-1} p_{11}], \\
P^{2\to 0} &= \frac{2}{5}\,\mathcal{P}^{\Pi}_{20-}\, p_{0-0-} + \frac{2}{5}\,[p_{11}\mathcal{P}^{\Sigma}_{10-}(1-p_{0-0-}) + (1-p_{11})\,\mathcal{P}^{\Pi}_{10-} p_{0-0-}] \\
&+ \frac{1}{5}\,[p_{0-0-}\,\mathcal{P}^{\Sigma}_{0-0-}(1-p_{0-0-}) + (1-p_{0-0-})\,\mathcal{P}^{\Pi}_{0-0-}\, p_{0-0-}], \\
P^{1\to 0} &= \frac{2}{3}\,[(1-p_{11})\,\mathcal{P}^{\Sigma}_{10-}(1-p_{0-0-}) + p_{11}\mathcal{P}^{\Pi}_{10-} p_{0-0-}].
\end{aligned}
\right\}
\tag{32}
$$

In Eqs. (31) we have retained terms of up to third order, since the probabilities $p$ and $\mathcal{P}^{\Sigma,\Pi}$ are not small. In order to keep the notation consistent in Eq. (31), for cases 2 and 3 we have used the notation $p_{0-0-}$ and $p_{11}$ for the probabilities of transitions from the states $|\Omega, j\rangle$ into the states $|^3\Pi_\Omega\rangle$ and $|^3\Sigma_\Omega\rangle$, so that in case 2 these are the probabilities of adiabatically passing through the region $R_1 = R_{\mathrm{exch}}$, and in case 3, the probabilities of nonadiabatic passing ($R_1 = R_{\mathrm{disp}}$). In the approximation of sudden perturbations, for the exponential model we obtain $p_{11} = 1/2$ and $p_{0-0-} = 2/3$.

The probabilities of the Coriolis transitions $\mathscr{P}_{\Omega\Omega'}{}^{\Pi}$ and $\mathscr{P}_{\Omega\Omega'}{}^{\Sigma}$ are determined by the angles $\Phi_{\Pi}$ and $\Phi_{\Sigma}$ of rotation of the molecular axis on passing through the region $R < R_1$ along the $^3\Pi$- and $^3\Sigma$- terms.

In the approximation of straight transit, the cross sections $\langle \sigma^{j\to j'} \rangle$ are given by

$$\left.\begin{aligned}
\langle\sigma^{2\to1}\rangle &= \frac{28}{90}\left(\frac{28}{90}\right)\pi R_1^2, & \langle\sigma^{1\to2}\rangle &= \frac{29}{54}\left(\frac{28}{54}\right)\pi R_1^2,\\
\langle\sigma^{2\to0}\rangle &= \frac{16}{135}\left(\frac{17}{135}\right)\pi R_1^2, & \langle\sigma^{0\to1}\rangle &= \frac{16}{27}\left(\frac{17}{27}\right)\pi R_1^2,\\
\langle\sigma^{1\to0}\rangle &= \frac{2}{27}\left(\frac{1}{27}\right)\pi R_1^2, & \langle\sigma^{0\to1}\rangle &= \frac{2}{9}\left(\frac{1}{9}\right)\pi R_1^2.
\end{aligned}\right\} \quad (33)$$

The first coefficients in Eqs. (33) correspond to the approximation of straight transit. (The coefficients in parentheses are discussed below.) In the approximation of straight transit, the cross sections $\langle \sigma^{j\to j'} \rangle$ are determined only by the radius $R_1$ at which the type of coupling changes and are independent of the type of interaction (dispersive or exchange), which disrupts the spin–orbit coupling in an M($^3P$) atom. The coefficients in Eq. (33) are found by expanding the wave function for type $e$ coupling in terms of functions for type $b$ coupling when the atoms approach one another to a distance $R_1$ and by taking the inverse expansion on passing through the neighborhood of $R_1$ as the atoms fly apart [36, 45, 46].

For the pairs Be, Mg + He, $R_1 = R_{exch} = 10$–$13a_0$ [36]. For the pairs Be + Ar, Kr, Xe, taking the hydrogenlike value for $\langle r^2 \rangle = 5.4a_0^2$ gives $R_1 = R_{disp} = 12.3a_0$, $13.2a_0$, and $14.4a_0$, respectively [38].

## 5.3. Experimental Data on Intramultiplet Transitions. Comparison with Computational Results

Experimental data on intramultiplet transitions during collisions of atoms of group II with inert gas atoms, as opposed to AM atoms, are fragmentary. The system Ca($4s4p\,^3P_j$) + He has been studied in the most detail.

Optical pumping with two lasers [47] has been used to measure the cross sections for transitions among all the fine structure components. The results of these measurements at an average energy of $E = (0.0556 \pm 0.0015)$ are listed in Table 2 along with the calculated [48] averages over the non-Maxwellian particle distribution $f(v)$ which existed in the experimental apparatus [47]. As a whole, the experimental cross sections are 3–4

**TABLE 2.** Cross Sections $\bar{\sigma}^{j\to j'}$ ($10^{-2}$ $nm^2$) for Ca($4^3P_j$) + He Collisions Averaged over the Velocities of the Colliding Particles. Average Collision Energy 0.0556 ± 0.0012 eV

| Cross section | Experimental data [47] | Computational result [48] |
|---|---|---|
| $\bar{\sigma}^{2\to1}$ | 31.9±4,2 | 11.2 |
| $\bar{\sigma}^{2\to0}$ | 5.5±1.6 | 0.8 |
| $\bar{\sigma}^{1\to0}$ | 2.0±5.0 | 2.5 |

**TABLE 3.** Cross Sections $\langle\sigma^{j\to j'}\rangle$ ($10^{-6}$ $nm^2$) for Sr($5^3P_j$) + Ar Collisions Averaged over a Maxwellian Distribution at $T$ = 700 K

| Cross section | Experimental data [49] | Computational result [39] |
|---|---|---|
| $\langle\sigma^{2\to1}\rangle$ | 16.8±2.6 | 16* |
| $\langle\sigma^{2\to0}\rangle$ | 0.8 | $5.0\cdot10^{-3}$ † |
| $\langle\sigma^{1\to0}\rangle$ | 5.4±1.0 | — |

*Calculated using the formula from [39] for $\langle\sigma_{rad}{}^{2\to1}\rangle = 9.0\cdot10^{-6}$ $nm^2$ and $\eta$ = 1.3.

†Only radiative transitions are included.

times smaller than the averaged theoretical values. Note that the relationship among the experimental cross sections, $\bar{\sigma}_{exp}{}^{2\to1} > \bar{\sigma}_{exp}{}^{2\to0} > \bar{\sigma}_{exp}{}^{1\to0}$, is consistent with that among the semiclassical cross sections, but not with a quantum mechanical calculation [48]. The difficulty in the comparison with the experimental data [47] is related to the known uncertainty in the form of $f(v)$. When $E \leq 0.1$ eV, the collisions have an almost adiabatic character, so that $\bar{\sigma}_{theor}$ is extremely sensitive to the form of $f(v)$ for $E < 0.1$ eV.

The reaction

$$\mathrm{Sr}\,(5^3P_j) + \mathrm{Ar}\,(^1S_0) \rightarrow \mathrm{Sr}\,(5^3P_{j'}) + \mathrm{Ar}\,(^1S_0) \tag{34}$$

has been studied [49] by exciting the $5^3P_1$ level with a laser based on a donor–acceptor mixture of two dyes in a gas cell. The experimental cross sections averaged over a Maxwellian distribution at $T = 700$ K are listed in Table 3. Also given there is the calculated value of $\langle\sigma^{2\to0}\rangle$. The cross section $\langle\sigma^{2\to1}\rangle$ has been evaluated for a reasonable value of the ratio $\eta = 1.3$ and $\langle\sigma_{rad}{}^{2\to1}\rangle = 9.0\cdot10^{-6}$ nm$^2$ [39].

In other experiments, either $^3P_1 \to {}^3P_0$ excitation transfer or quenching of the $^3P_{0,2}$ metastable states have been measured. Thus, fluorescence spectroscopy has been used [50] to study the reaction

$$\mathrm{Zn}\,(4\,{}^3P_1) + \mathrm{Ar} \to \mathrm{Zn}\,(4\,{}^3P_0) + \mathrm{Ar}. \tag{35}$$

At $T = 613$ K the cross section averaged over a Maxwellian distribution is $\langle\sigma_{exp}{}^{1\to0}\rangle = 5\cdot10^{-6}$ nm$^2$. Using the experimental value, we can also estimate the cross section $\langle\sigma^{2\to1}\rangle$. In fact, a calculation yields $\langle\sigma_{rad}{}^{2\to1}\rangle =$ $5.6\cdot10^{-5}$ nm$^2$, so that this estimate yields $\langle\sigma^{2\to1}$ (613 K)$\rangle = 6.4\cdot10^{-5}$ nm$^2$ and the cross section is determined primarily by the contribution from the radial mechanism.

Note that the experimentally obtained cross sections $\langle\sigma_{exp}{}^{1\to0}\rangle$ for Sr($5^3P_j$), Zn($4^3P_j$) + Ar collisions [49, 50] are in agreement with the theoretical cross sections for ordinary values of the parameters, e.g., when $\alpha_2 = 2.7$, $R_2 = 6$, and $U_0{}^{(4)} = 1\cdot10^{-3}$ (all quantities in atomic units).

Estimates based on the formulas given above show that the cross sections for intramultiplet transitions in Cd($5^3P_j$), Hg($6^3P_j$) + X($^1S$) collisions are many orders of magnitude smaller than the experimental cross sections for quenching of excited states [51–57]. One of the actual mechanisms for quenching may be deexcitation of metastable states in collisions. The cross sections for such processes in Cd, Hg + X are orders of magnitude greater than those for intramultiplet transitions [58], but are nevertheless an order of magnitude smaller than the experimental quenching cross sections. Perhaps the quenching of the $^3P_2$ states of Cd and Hg atoms in collisions with Ar, Kr, and Xe atoms is associated with transitions through intermediate terms of the $M^-$–$X^+$ configuration with $R \sim 0.2$ nm. Using the terms established experimentally in [49, 59], we can show that $U_2$(R ~ 0.2 nm) ~ 0.05 eV for Hg–Ar, Kr, Xe.

## 6. CONCLUSION

With its relative simplicity, the semiclassical theory provides a description of intramultiplet mixing processes for a wide range of collision partners M* + X. Invoking the asymptotic theory of atomic interactions [10, 11, 15] and semiclassical models in the theory of slow atomic collisions [9, 10, 16] makes it possible to: 1) establish the basic mechanisms for nonadiabatic transitions in slow $M(^2P) + X$ and $M(^3P) + X$ collisions; 2) determine the character of the temperature dependences of the rate constants $K(T) = \bar{v}\langle\sigma(T)\rangle$ and the energy dependences of the cross sections $\sigma(E)$ for elementary processes; 3) trace the character of the changes in the cross sections as the collision partners are switched; and 4) evaluate the cross sections numerically for many pairs M + X. This approach appears to be suitable for a preliminary analysis of elementary collision processes involving a wide circle of different pairs of atoms with similar structure in their outer electronic shells. In particular, simple theoretical estimates of the cross sections are necessary where reliable calculations of the interaction potentials are lacking and the accuracy of a quantum mechanical solution to the collision problem is nullified by the errors in the interaction potentials that have been used [39, 49, 59].

As more reliable calculations of the potential curves and experimental data on differential cross sections become available, it becomes necessary to invoke more exact versions of the semiclassical theory, for example, the multitrajectory semiclassical approximation [60], or a complete strong-coupling quantum mechanical calculation. This is precisely the situation that has arisen now in the theory of intramultiplet transitions during collisions of resonantly excited AM atoms with inert gas atoms [14]; however, these data are currently unavailable for the processes (1) that have been examined in this article.

## REFERENCES

1. H. Massey and E. Burhop, *Electronic and Ionic Impact Phenomena*, Oxford University Press, Oxford (1952).
2. N. Mott and H. Massey, *Theory of Atomic Collisions*, Oxford University Press, Oxford (1965).
3. C. Th. Alkemade and P. J. Zeegers, *in*: *Spectrochemical Methods of Quantitative Analysis of Atoms and Molecules*, J. D. Winefordner (ed.), Wiley (1971), pp. 3–21.
4. P. L. Lijnse, "Review of literature on quenching, excitation and mixing collision cross sections for the first resonance doublets of the alkalies," Report 398, Rijksuniversiteit, Utrecht (1972).
5. L. Krause, *in*: *The Physics of Electronic and Atomic Collisions*. 7th ICPEAC. Invited Papers and Progress Reports, T. R. Govers and F. G. de Heer (eds.), North Holland (1972), pp. 65–72.
6. L. Krause, *Adv. Chem. Phys.* **28**, 267–274 (1975).

7. E. E. Nikitin, *Adv. Chem. Phys.* **28**, 317–377 (1975).
8. E. I. Dashevskaya, "Relaxation processes in alkali metal atoms during optical pumping and sensitized fluorescence," Doctoral Dissertation, Phys.-Math. Sciences, Moscow (1980).
9. E. E. Nikitin and S. Ya. Umanskii, *Nonadiabatic Transitions in Slow Atomic Collisions* [in Russian], Atomizdat, Moscow (1979).
10. V. M. Galitskii, E. E. Nikitin, and B. M. Smirnov, *The Theory of Collisions of Atomic Particles* [in Russian], Nauka, Moscow (1981).
11. B. M. Smirnov, *Asymptotic Methods in the Theory of Atomic Collisions* [in Russian], Atomizdat, Moscow (1973).
12. E. E. Nikitin, *Opt. Spektrosk.* **22**, 689–698 (1967).
13. E. I. Dashevskaya, E. E. Nikitin, and A. I. Reznikov, *J. Chem. Phys.* **53**, 1175–1180 (1970).
14. E. I. Dashevskaya and A. I. Reznikov, *Opt. Spektrosk.* **48**, 644–650 (1980).
15. E. E. Nikitin and S. Ya. Umanskii, *Semiempirical Methods for Calculating Atomic Interactions* [in Russian], VINITI, Moscow (1980).
16. E. E. Nikitin and S. Ya. Umanskii, *Theory of Slow Atomic Collisions*, Springer Verlag, Heidelberg–New York (1984).
17. M. Krauss, P. Maldonado, and A. V. Wahl, *J. Chem. Phys.* **54**, 4944–4952 (1971).
18. F. Masnou-Seeuws, M. Philippe, and P. Valiron, *Phys. Rev. Lett.* **41**, 395–399 (1978).
19. W. E. Baylis, *J. Chem. Phys.* **51**, 2665–2678 (1969).
20. J. Hanssen, R. McCarroll, and P. Valiron, *J. Phys. (Paris)* **54**, 4944–4951 (1979).
21. G. K. Ivanov, *Teor. Éksp. Khim.* **14**, 610–617 (1978).
22. A. I. Reznikov, *Chem. Phys. Lett.* **44**, 41–45 (1976).
23. J. Pascale and J. Vandeplanque, *J. Chem. Phys.* **60**, 2278–2283 (1974).
24. A. Z. Devdariani and A. L. Zegrebin, *Khim. Fiz.*, No. 7, pp. 947–956 (1982).
25. C. Bottcher, K. K. Docken, and A. Dalgarno, *J. Phys.* **B8**, 1756–1764 (1975).
26. A. R. Malvern, *J. Phys.* **B11**, 831–845 (1978).
27. E. Czuchaj and J. Sienkievicz, *J. Phys.* **B15**, 2251–2267 (1984).
28. I. N. Demetropoulos and K. P. Lawley, *J. Phys.* **B15**, 1855–1861 (1982).
29. B. Pouilly, B. H. Lengsfield, and D. R. Yarkony, *J. Chem. Phys.* **80**, 5089–5094 (1984).
30. E. I. Dashevskaya and E. E. Nikitin, *Opt. Spektrosk.* **22**, 866–871 (1967).
31. L. Krause, *Appl. Opt.* **5**, 1375–1381 (1986).
32. J. C. Gay and W. B. Schneider, *Z. Phys.* **A278**, 211–220 (1976).
33. W. B. Schnieder, *Z. Phys.* **A255**, 14–19 (1972).
34. J. Pitre and L. Krause, *Can. J. Phys.* **45**, 2671–2682 (1967).
35. A. Gallagher, *Phys. Rev.* **172**, 88–91 (1968).
36. A. Z. Devdariani and A. L. Zagrebin, *Khim. Fiz.*, No. 8, 1141–1143 (1982).
37. A. Z. Devdariani and A. L. Zagrebin, *Khim. Fiz.*, No. 2, 163–167 (1983).
38. A. Z. Devdariani and A. L. Zagrebin, *Energy Transport Processes in Metal Vapors* [in Russian], Izd. Latv. Gos. Univ. im. P. Stuchki, Riga (1985), pp. 29–38.
39. A. Z. Devdariani and A. L. Zagrebin, *Khim. Fiz.*, **5**, No. 2, 147–155 (1986).
40. E. E. Nikitin, *Adv. Quant. Chem.* **5**, 135–184 (1970).
41. A. K. Belyaev, A. Z. Devdariani, and A. L. Zagrebin, *Opt. Spektrosk.*, **55**, 807–811 (1982).
42. S. Ja. Umanski and E. E. Nikitin, *Theor. Chim. Acta (Berlin)* **13**, 91–105 (1969).

43. S. Ja. Umanski and A. I. Voronin, *Theor. Chim. Acta (Berlin)* **12**, 166–174 (1968).
44. A. I. Reznikov, *J. Phys.* **B15**, L157–L161 (1982).
45. E. E. Nikitin, *Atomic Physics*, Vol. 4, Plenum, New York (1975), pp. 529–536.
46. E. I. Dashevskaya, F. Masny, R. McCarroll, and E. E. Nikitin, *Opt. Spektrosk.* **37**, 209–215 (1974).
47. H.-J. Yuh and P. J. Dagdigian, *Phys. Rev.* **A28**, 63–72 (1983).
48. M. H. Alexander, T. Orlikowski, and J. E. Straub, *Phys. Rev.* **A28**, 73–82 (1983).
49. E. N. Borisov, N. P. Penkin, and T. P. Red'ko, *Khim. Fiz.* **5**, 605–609 (1980).
50. M. Czajkowski, E. Walentynowicz, and L. Krause, *J. Quant. Spectrosc. Radiat. Transfer* **28**, 493–501 (1982).
51. N. A. Kryukov, N. P. Penkin, and T. P. Red'ko, *Opt. Spektrosk.* **51**, 756–761 (1981).
52. N. P. Kryukov, N. P. Penkin, and T. P. Red'ko, *Opt. Spektrosk.* **42**, 33–41 (1977).
53. M. Czajkowski, E. Walentynowicz, and L. Krause, *J. Quant. Spectrosc. Radiat. Transfer* **28**, 13–20 (1982).
54. H. Umemoto, S. Tsunashima, and S. Sato, *Chem. Phys.* **47**, 257–269 (1980).
55. N. P. Penkin and T. P. Red'ko, *Opt. Spektrosk.* **57**, 979–982 (1984).
56. W. H. Breckenridge and H. Umemoto, *Adv. Chem. Phys.* **50**, 325–394 (1982).
57. W. H. Breckenridge, T. W. Broadbent, and D. S. Moore, *J. Phys. Chem.* **79**, 1233–1241 (1975).
58. A. Z. Devdariani and A. L. Zagrebin, *Opt. Spektrosk.* **58**, 1223–1227 (1985).
59. N. A. Kryukov, N. P. Penkin, and T. P. Red'ko, *Opt. Spektrosk.* **51**, 756–761 (1981).
60. E. E. Nikitin, *Khim. Fiz.*, No. 7, pp. 867–882 (1982).

# LASER MAGNETIC RESONANCE STUDIES OF PROCESSES INVOLVING FREE RADICALS IN THE GASEOUS PHASE

L. N. Krasnoperov

## 1. INTRODUCTION

The chemical behavior of a majority of reacting systems in the gaseous phase is determined by active intermediate particles: atoms and free radicals. The enhanced reactivity owing to the presence of free valence leads to a short lifetime and a low concentration for these particles. Thus, highly sensitive techniques are needed to detect and study the chemical and physical properties of atoms and free radicals. Over the last few decades, electron paramagnetic resonance (EPR) has been used extremely successfully for these purposes [1–3]. Unfortunately, the applicability of this quite universal method is restricted (in the gaseous phase) to atoms and a few diatomic free radicals [1–3]. The large number of populated rotational levels in polyatomic free radicals causes the intensity of the individual Zeeman transitions to fall below the sensitivity limits of EPR spectrometers.

Over the last 15 years, the evolution of laser technology has led to the development of a series of new, highly sensitive methods for detecting particles in the gaseous phase. These include laser-induced fluorescence (LIF) [4], intracavity laser spectroscopy (ICLS) [5–7], and laser magnetic resonance (LMR) [8–16]. While LIF and ICLS are also sensitive to stable molecules, the LMR method employs the paramagnetism intrinsic to free radicals, as does EPR.

In its essence, LMR is a natural modification of EPR. It is based on tuning a transition of a free radical to a laser line with a fixed frequency by means of the Zeeman effect.

Let us examine the behavior of the energy levels of a polyatomic free radical with spin 1/2 in a gaseous medium in a magnetic field [17, 18]. For a first-order description of the behavior of the energy levels of a free radical in a magnetic field, it is sufficient to consider the energy of the spin–rotational interaction and the Zeeman interaction of the unpaired electron spin with the external magnetic field. The Hamiltonian can then be written in the form [17, 18]

$$H = H_0 + \gamma\,(\mathbf{SN}) + g\beta\,(\mathbf{BS}). \tag{1}$$

Here $H_0$ is the vibrational–rotational Hamiltonian, the second term describes the spin–rotational interaction, and the third, the Zeeman interaction of the spin with the external magnetic field; $\gamma$ is the effective spin–rotational interaction constant; **N** is the rotational moment; $g$ is the spin $g$-factor; $\beta$ is the Bohr magneton; and **B** is the magnetic induction. In zero-magnetic field, an individual vibrational–rotational level $N, S$ is split by the spin–rotational interaction into two sublevels with total angular momenta of $J = N + S = N + 1/2$ and $J = N - S = N - 1/2$. The level with $J = N + 1/2$ is denoted by $F_1$ and that with $J = N - 1/2$, by $F_2$. The energy shifts in the energies of the $F_1$ and $F_2$ levels owing to the spin–rotational interaction are given by $\gamma(\mathbf{NS}) = (\gamma/2)(\mathbf{J}^2 - \mathbf{N}^2 - \mathbf{S}^2)$, which yields $E_{F1} = \gamma N/2$ and $E_{F2} = -\gamma(N + 1)/2$, so that the splitting of the vibrational–rotational level in zero field is $\Delta = E_{F1} - E_{F2} = \gamma(N + 1/2)$. The constant $\gamma$ generally depends on the vibrational–rotational state [18].

In weak magnetic fields ($g\beta B \ll \Delta$), both of these levels are split into $2J + 1$ components ($J = N \pm 1/2$). The energies, relative to the energy in zero field, are given by

$$E_z = g\beta B\,\langle S_z\rangle = g_{\text{eff}}\,\beta B M_J, \tag{2}$$

where the effective $g$-factor is

$$g_{\text{eff}} = g\,\frac{(\mathbf{SJ})}{\mathbf{J}^2} = g\,\frac{\mathbf{J}^2 + \mathbf{S}^2 - \mathbf{N}^2}{2\mathbf{J}^2} = g\,\frac{J(J+1) + S(S+1) - N(N+1)}{2J(J+1)}. \tag{3}$$

For a spin $S = 1/2$, this yields $g_{\text{eff}} = \pm g/(2N + 1)$, where the + and – signs refer to the $F_1$ and $F_2$ levels, respectively.

In strong magnetic fields ($g\beta B \gg \Delta$), the coupling between the spin and the rotation is broken and there are two levels $S_z = \pm 1/2$ with energies $E_z = \pm g\beta B/2$, respectively, which are split by the spin–rotational interaction into $2N + 1$ components. The energies are given by $E = E_z + \gamma\langle(\mathbf{SN})\rangle = E_z + \gamma S_z N_z$. In the intermediate magnetic field range, the energies of the levels are given by the roots of a quadratic equation [18].

Figure 1 is a diagram of the energy levels of a free radical with spin 1/2 in a magnetic field, neglecting the hyperfine and nuclear Zeeman interactions. This diagram shows that two types of transition are possible [17, 18]. The first type includes transitions between levels that correlate in high fields to levels with the same projection of the spin in the direction of the external magnetic field. The frequency of these transitions varies with the field only when the field is weak and ceases to vary for strong fields. Then the characteristic scale of the frequency changes is determined by the spin–rotational interaction. The second type of transition includes those between levels that correlate in strong fields to states with different projections of the spin on the external field. For transitions of this type, frequency tuning occurs for all values of the magnetic field, and in strong fields the rate of tuning is the same as the rate of tuning of the transition between the Zeeman spin sublevels of a free electron $d(\Delta E)/dB = g\beta$. While the transitions of the first type are tunable over a range of order $\gamma$, they continue to be electric dipole transitions for all values of the field. Transitions of the second type, on the other hand, rapidly (as $1/B^2$) lose the electric dipole component of their intensity when the field is raised and the Zeeman energy exceeds the energy of the spin–rotational interaction. Thus, with the aid of a magnetic field it is possible, in effect, to tune the

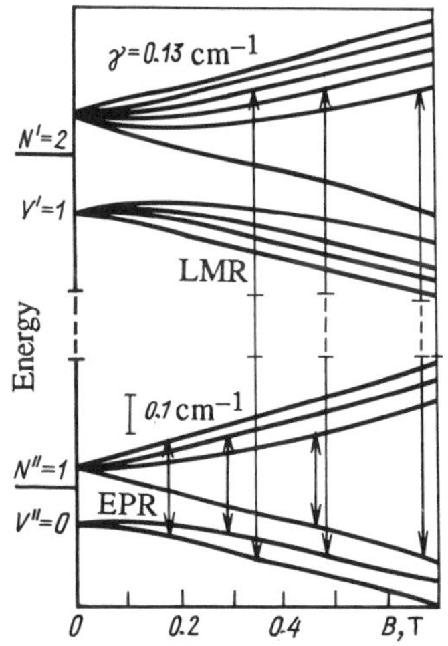


**Fig. 1.** Energy level diagram for a free radical with a spin of 1/2 in a magnetic field.

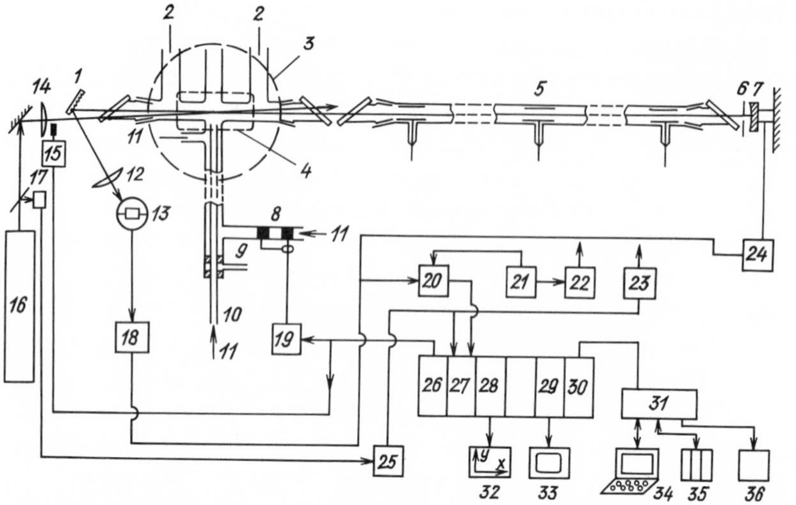


**Fig. 2.** Block diagram of an LMR spectrometer [23-25]: 1) diffraction grating; 2) pumpout for streaming reactor; 3) poles of electromagnet; 4) region in which the magnetic field is modulated; 5) $CO_2$ laser discharge tube; 6) iris diaphragm; 7) spherical mirror on a piezoelectric ceramic; 8) electrodes for rf discharge; 9) evacuable chamber for varying location of reagent delivery; 10) movable reagent inlet; 11) reagent feed; 12) NaCl lens; 13) Ge–Au photoconductive detector, 77 K (FSG–22–3A2) or Ge–Hg, 63 K (FSG–28–RTA); 14) LiF lens; 15) controllable beam chopper; 16) pulsed laser (LTIPCh-7.8); 17) photomultiplier for triggering the measurement circuit; 18) emitter follower; 19) rf oscillator (20 MHz); 20) synchronous detector (UPI-1); 21) master oscillator (GZ-33); 22) amplifier for driving the modulation coils (200 W); 23) circuit for producing a jump in the magnetic field; 24) laser frequency stabilizer; 25) pulse generator; 26) controller for discharge and beam chopper; 27) analog-to-digital converter (256 channels, 6 bits, up to 50 nsec/channel); 28) driver for XY recorder; 29) driver for graphics display; 30) crate controller; 31) Élektronika-60 minicomputer; 32) XY recorder (N-306); 33) graphics display; 34) alphanumeric display; 35) magnetic disc storage; 36) alphanumeric printer.

frequency of the vibrational–rotational transition by an amount on the order of the spin–rotational interaction, which is of order 0.1 $cm^{-1}$. Even for radicals with a small spin–rotational interaction, however, LMR spectra may appear because of anticrossing of the Zeeman levels [19–22].

In EPR spectroscopy, transitions between Zeeman sublevels of a single vibrational–rotational state are used, but in LMR spectroscopy, transitions between the Zeeman sublevels of different rotational or vibrational–rotational states are used. Lasers with submillimeter wavelengths are used in the first case and infrared lasers, in the second. The main factors responsible for the greater sensitivity of LMR compared to EPR are that electric dipole transitions are observed and there is a greater difference in the equilibrium populations of the levels involved in the transition. In the case of vibrational–rotational transitions, there is some loss of sensitivity because of the smallness of the dipole moment of the vibrational transition, but on the whole there is still a substantial advantage.

## 2. EXPERIMENTAL TECHNIQUES

Figure 2 shows a block diagram of an LMR spectrometer. The main components of the spectrometer are a laser, a source of constant magnetic field, a source of magnetic modulation, and a radiation detector. Depending on the type of transitions being observed, rotational or vibrational–rotational, LMR spectrometers can be divided into two classes. In the first case, far IR (FIR) lasers (30–1000 μm) are used. Electrical discharge HCN (311 and 337 μm) [26] and water vapor (78.8 and 119 μm) [27] lasers were used in the first experiments. Later on, lasers that are optically pumped by $CO_2$ laser radiation (optically pumped lasers, OPL) were used most widely [9]. This type of laser is much more versatile than an electrical discharge laser. By simply replacing the working medium in an OPL, it is possible to cover the range from 30 to 1000 μm.

$CO_2$ lasers are most often used in LMR spectrometers in the IR [10, 13, 14, 28]. Isotopically substituted $CO_2$ molecules with the carbon isotopes $^{12}C$ and $^{13}C$ and the oxygen isotopes $^{16}O$ and $^{18}O$ can be used to obtain lasing on several hundred lines in the range from 875–1110 $cm^{-1}$. A number of experiments using a CO laser in the 1200–2000 $cm^{-1}$ range have been reported [29–31]. A color center laser has been used to observe the OH radical at 3670 $cm^{-1}$ [32] and an InSb crystal spin–flip laser pumped by a CO laser has been used to record spectra of NO with Zeeman modulation [33].

Solenoids and electromagnets are used to create a magnetic field. Because of the impossibility of obtaining the π-component of the spectra in the ordinary configuration (laser beam along the solenoid), the small magnetic fields attainable with conventional solenoids ($B \leq 0.2$ T), and the difficulties of obtaining substantial amplitude modulation, electromagnets have been used in the overwhelming majority of this work. A small modulation of the magnetic field is achieved with the aid of coils mounted in the gap of the electromagnet and driven by an alternating current. The choice of modulation frequency is extremely important and will be examined separately.

As a rule, the measured quantity is the variable component of the laser output intensity, although the power absorbed in the sample (acousto-optical method) [34], the rotation of the plane of polarization [35], and optical detection [36] can also be used. In FIR LMR spectrometers, Golay cells, bolometers, and semiconductor photoconductive detectors are used to detect the radiation [37]. In the first designs, a Golay cell was used as a detector [26]. The slow response time of this detector means that magnetic field modulation frequencies of 10–100 Hz must be used. Semiconductor bolometers based on gold- or gallium-doped germanium cooled by liquid helium were more widely used subsequently [38]. These detectors have a higher detectivity and shorter response time, so that modulation frequencies of roughly 1 kHz can be used. Recently, extremely fast ($\tau < 1$ μsec) semiconductor photoconductive detectors, Ge–B, Ge–In, Ge–Ga, Ga–As, and *n*In–Sb cooled to 4 K, have been used. These detectors allow modulation frequencies on the order of hundreds of kilohertz to be used [37].

Cooled semiconductor photoconductive detectors are used exclusively in IR LMR spectrometers. Photoconductors based on liquid-nitrogen cooled Ge–Au and Hg–Cd–Te, as well as Ge–Hg and Ge–(Zn, Sb), which require cooling to the melting temperature of nitrogen, are used for detecting the light from $CO_2$ lasers, which are most widely used in LMR spectrometers in this range. Detectors based on liquid-nitrogen cooled Ge–Au and In–Sb are used to detect the emission from CO lasers.

The correct choice of the modulation frequency for the magnetic field is of great importance in obtaining the maximum sensitivity. Here several factors must be taken into account. On one hand, it is necessary to get as high a modulation frequency as possible in order to reduce the contribution from mechanical vibrations of the spectrometer. On the other, it is easy to show that the amplitude of the modulation in the magnetic field is given by the formula

$$B_M = \left(\frac{2\mu_0 QP}{\omega V}\right)^{1/2} \quad \text{(SI)}$$

or

$$B_M = 10^{-4}\left(\frac{4\cdot 10^7 QP}{fV}\right)^{1/2}, \tag{4}$$

where in the second formula $B_M$ is the amplitude of the modulation (T), $P$ is the power of the generator that drives the modulation coils (W), $V$ is the volume in which the modulation field must be maintained (cm$^3$), and $f$ is the modulation frequency (Hz). It is clear from this formula that the amplitude of the modulations decreases as $1/\sqrt{f}$ for a fixed generator power. For extracavity LMR spectrometers, the optimum modulation frequency is $f = 15$–$20$ kHz, since at these frequencies the noise resulting from vibrations of the spectrometer ceases to be important [39]. Placing the cuvette with the gas sample inside the cavity of the spectrometer will increase the sensitivity. This question has been examined elsewhere [8, 37, 40, 41]. Two mechanisms are responsible for the increased sensitivity of an intracavity spectrometer: an increase in the effective number of passes of the beam through the sample [42] and using modulations at the frequency of the relaxation oscillations in the transient response of the laser (the latter effect has been used to raise the sensitivity in experiments with intracavity absorption in solid state lasers [43] and it has been proposed [40] that this effect be used to increase the sensitivity of LMR spectrometers). For homogeneously broadened laser lines in the approximation of a small steady-state gain in each pass, with single-exponential relaxation of the inverted population, and the approximation of kinetic equations for the frequency characteristic of the spectrometer $K(\omega)$ [where $K(\omega)$ is the ratio of the complex amplitude of the variable component of the laser output when the sample is placed in the cavity to the amplitude of this component with single-pass absorption], the following expression can be obtained [40, 41]:

$$K(\omega) = K_a \frac{1 - i\,\omega\tau_0 K_r}{1 - (\omega\tau_0)^2 K_a K_r - i\,\omega\tau_0 K_a}. \tag{5}$$

Here $K_a$ is the increase in the sensitivity with adiabatic modulation ($\omega \to 0$), $K_r$ is the increase in the sensitivity at the resonant modulation frequency $\omega_r = (\tau_0\sqrt{K_a K_r})^{-1}$, and $\tau_0$ is the time of flight for a photon along the cavity length. In this model, these quantities are given in terms of the laser parameters by [40, 41]

$$K_a = \frac{\varkappa_e}{\beta(\varkappa_e - \beta)}, \quad K_r = \frac{\tau}{\tau_0}\,\frac{\beta}{\varkappa_e}, \tag{6}$$

where $\varkappa_e$ and $\beta$ are the unsaturated gain and loss, respectively, in a single pass, and $\tau$ is the relaxation time for the population inversion.

Equation (5) predicts a resonant form for $|K(\omega)|$. The conclusions of this theory have been verified experimentally [41]. Figure 3 shows the frequency characteristics of a spectrometer based on a $CO_2$ laser for different cavity losses. The curves have a true resonant character, and a substantial gain in sensitivity can be obtained at a resonant modulation frequency (about 150 kHz). It was shown that placing the cuvette with the sample radicals inside the laser cavity results in a sensitivity gain of 100–

1000 times at the resonant modulation frequency. It was also found that the theoretical model described above agrees qualitatively with the experiment. An analysis of Eq. (5) at low modulation frequencies also reveals some interesting features [37, 44, 45]. In terms of the same model (homogeneously broadened line, exponential relaxation of the population inversion, low steady-state gain per pass), the output intensity of the laser is given by

$$J = t\frac{J_s}{2}\left(\frac{\varkappa_e}{\beta} - 1\right), \tag{7}$$

where $t$ is the coupling coefficient and $J_s$ is the saturation intensity. Using Eqs. (5) and (6), we find the amplitude of the variable signal in the laser output during adiabatic modulation to be

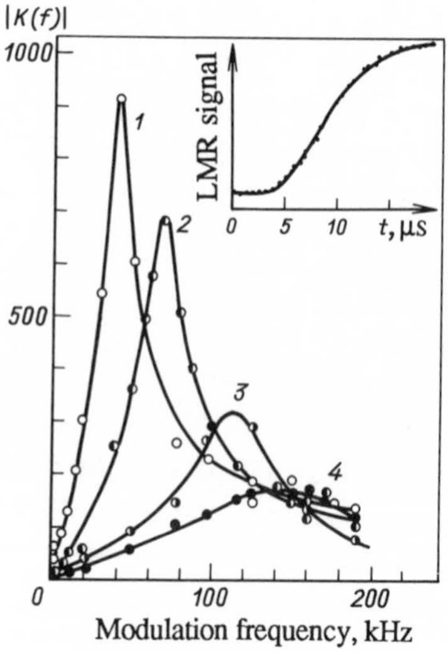


**Fig. 3.** Frequency characteristics of a $CO_2$ laser LMR spectrometer for various cavity losses [41]: the losses decrease from curve 1 to curve 4. The values of the frequency characteristic with a zero modulation frequency are $K(0) = 33, 20, 13$, and 12, respectively, for curves 1–4. The insert at the upper right shows the transient characteristic of the spectrometer obtained during pulsed photodissociation of molecular chlorine by the third harmonic of a neodymium laser [23]. The risetime $\tau_{0.1-0.9} = 9$ μsec is equivalent to a time constant of 4 μsec.

$$J_{\sim} \propto K_a J = \frac{tJ_s}{2} \frac{\varkappa_e}{\beta^2}. \quad (8)$$

This expression implies that, as the laser approaches the lasing threshold ($\beta \to \varkappa_e$), the signal ceases to depend on the output power of the laser [8, 37, 44]. This case is apparently realized for FIR LMR spectrometers [8]. Placing the sample inside the laser cavity and using resonance modulation are now used extensively to increase the sensitivity of LMR spectrometers.

Several other ways of increasing the sensitivity have been proposed. When the noise of the spectrometer is limited by detector noise (as is true for a $CO_2$ laser), a further increase in the sensitivity can be achieved when the response of the laser at the resonant modulation frequency is increased by placing a bleachable gas inside the laser cavity (thereby approaching a $Q$-switched regime) [41, 46]. This approach has been studied theoretically and experimentally [41]. In this way it is easy to obtain an additional sensitivity increase of an order of magnitude. Further increases in the gain by this method will impose more stringent demands on the stability of the laser parameters.

When the output power of the laser is much greater than the power needed to saturate the detector, it is possible to increase the fraction of modulated signal by passing the light through a saturating absorber gas [37, 45]. An analysis of this situation shows that the variable component of the intensity is attenuated less than the dc component. This approach can be used for low-frequency modulation, and theoretical estimates show that the sensitivity gain may be two orders of magnitude [37, 45]. With high-frequency modulation, placing the saturating absorber gas in the path of the laser output yields no gain in sensitivity [37, 45].

If the output power greatly exceeds the power that would saturate the detector and the laser noise is predominant, then an increase in the sensitivity may be achieved by observing the rotation of the plane of polarization of the laser light passing through the sample [35]. A gain of approximately a factor of 20 in the sensitivity of an intracavity LMR spectrometer based on a CO laser has been obtained in this way [35].

## 3. TIME-RESOLVED LASER MAGNETIC RESONANCE

LMR in its traditional variant (modulated magnetic field, synchronous detection of the signal followed by averaging with a time constant of 1–10 sec) is most suitable for stationary measurements; hence, a combination of LMR techniques with a discharge-stream method has gained the widest acceptance. In a number of applications (studies of fast processes, detection of LMR spectra of excited particles, studies of elementary processes involving excited particles, etc.), however, high time resolution is required. The first experiments on the use of LMR for observing processes with time resolution were done by Krasnoperov and Panfilov [47], who observed a population inversion in the fine structure levels of the chlorine atom during photolysis of ICl by the second harmonic of neodymium laser light. In this experiment an extracavity scheme without magnetic field modulation was used. The time resolution was 70 nsec, and the sensitivity for chlorine atoms was about $10^{14}$ cm$^{-3}$. The low sensitivity, as well as the parasitic signal caused by acoustic oscillations of the gas in the cuvette after absorption of the photolytic pulse, made it impossible to perform detailed kinetic studies. Later on, placing the cuvette inside the cavity of the spectrometer and modulating the magnetic field at a frequency of 20 kHz with subsequent narrow-band amplification of the signal made it possible to raise the sensitivity substantially (to $3 \cdot 10^{12}$ cm$^{-3}$) and to study the heterogeneous recombination of chlorine atoms after pulsed photolysis of molecular chlorine [48]. The time resolution in this experiment was determined by the risetime of the amplifier and was about 100 μsec. LMR in the FIR has been used [49] for observing short-lived particles during the multiphoton dissociation of $SF_6$ in the presence of $H_2O$ and NO. An intracavity scheme was used, but the magnetic field was not modulated. The time resolution was roughly 1 μsec and the sensitivity (estimated from the published data [49]) for OH radicals was on the order of $10^{12}$ cm$^{-3}$.

It should be noted that there are several advantages to using magnetic modulation in time-resolved LMR spectroscopy experiments. First, magnetic modulation makes it possible to avoid the parasitic signal caused by acoustic oscillations of the gas in the cuvette after the light signal is absorbed. Second, using modulation at the frequency of the relaxation oscillations in the transient response of the laser [40, 41] makes it possible to increase the sensitivity of the apparatus substantially. Time-resolved LMR has been used to study several elementary reactions of chlorine atoms with the $SiH_3$ radical [23]. An intracavity scheme and resonant modulation of the magnetic field at a frequency of 150 kHz, in combination with a signal sampling and averaging technique, made it possible to attain high sensitivity ($4 \cdot 10^{10}$ cm$^{-3}$ for chlorine atoms) with a time resolution of 4 μsec, which is quite sufficient for measurements of this type. The main principles of operation of time-resolved LMR spectrometers have been discussed by Krasnoperov et al. [23] The pulse characteristic of the spectrometer, i.e., the response of the radiation to a δ-function absorbed pulse inside the cavity with magnetic field modulation at the frequency $\omega_r$ of the relaxation oscillations and synchronous detection of the signal with averaging over the initial phase of the modulation by multiple sampling of the signal, is given by [23]

$$h^*(t) = \frac{2}{\pi\tau_0} \exp(-\Omega_1 t/2) \cos^2(\omega_r t), \tag{9}$$

where the parameter $\Omega_1 = \varkappa_e/\beta\tau$ characterizes the width of the resonance at the frequency characteristic. Figure 3 shows that the width of the resonance peak is comparable to the resonance frequency; hence, we may expect a time resolution comparable to the period of the magnetic field modulations. Thus, this type of spectrometer can be used to study processes taking place over times $\geq$10 μsec.

## 4. RADICALS DETECTED BY LMR. THE IDENTIFICATION PROBLEM

So far LMR studies have been devoted mainly to detecting and interpreting the LMR spectra of different radicals. The LMR method is highly sensitive ($10^8$–$10^{10}$ cm$^{-3}$) and has high resolving power ($10^6$–$10^7$), so that particles can be detected in very low concentrations and their structural parameters can be determined with very high accuracy. The interpretation of LMR spectra makes it possible to determine with high accuracy such spectroscopic parameters as vibrational frequencies, rotational constants, spin–rotational and hyperfine interaction constants, etc. The spectroscopic problems have been examined in a number of reviews [8–12, 15, 16], so that we shall not consider them here.

Table 1 is a list of the particles that have been detected with the aid of laser magnetic resonance spectrometry.

The table shows that, so far, about 70 particles have been detected by LMR and the list of these particles is growing continuously. The initial rapid growth in the number of publications on the detection and interpretation of LMR spectra of short-lived particles was apparently a consequence of the fact that most of them are simple and supplementary spectroscopic information is available for several of them. This greatly simplified the interpretation of the corresponding LMR spectra, which is generally not a simple problem. The difficulty arises because the LMR spectra may ap-

**TABLE 1.** Particles Detected by LMR in the Gaseous Phase

| Particle | Spectral range | References |
|---|---|---|
| | Stable particles | |
| $O_2$ | FIR | [26, 50—55] |
| NO | FIR | [56] |
| | IR | [29—31, 57—60] |
| $NO_2$ | FIR | [61] |
| | IR | [62, 63] |
| $ClO_2$ | IR | [19—22] |
| | Atoms | |
| Cl | IR | [23, 47, 48, 64—70] |
| O | FIR | [71—73] |
| C | FIR | [74] |
| | Diatomic radicals | |
| OH | FIR | [27, 49, 73, 75—93] |
| | IR | [32] |
| OD | FIR | [79, 93] |
| CH | FIR | [94—96] |
| CF | FIR | [97] |
| | IR | [98, 237] |
| NH | FIR | [99, 100] |
| ND | FIR | [100] |
| NF | — | [15] |
| NS | FIR | [101] |
| NSe | — | [12] |
| SH | IR | [102] |
| SD | IR | [103, 104, 239] |
| SO | IR | [105] |
| $S_2$ | — | [12] |
| SeH | FIR | [106] |
| | IR | [107] |
| SeD | FIR | [108] |
| | IR | [109, 110] |
| SeO | IR | [111] |
| SeN | IR | [240] |
| PH | FIR | [112—114] |
| PD | FIR | [114] |
| | IR | [115] |
| PO | FIR | [116] |
| SiH | FIR | [117] |
| | IR | [118] |
| SiD | IR | [241] |
| FO | IR | [119] |
| ClO | FIR | [100, 120, 121] |
| | IR | [122] |
| BrO | IR | [123] |
| AsO | IR | [124] |
| GeH | FIR, IR | [125] |
| | Triatomic radicals | |
| $HO_2$ | FIR | [38, 73, 86—92, 120—124, 126—129, 130—134] |
| | IR | [60, 135—139 237] |
| $DO_2$ | FIR | [140] |
| | IR | [141, 142] |
| $FO_2$ | FIR | [143] |
| | IR | [144, 145] |
| $CH_2$ | FIR | [146—151] |
| | IR | [152—153] |
| $CD_2$ | FIR | [154] |
| HCO | FIR | [155] |
| | IR | [156—159] |
| DCO | IR | [160] |
| FCO | IR | [16, 161] |
| NCO | IR | [162, 238] |
| $NH_2$ | FIR | [163—165] |
| $ND_2$ | IR | [156, 157, 166, 167] |
| NHD | FIR | [168] |
| | IR | [24, 169] |
| $NF_2$ | IR | [28, 170—173] |
| $PH_2$ | FIR | [174, 175] |
| | IR | [176] |
| $PO_2$ | FIR | [177] |
| HSO | IR | [12, 178] |
| ClSO | FIR | [179] |
| FSO | FIR | [179] |
| $C_2H$ | FIR | [180] |
| $N_2H$ | — | [15] |
| | Tetraatomic radicals | |
| $CH_2F$ | FIR | [181] |
| $NH_2O$ | FIR | [182, 242] |
| $SiH_3$ | IR | [23, 66, 67, 69, 183, 184] |
| $CH_3$ | — | [12] |
| | Pentaatomic radicals | |
| $CH_3O$ | FIR | [130, 185, 186] |
| $CD_3O$ | FIR | [240] |
| $CH_2OH$ | FIR | [187] |

**Table 1.** Continued

| Particle | Spectral range | References | Particle | Spectral range | References |
|---|---|---|---|---|---|
| Excited states | | | $CO^*$ $(^3\Pi)$ | — | [9, 15] |
| $Hg^*$ $(6^3P_0)$ | IR | [188] | $HO_2^{\bullet}$ | — | [9, 193] |
| $Kr^*$ | IR | [189] | $O_2^{\bullet}$ $(a_1{}^1\Delta_g)$ | FIR | [194] |
| $Xe^*$ | IR | [189] | $NF^*$ $(a^1\Delta)$ | FIR | [195, 196] |
| $He^*$ | IR | [36] | Ions | | |
| $Mg^*$ $(3P)$ | — | [190] | $HBr^+$ | FIR | [197] |
| $PH^*$ $(^1\Delta)$ | FIR | [112[ | $HCl^+$ | FIR | [198] |
| $SO^*$ $(a^1\Delta)$ | IR | [145, 191] | $DCl^+$ | IR | [199, 236] |
| $CH_2^{\bullet}$ $(a^1A_1)$ | FIR | [192] | | | |

pear at arbitrary distances between the frequency of a given laser line and that of a rotational or vibrational–rotational transition, so that they are, in principle, incomplete. Although a number of methods for interpreting LMR spectra have been developed [8–12, 15, 18, 200], when no additional spectroscopic information is available, the analysis of spectra is an extremely complicated problem, especially when the spectra are not well resolved. At the same time, for a number of kinetic applications, such as the detection of active particles in complicated chemical reactions and the study of elementary chemical processes involving such particles, it is not necessary to interpret the LMR spectra of these particles, although that is extremely desirable. For kinetic applications of this type, reliable identification of the particles responsible for the observed LMR spectra is necessary. Although the most reliable identification, of course, involves a full interpretation of the LMR spectra and a determination of the structural parameters, in many cases a sufficiently reliable identification may be based on various other physical and chemical considerations.

Let us examine this question using the example of the $SiH_3$ and FCO radicals, whose spectra have not yet been interpreted. LMR spectra of the $SiH_3$ radical were first detected by Krasnoperov et al. [183]. (This radical had not been detected in the gaseous phase before.) Although spectra could be generated by a score of $CO_2$ laser lines, most of them were poorly resolved. The interpretation was made more complicated by the fact that two vibrational frequencies of the $SiH_3$ radical lie in the output range of the $CO_2$ laser. In that case, the spectra were attributed to this radical on the basis of the following spectroscopic and chemical considerations:

a) The $SiH_3$ radical has two vibrations: $\nu_2$ (996 $cm^{-1}$) and $\nu_1$ (925 $cm^{-1}$) [201], whose frequencies lie in the output range of the $CO_2$ laser and, therefore, can produce LMR spectra.

b) The $SiH_2$ radical has a vibrational frequency in the output range of the $CO_2$ laser ($\nu_2$, 1004 $cm^{-1}$) [201], but this radical has a singlet ground state. The singlet–triplet splitting is large (~1 eV) [202], so that $SiH_2$ in its ground state should not produce an LMR spectrum.

c) The SiH radical has a vibrational frequency of 2041 $cm^{-1}$ [201], which is far from the output range of a $CO_2$ laser (~1000 $cm^{-1}$).

d) Identical spectra were obtained from the products of the reactions Cl + $SiH_4$ and F + $SiH_4$, where the chlorine atoms in the first reaction were obtained from an electrical discharge in $Cl_2$ and the fluorine atoms in the second, from a discharge in $CF_4$. It follows immediately that the source of the observed LMR spectra may consist exclusively of silicon and hydrogen atoms. These considerations show that the most probable source of the LMR spectra is the $SiH_3$ radical, although they cannot exclude the possibility that the spectra in these systems are caused by radicals of the form $Si_nH_m$. Further confirmation of this identification was based on kinetic considerations.

e) A time-resolved LMR spectrum has been observed at the 11.4 μm P(16) transition of a $^{13}C^{16}O_2$ laser during pulsed production of chlorine atoms by laser photolysis of $S_2Cl_2$ in the presence of $SiH_4$. The kinetics of the disappearance of the chlorine atom was also monitored through the LMR signal [23, 67]. It was found that the LMR signal at the P(16) line appears with no delay relative to the time the signal from the chlorine atom [the P(36) line of the $^{13}C^{16}O_2$ laser] disappears. The conditions in these experiments were such that if the particle responsible for the LMR spectrum had appeared in a secondary processes, then the characteristic risetime of the signal would have been at least 1 msec [the density of the radicals was $(2\text{–}10)\cdot 10^{12}$ $cm^{-3}$], while the (actual) risetime of the signal was about 15 μsec. This indicates unambiguously that the particle responsible for the spectrum appears in the first step of the reaction of a chlorine atom with a silane molecule, and therefore cannot contain more than one atom of silicon. The attribution of the observed LMR spectra to the $SiH_3$ radical on the basis of all these considerations is, thus, quite reliable. Somewhat indirect confirmation is provided by observations of identical spectra during chain branching oxidation and chlorination reactions of silane [66–69, 184], as well as during pyrolysis of silane. An example of an LMR spectrum of the $SiH_3$ radical is shown in Fig. 4.

The attribution of LMR spectra to the FCO radical was based on chemical considerations and isotope substitution experiments [161]. Three chemical reactions capable of generating FCO radicals were used [16, 161]:

$$O + C_2F_4 \rightarrow \begin{cases} \xrightarrow{a} CF_2^{\bullet}\,(T) + CF_2O, \\ \xrightarrow{b} CF_2\,(S) + CF_2O, \\ \xrightarrow{c} FCO + CF_3, \end{cases} \quad (10)$$

$$F + HFCO \rightarrow FCO + HF, \quad (11)$$

and

$$F + CO + M \rightarrow FCO + M. \quad (12)$$

Identical LMR spectra were observed with these reactions at about thirty $CO_2$ laser lines and could be attributed to the FCO radical, which has a vibrational frequency ($\nu_2 = 1018$ cm$^{-1}$ [201]) in the output range of a $CO_2$ laser, on the basis of isotope substitution experiments. Isotope separation techniques have long been known as a powerful instrument for identifying unknown spectra of short-lived particles [203]. Chasovnikov et al. [161] have used a simple isotope substitution technique which does not require a large amount of the isotope-enriched substance but yields information on the amount of one or another atom in the radical responsible for the spectrum. The arrangements for such an experiment are sketched in Fig. 5. A loop containing one of the reagents, but in isotope-enriched form, was inserted into the line through which that reagent was fed to a valve which released the reagent into a streaming reactor. The spectrometer was tuned to the peak of one of the LMR spectrum lines. In the case of reaction (10), oxygen enriched in the isotope $^{18}O$ was used. If the radical responsible for a given LMR spectrum contains $n$ oxygen atoms, then the signal should drop to $(1-\alpha)^n$ times its initial value when enriched oxygen is fed into the cuvette, where $\alpha$ is the degree of enrichment. (The isotope shifts for the free radicals are so large on the scale of LMR spectroscopy that this clearly shifts the line out of resonance. The appearance of a resonance at a given magnetic field for the isotopically enriched radical is extremely unlikely and is easily checked experimentally.) It is easy to determine the number $n$ of oxygen atoms in the radical responsible for the spectrum by measuring the drop in the signal in this sort of experiment with a known enrichment of the reagent. The result of such an experiment is shown in Fig. 5. When oxygen enriched to $(57 \pm 1)\%$ in $^{18}O$ was used, the signal

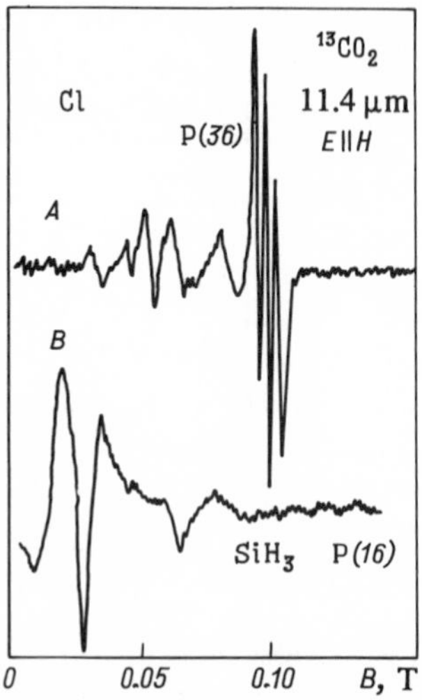


**Fig. 4.** LMR spectra of the chlorine atom (A) and $SiH_3$ radical (B) recorded during spontaneous chain branching in the chlorination of silane [66, 69].

fell by 55 ± 3%, which immediately shows that the radical responsible for the spectrum contains a single oxygen atom. Similar experiments have been done with CO enriched in $^{13}C$ in reaction (12). They showed that the radical responsible for the spectrum has a single oxygen atom. These experiments made it possible to attribute a number of the LMR spectrum lines unambiguously to the FCO radical. In order to obtain correct results in experiments of this type, however, a number of precautions must be observed [161]:

1) The radical transition must not be saturated, since impurities contained in the isotopically enriched substance may include strong relaxants (and vice versa).

2) The reactor must not contain significant amounts of molecular gases which absorb the laser light, since this can lead to a large change in the sensitivity of the spectrometer as it approaches a *Q*-switched regime [41, 46]. Then the sensitivity might be changed when an isotopically mixed reagent is added, if it contained another impurity.

3) And, finally, neither the standard nor the isotopically enriched reagent can contain impurities which react with the observed radical.

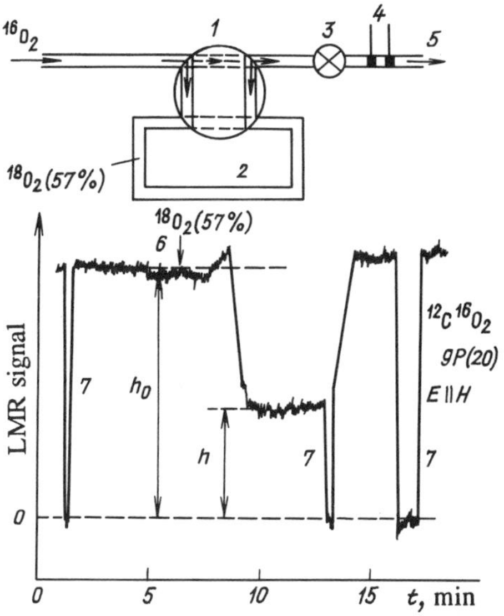


**Fig. 5.** A sketch of the experiment with isotopic substitution of the reagent and the result of an experiment with the isotopic substitution of $^{16}O_2$ for $^{18}O_2$ (57%) in the reaction $O + C_2F_4 \rightarrow FCO + CF_3$. The spectrometer is tuned to the maximum of the LMR line at the $9P(20)$ line of a $^{12}C^{16}O_2$-laser with E ∥ H: 1) valve; 2) loop filled with isotopically enriched reagent; 3) needle valve; 4) rf discharge electrodes; 5) to the streaming reactor; 6) the time the loop is switched in; 7) zero check [161].

In a number of cases, therefore, reliable attribution of LMR spectra may be achieved without having to interpret them. Although the amount of work required to do this may seem large, it more than makes up for that by offering the possibility of directly detecting a number of chemically interesting particles with high sensitivity, especially if they are not detected by other methods. It may be hoped that the low rate at which the list of particles detected by LMR is being extended (which appears to be caused by a lack of cases accessible to interpretation) will be increased as the contribution of kinetics studies to LMR studies becomes greater, although even now the prospects for applications of LMR spectroscopy are very extensive (Table 1).

The following estimate is useful in this connection: let one of the vibrational frequencies of a radical lie in the output range of a $CO_2$ laser. We now evaluate the probability that a vibrational–rotational line of the free radical is close enough to one of the laser lines that an LMR spectrum can be recorded. Assuming that the effective tuning width of the line is on the order of the spin–rotational interaction (~0.1 $cm^{-1}$), and given that the separation between laser lines is ~2 $cm^{-1}$, we find that the probability that an LMR spectrum can be recorded at one of the laser lines is ≈1/20. Given the large number of intense transitions of a polyatomic radical (more than 200), we find that the probability of detecting a spectrum at some laser line is 200/20 ≥ 10; that is, on the average, a spectrum should be observed at more than ten laser lines. This estimate, although it is not rigorous, is in agreement with experimental observations and shows that many radicals should be observable by $CO_2$ laser LMR spectroscopy, for which it is sufficient that the frequency of one of the vibrations lie within the output range of the laser, the transition not be too weak, and the spin–rotational interaction not be too small.

## 5. STUDIES OF ELEMENTARY CHEMICAL PROCESSES

The high sensitivity of LMR techniques means that they can be used to study elementary chemical processes in the gaseous phase and measure their rate constants. It was first used for these purposes by Howard and Evenson [81] in 1974 in a study of several elementary processes involving the hydroxyl radical OH. A combination of an FIR LMR spectrometer based on an $H_2O$ laser and a discharge–stream system was used. The high sensitivity of the method was demonstrated ($2 \cdot 10^8$ OH radicals $cm^{-3}$) and the rate constants of elementary reactions of OH with CO, NO, and $NO_2$ molecules were measured. A very important characteristic of the spectrometer from the standpoint of kinetic applications was also verified [81]: the linear dependence of the LMR signal amplitude on the density of the radicals being studied. It was shown that the signal amplitude is proportional to the density of OH radicals over a very wide range of variation. Subsequently, a group of other processes involving OH were studied by this method [73, 81–85, 87, 91, 93], including several which are important in atmospheric photochemistry. Laser magnetic resonance in the IR ($CO_2$ laser) was first used to study short-lived particles by Gershenzon et al. [28], who recorded the LMR spectrum of $NF_2$ radicals produced by thermal dissociation of tetrafluorohydrazine. Some kinetic studies using LMR in the IR were made in 1978 [171]. The dissociation of tetrafluorohydrazine was studied by observing the density of product $NF_2$ radicals in a static system with the aid of the LMR spectrum [171]. Subsequently, the spectrometer was combined with a discharge-stream system and a number of radical–radical processes involving $NF_2$ radicals were studied [172, 173]. As can be seen from Table 1, although the number of free radicals that can be detected by LMR techniques is large, the number of atoms accessible to detection by this method is extremely small. At the same time, the reactions of radicals with atoms are of considerable interest. Detection of a variety of atoms, such as H, O, and N, is also desirable because of a need to monitor their densities in discharge-stream experiments, since they can be produced in significant amounts through dissociation, in an electrical discharge, of impurities contained in the inert gases used as a carrier gas. In order to overcome this shortcoming of the LMR technique, it has been proposed [88, 172] that an LMR spectrometer be combined with an EPR spectrometer that uses the same magnet. The range of test objects can then be greatly extended; one can detect a large number of atoms and a variety of diatomic radicals by EPR [1–3]. This sort of two-band EPR-LMR spectrometer fits the discharge-stream technique well and can be used to study a broad class of processes. Combined EPR-LMR spectrometers are especially convenient for studying reactions between radicals, since that requires calibration of the absolute sensitivity, which is easily done using the EPR method [88, 139, 172, 173].

At about the same time (1978), the first experiments using time-resolved LMR to investigate the photolysis of $Cl_2$ and ICl molecules were undertaken [47, 48]. A population inversion was observed in the fine structure levels of the chlorine atom during pulsed photolysis of the ICl molecule by the second harmonic of a neodymium laser [47] and the efficiency of heterogeneous recombination on a quartz surface of chlorine atoms formed by photolysis of molecular chlorine with a pulsed lamp was determined [48]. Subsequently, the sensitivity of the method was greatly increased by using an intracavity spectrometer with resonance modulation of the magnetic field and using signal averaging techniques. This made it possible to study a variety of very fast processes involving chlorine atoms and $SiH_3$ radicals [23, 67]. The feasibility of using LMR in the far infrared for time-resolved experiments was demonstrated [49] and the first kinetic studies were completed [204].

Most of the methods currently in use for determining the rate constants of elementary processes can be divided arbitrarily into two groups: stream and pulsed [4, 6, 205, 206]. In the stream methods, as a rule, the reaction kinetics are measured by observing the density of radicals at some point in the vessel while varying the distance from the point where the reagents are mixed to the observation zone. Then the reaction time is related to the length by the simple formula $t = x/\bar{v}$, where $\bar{v}$ is the average velocity of the flow over the cross section of the vessel. Generally, electrodeless rf or microwave discharges in suitable molecules or reactions of the discharge products with molecules are used to obtain atoms and free radicals, but other sources may be used (thermal, photochemical, etc.) [205]. In the pulsed methods, the drop (or rise) in the density of the particles is observed in real time after they are created in a pulse. For this purpose, pulsed photolysis of suitable molecules by visible or UV light is often used, but other initiation sources can also be used (pulsed discharge, electron beam, etc.). Recently, yet another powerful source for pulsed generation of short-lived particles has appeared: multiphoton dissociation of molecules by pulsed IR laser light [4, 207]. Although a stream system is often used in pulsed experiments to establish the conditions in the reaction vessel, it is of subsidiary importance, and over the time that the reaction takes place, the flow can be regarded as motionless. Both the stream and pulsed techniques have a number of advantages and disadvantages, regardless of the method used to detect the particles under investigation [205, 206]. The basic advantages of the stream method are the ease of generating many free radicals, the possibility of obtaining them in independent sources after which they can be mixed in the reaction volume, moderate requirements on the response time of the detection system, and greater simplicity of the experimental apparatus. The disadvantages of this method include the following: a limited pressure range (20–2000 Pa), stiffer requirements on the purity of the carrier gas, and special difficulty in taking account of heterogeneous processes in which the particles under study may participate. The last difficulty is fundamental, since meeting the conditions of flow uniformity requires rapid radial diffusion, which assumes *a priori* that the particles will undergo many collisions with the walls during the time of the reaction being studied. The standard procedure, in which the reaction time is measured with different concentrations of the reagent, does not eliminate this problem, since the reaction being studied may take place on the walls of the reaction vessel. A more definite result can be obtained by measuring the same process in vessels with different diameters, but this procedure is extremely tedious and is used quite rarely. The problem of heterogeneous reactions is almost completely overcome in a modification of the stream method by studying the reaction in a diffusion cloud with the flow [208, 209], but this technique is not as widely used as the standard stream method.

**TABLE 2.** Comparison of Stationary LMR in Combination with the Stream Method and Time-Resolved LMR

| Parameter | Stationary LMR + stream | Time-resolved LMR |
|---|---|---|
| Pressure range | 20-2000 Pa | 20-5000 Pa |
| Time resolution $\tau$ | 0.3 msec (determined by the spatial resolution along the stream) | 4 μsec |
| Sensitivity $n_{min}$ | $2 \cdot 10^9$ cm$^{-3}$ (for Cl atoms) | $4 \cdot 10^{10}$ cm$^{-3}$ (for Cl atoms [23]) |
| The product $n_{min}\tau$ | $6 \cdot 10^5$ cm$^{-3}$sec | $1.6 \cdot 10^5$ cm$^{-3}$sec |
| Range of measurable rate constants | $3 \cdot 10^{-10}$–$10^{-16}$ cm$^3$sec$^{-1}$ | $3 \cdot 10^{-10}$–$10^{-17}$ cm$^3$sec$^{-1}$ |
| Absolute calibration | Relatively simple | More complicated |
| Possibility of studying radical processes | Yes | Limited |
| Possibility of studying processes involving excited particles | Limited | Yes |
| Problem with heterogeneous processes | Yes | No |
| Problem in studies with adsorbed molecules | Yes | Yes, but to a much smaller extent |

The basic advantages of the pulsed methods include a complete absence of problems with heterogeneous processes, less rigid requirements on monitoring the gas composition in the reaction cuvette, the ability to study processes involving excited particles, and a wider range of pressures. The disadvantages of this method are a greater difficulty of generating the particles to be studied, especially if a radical–radical process is under investigation, high specifications for the response time of the detectors, and a relatively complicated experimental apparatus (now, digital data acquisition and processing techniques are generally employed).

As a rule, the rate constants of elementary bimolecular reactions are measured, if possible, under conditions such that the reaction is of pseudo-first order, i.e., such that the concentration of one reagent is much greater than that of the other. Then the absolute concentration of only one reagent has to be known: the one that is in excess. This simplifies the measurements considerably, especially when stable molecules are in excess, since then their concentrations in the reactor can be calculated from the flow rates for the reagents. The greatest difficulties arise in measuring the rates of very fast processes, since small concentrations of the reagents must then be used and this imposes rigid conditions on the sensitivity of the detection apparatus. We now estimate the sensitivity required to measure the rate constants of the fastest processes under pseudofirst order reaction conditions. Let the apparatus have a time resolution $\tau$. A reliable determination of the reaction time requires that $t_r \geq 3\tau$. On the other hand, $t_r^{-1} = k[A]$, where [A] is the concentration of the reagent that is in excess. Requiring that the concentration exceed the initial concentration $n_0$ of the radicals under study by an order of magnitude, i.e., $[A] \geq 10n_0$ (the condition for a pseudofirst-order reaction), while $n_0$ is an order of magnitude greater than the minimum detectable density $n_{min}$ (for a signal-to-noise ratio of at least 10), we obtain $(100k \cdot n_{min})^{-1} \geq 3\tau$. Then, for the fastest processes ($k \approx 3 \cdot 10^{-10}$ cm³/sec), we obtain the following requirement for the product of the minimum detectable density and the time resolution of the apparatus:

$$n_{min}\tau \leqslant 10^7 \text{ cm}^3\text{/sec.} \tag{13}$$

A relative comparison of the stationary LMR method in combination with the stream technique and time-resolved LMR is shown in Table 2. It can be seen from the table that criterion (13) is satisfied for both methods with room to spare. Pulsed measurements are of some advantage when it is necessary to study fast reactions involving easily adsorbed molecules, such as NOCl or ICl. In this case, it becomes necessary to work with very low concentrations of these molecules in the stream method, and they may be significantly disturbed by adsorption processes. With time-resolved LMR, the concentrations can be raised by two orders of magnitude, so that the adsorption problem naturally becomes less significant. A comparison of the two methods shows that at the present stage, each is more suitable for its own class of problems. In particular, reactions of free radicals with stable molecules and processes involving excited particles are best studied with the aid of time-resolved LMR, while processes involving two radicals are more conveniently studied with the aid of the stationary LMR method in combination with a stream system.

As an illustration of the prospects for time-resolved LMR, we now discuss some experiments used to study several reactions of chlorine atoms with the $SiH_3$ radical [23, 67]. The experimental apparatus used in this work is basically (except for the data analysis procedure) the same as that shown in Fig. 2. Light at the fourth harmonic of a neodymium laser was aimed into a streaming reactor at a small angle (≈2°) to the axis. The reactor was located in the cavity of a $CO_2$ laser and between the poles of a 20-cm electromagnet. The modulation coils, driven by an oscillator through an amplifier (200 W), modulated the magnetic field at the frequency of the relaxation oscillations in the output radiation, 150 kHz, with a peak-to-peak amplitude of 8 mT over the 12-cm length of the reactor. Chlorine atoms were obtained by photolysis of $S_2Cl_2$ at 265 nm [210]. Kinetic plots of the disappearance of the chlorine atoms and of the appearance and disappearance of $SiH_3$ radicals were obtained by tuning to the peak of one of the lines of the LMR spectrum of these particles, which are shown in Fig. 4. Figure 2 shows a block diagram of the detection system used in a number of later papers [25, 211, 212]. The signal from the photoconductor is detected synchronously, averaged with a time constant of 3 μsec, sent to a fast analog-to-digital converter, and stored in the memory of a minicomputer. In order to improve the quality, 1000 kinetic curves were collected. The final kinetic curve was processed (as an expansion in one or two exponentials) by the method of least squares. Figure 6 shows an example of the kinetic curves for disappearance of chlorine atoms at different pressures of $S_2Cl_2$. When silane is added to the cuvette, it is possible to observe the appearance of $SiH_3$ radicals through the very rapid reaction

$$Cl + SiH_4 \rightarrow HCl + SiH_3, \tag{14}$$

and their subsequent disappearance in a reaction with $S_2Cl_2$. Some examples of the kinetic curves for the $SiH_3$ radical are shown in Fig. 6. By using the standard procedure for analyzing kinetic data, it is easy to obtain the rate constants of the corresponding processes from these experiments. Thus, for reaction (14), the rate constant was close to the number of gas kinetic collisions [$(2.3 \pm 0.5)\cdot 10^{-10}$ $cm^3sec^{-1}$] and is in good agreement with IR chemiluminescence data [213]. The rate constants for the reaction of a chlorine atom and $SiH_3$ radical with the $S_2Cl_2$ molecule were also measured. In these experiments the limiting sensitivity of the method (signal/noise = 1; 1000 samples) was determined for chlorine atoms ($4\cdot 10^{10}$ $cm^{-3}$) and for $SiH_3$ radicals ($10^{11}$ $cm^{-3}$).

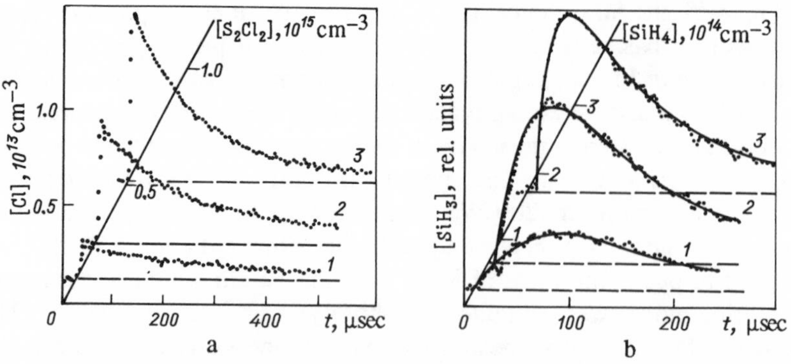


**Fig. 6.** Kinetic curves for the formation and consumption of chlorine atoms (a) and $SiH_3$ radicals (b) during pulsed photolysis of $S_2Cl_2$ in the presence of $SiH_4$ [23].

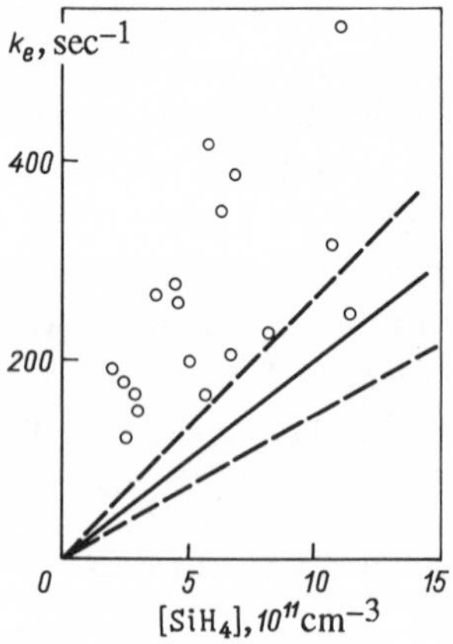


**Fig. 7.** Rate constants for the reaction $Cl + SiH_4 \rightarrow HCl + SiH_3$ measured with a stream system [70]: the smooth curve and the dashed curves are the results of time-resolved measurements, $(2 \pm 0.5) \cdot 10^{-10}$ [23, 213]. The dashed lines represent the errors in these measurements.

Reaction (14) is a good illustration of the difficulties associated with heterogeneous processes in a stream system. The initial attempts to measure the rate constant for this reaction with the aid of a fast stream system with a movable reagent feed in combination with an LMR spectrometer for detecting the chlorine atoms yielded the following results [70]: first, a series of experiments conducted at different times was found to be highly irreproducible. Second, the average value of the rate constant measured in

TABLE 3. Elementary Reactions in the Gas Phase Studied by Laser Magnetic Resonance

| Radical | Reagent | References |
|---|---|---|
| OH | CO, NO, $NO_2$ | [81] |
| OH | $C_2H_4$, $C_2H_3Cl$, $CH_2CH_2$, $C_2HCl_3$, $C_2F_2Cl_2$, $C_2Cl_4$ | [82] |
| OH | $CH_4$, $CH_nF_mCl_kBr_l$ (15 halogen substitutions in methane) | [83] |
| OH | $CX_4$ (X – various halogens), $C_2X_6$, $C_2X_4$ (19 reactions) | [84] |
| OH | $C_2H_6$, $C_2X_6$ (12 different halogen substitutions in ethane) | [85] |
| OH | $HO_2$ | [87] |
| OH | $NO_2$ | [93] |
| OH | NO | [91] |
| OH | $HO_2$ | [73] |
| $HO_2$ | NO | [86] |
| $HO_2$ | $NO_2$ | [126] |
| $HO_2$ | OH | [87] |
| $HO_2$ | O | [88] |
| $HO_2$ | H | [127] |
| $HO_2$ | ClO | [120] |
| $HO_2$ | NO | [89] |
| $HO_2$ | O, OH, NO, Cl, $SO_2$, CO | [128] |
| $HO_2$ | $HO_2$ | [129] |
| $HO_2$ | NO | [90] |
| $HO_2$ | NO | [82] |
| $HO_2$ | $O_3$ | [132] |
| $HO_2$ | OH, NO | [91] |
| $HO_2$ | H | [133] |
| $HO_2$ | Cl | [92] |
| $HO_2$ | OH | [73] |
| $HO_2$ | $HO_2$, NO, OH | [139] |
| $HO_2$ | Heterogeneous recombination | [243] |
| ClO | $HO_2$ | [120] |
| ClO | NO, $NO_2$ | [121] |
| $CH_2$ | H | [148] |
| $CH_2$ | $O_2$ | [149] |
| $CH_2$ | Hydrocarbons | [150] |
| $NF_2$ | $NF_2$ | [171] |
| $NF_2$ | O | [172] |
| $NF_2$ | O, N | [173] |
| $CH_2OH$ | $O_2$ | [130] |
| $SiH_3$ | $S_2Cl_2$ | [23, 67] |
| $SiH_3$ | $O_2$ | — |
| $SiH_3$ | NOCl | — |
| Cl | Heterogeneous recombination | [48] |
| Cl | ICl | [47] |
| Cl | $SiH_4$, $S_2Cl_2$ | [23, 67] |
| Cl | Heterogeneous recombination | [68] |
| Cl | $SiH_4$ | [70] |
| Cl | NOCl | [25] |
| $Cl(^2P_{1/2})$, $Cl(^2P_{3/2})$ | ICl | [211, 212] |
| $Cl(^2P_{1/2})$, $Cl(^2P_{3/2})$ | NOCl | [214] |

this way was substantially (roughly a factor of 2) higher than that obtained by time-resolved methods (LMR and IR chemiluminescence). Third, in some sets of measurements a value of $7 \cdot 10^{-10}$ $cm^3sec^{-1}$ for the rate constant for this reaction was obtained; this exceeds the number of gas-kinetic collisions for these particles. The results of these experiments are shown in Fig. 7. This behavior was interpreted [70] in terms of the unusual properties of heterogeneous recombination observed at low concentrations of chlorine atoms [68]; however, it is possible that the reaction being studied took place directly on the walls of the stream reactor.

The development of the time-resolved laser magnetic resonance technique is opening up broad possibilities for studies of reactions involving chlorine atoms that have been excited to the upper fine structure state $^2P_{1/2}$. A large number of chlorine atoms in this state (up to a population inversion) can be obtained during photolysis of such molecules as ICl [47, 211], NOCl [214], and several others [215]. By observing the kinetics of the gain on the $^2P_{1/2}$–$^2P_{3/2}$ transition, it is possible to determine the sum of the rate constants for the reaction and for relaxation of the upper state $^2P_{1/2}$ and, in a number of cases, to reach some conclusions about its reactivity. This question will be examined in more detail in a later section.

To conclude this section, Table 3 shows a list of the elementary chemical processes that have been studied with the aid of laser magnetic resonance.

## 6. STUDIES OF COMPLEX CHEMICAL PROCESSES

Another kinetic application of LMR is the detection of active intermediate particles in complex chemical reactions. Finding the active intermediate particles in a chemical reaction and, especially, measuring their concentrations provide valuable information for a determination of the mechanism of the reaction. Among spontaneous reactions, a special place is occupied by branching chain reactions, in which the concentrations of the active particle can be many orders of magnitude greater than their equilibrium values and, in a number of cases, may be comparable to the concentrations of the initial substances [216]. Reactions of this type include the oxidation of hydrocarbons, certain fluoridation reactions, and a number of other processes [216, 217]. Observations of atoms and free radicals yield information on the mechanism of the process; measuring their concentrations makes it possible to determine an important kinetic characteristic of these processes, namely the probability of branching into a link of the chain [216]. It is this quantity which determines the fraction of the initial reagents that can be converted into active particles during a chain branching process and is directly related to the concentrations of the active particles. This same quantity, together with the rate constant of the limiting stage and the rate constant for termination of the chain, determines the limits of the spontaneous ignition region [216].

The first publication in which LMR was used to detect a free radical produced during a spontaneous branching chain reaction is the paper by Evenson and colleagues [94], who recorded LMR spectra of the extremely short-lived CH radical formed in an oxygen–acetylene flame. The flame was used as a source of radicals for the spectroscopy and no chemical studies of the process itself were made.

LMR has been used to study the oxidation of silane [184], a reaction whose branching chain character was established in the mid 1930s [218]. In the hypothetical mechanism for this reaction, $SiH_2$ radicals and oxygen atoms were considered for the role of active particles. A reactor with a series of pipes for introducing the reagents, which is kinetically equivalent to an ideal mixing reactor, was used to study this reaction [69]. Figure 8 is a sketch of this reactor. The reagents were fed in at equal separations through a series of inlets positioned along the reactor. Figure 9 shows an LMR spectrum of the $SiH_3$ radical recorded when flows of oxygen and silane were mixed in this reactor. A study of the dependence of the intensity of the LMR spectrum on the pressures of the reagents showed that there is a threshold pressure of silane at which radicals begin to appear. No threshold in the oxygen pressure was observed. The concentration of $SiH_3$ radicals was also estimated in these experiments and found to be on the order of $10^{-3}$–$10^{-2}$ times the concentrations of the reagents.

These measurements, together with an earlier determination of the oxygen atoms and OH radicals in this system by means of EPR [219]. can serve as a basis for determining the previously unknown mechanism of this reaction.

Azatyan et al. [138] have used LMR to detect the $HO_2$ radical formed in a rarefied hydrogen–oxygen flame.

This experimental method has also been used to study the chlorination of silane, a reaction whose chain branching character was recently established [220]. When flows of chlorine and silane were mixed at very low partial pressures (on the order of a few Pascals), LMR spectra of chlorine atoms and $SiH_3$ radicals were detected [66] (see Fig. 4). The dependences of the LMR signals from Cl atoms and $SiH_3$ radicals on the partial

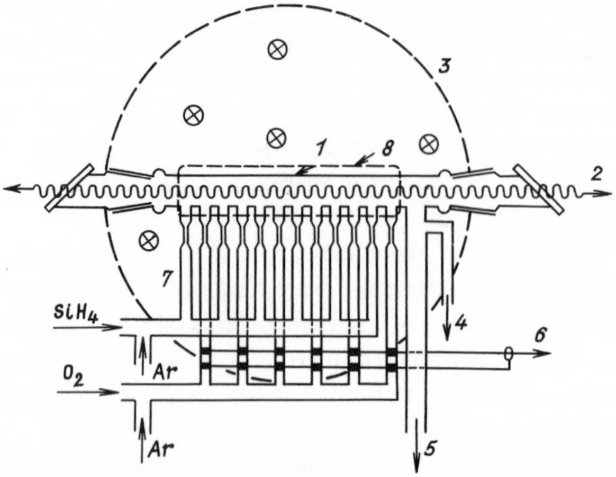


**Fig. 8.** The streaming reactor equivalent of an ideal mixing reactor used for studying branching chain reactions of oxidation and chlorination of silane [66, 69, 184]: 1) Pyrex reactor; 2) emission in the cavity of a $CO_2$ laser; 3) poles of the electromagnet; 4) pressure measurement; 5) pumpout; 6) rf discharge power; 7) set of tubes with capillaries for supply of reagents; 8) region where the magnetic field is modulated.

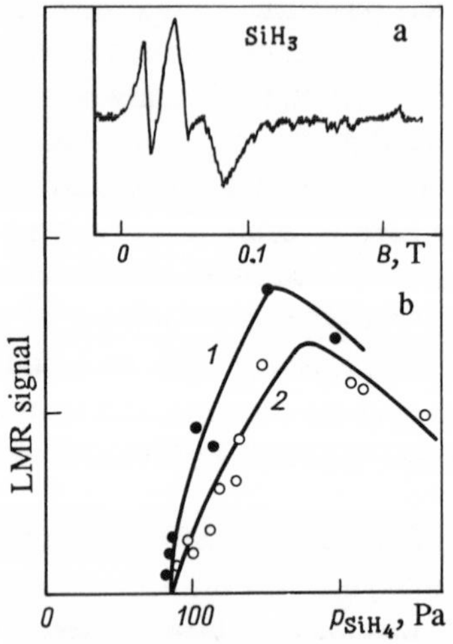


**Fig. 9.** LMR spectrum of the $SiH_3$ radical recorded during oxidation of silane (a) and the dependence of the amplitude of the LMR spectrum on the fluxes of reagents (b): curves 1 and 2 correspond to different oxygen pressures ($p_{O_2}$ = 120 Pa for curve 1 and 200 Pa for curve 2) [69, 184].

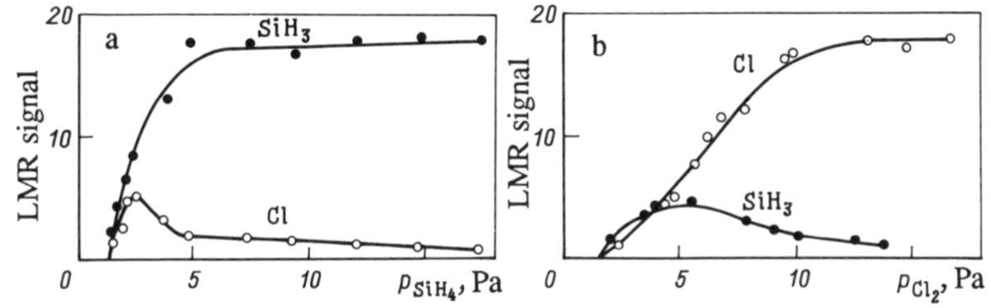


**Fig. 10.** The dependence of the LMR signals from chlorine atoms and $SiH_3$ radicals on the silane pressure for $p_{Cl_2} = 6$ Pa (a) and on the chlorine pressure for $p_{SiH_4} = 2.5$ Pa (b). $p_{Ar} = 700$ Pa [66, 69].

pressures of the reagents are shown in Fig. 10. Threshold effects are observed, as well as the "conversion" of the concentration of one of the radicals into the other as the composition of the mixture is changed, which is typical of chain processes. A study of the ignition region for this system, together with the measured radical concentrations, made it possible to estimate the branching probability into a link of the chain in this reaction as $(2–16)\cdot 10^{-4}$. As mentioned above, both the concentration limits and the absolute concentrations of the radicals are determined by the branching probability into a link of the chain [216]. However, this parameter appears in combination with a very poorly reproducible and controllable quantity, the rate constant for recombination of the active centers at the wall of the reaction vessel. Thus, only semiquantitative information can be obtained from such measurements.

In order to obtain quantitative kinetic information, a method for pulsed studies of chain branching reactions has been developed [25]. The basic principle of the method is illustrated in Fig. 11. If a chain branching reaction is initiated in a pulse in the stability region of the mixture and the loss kinetics of the radicals are measured, then the effective lifetime of the radicals should increase on approaching the self-ignition threshold because of the increased contribution of branching processes. The self-ignition threshold corresponds to an infinite lifetime for the active particles. When certain conditions are met (specifically, when a quasisteady state is established over times short compared to the lifetime of the radicals, one of the stages of propagation of the chain is limiting, and various other easily fulfilled conditions), an extremely simple expression can be obtained for the inverse lifetime of the radicals [25]:

$$\tau^{-1} = k_0 - k\,[A]\,\delta, \qquad (15)$$

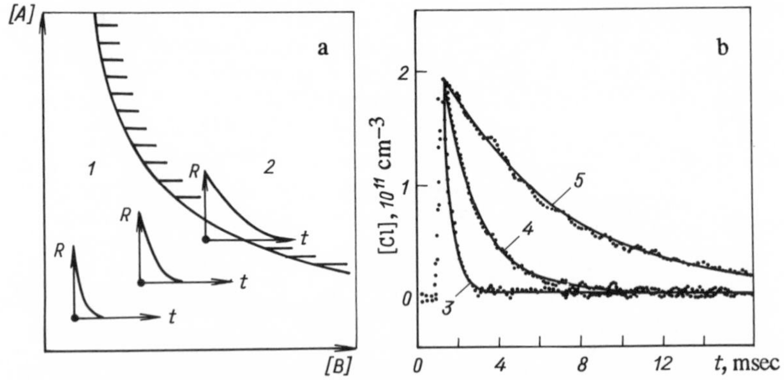


**Fig. 11.** The kinetics of the disappearance of *R* radicals as the region of self-ignition is approached during pulsed initiation of a branching chain reaction between reagents A and B in the region of stable mixture (a) and an illustration of this behavior for the disappearance of chlorine atoms during pulsed initiation of the branching chain chlorination of silane (b): 1) the stability region; 2) the self-ignition region; curves (3–5) correspond to different concentrations of silane ($[SiH_4] = 0, 2.1 \cdot 10^{15}$, and $2.8 \cdot 10^{15}$ cm$^{-3}$, respectively) for a constant concentration of inhibitor ($[NOCl] = 5.6 \cdot 10^{13}$ cm$^{-3}$) and pressure of chlorine ($p_{Cl_2} = 560$ Pa) [25].

where $k_0$ is the rate constant for loss of the radicals participating in the limiting stage of chain propagation, $k$ is the rate constant for that stage, $[A]$ is the concentration of the reagent that takes part in the limiting stage of chain propagation, and $\delta$ is the probability of branching into a link of the chain. This expression predicts that the inverse lifetime of the radicals should depend linearly on the concentration of the reagent and that the product $k\delta$ can be found from the slope of this function. If the rate constant for the limiting stage of the process is measured reliably by some other independent method, then it is possible to determine $\delta$.

This method has been used to study the chlorination of silane with the aid of time-resolved LMR [25]. Figure 11 shows the resulting kinetic curves for the disappearance of chlorine atoms formed by photolysis of molecular chlorine when different amounts of silane have been added. The lifetime of the chlorine atoms increases as the silane pressure is raised, unlike the case in which the elementary reaction is measured, when the lifetime decreases as reagents are added. On plotting the inverse lifetime as a function of the silane concentration, we should obtain a straight line whose slope gives the product $k\delta$. The value of $\delta$ determined in this fashion for the chlorination of silane was equal to $(4.6 \pm 0.6) \cdot 10^{-3}$. This method makes it possible to determine this important characteristic of branching chain reactions with a kinetic accuracy (±15%) and will be useful for studies of other chain branching processes.

LMR has also been used to study chain branching processes in the above-threshold region in an ideal displacement stream reactor. A stream reactor has been used [221] to investigate chain branching in the reaction of the $NF_2$ radical with $H_2O_2$. This method makes it possible to follow the exponential time evolution of the reaction (along the flow) and to determine its characteristics.

## 7. STUDIES OF PROCESSES INVOLVING EXCITED PARTICLES

The development of laser techniques for studying and stimulating chemical reactions has sharply increased the prospects for research on processes involving particles in definite quantum states [6, 207, 222]. These processes include the relaxation of internal energy and chemical reactions of particles in excited states. Internal energy relaxation in stable particles has been studied quite well by now [223], but information on relaxation processes involving short-lived particles is extremely limited. There is great interest in research on vibrational relaxation in free radicals, both from the standpoint of the kinetics of reactions under nonequilibrium conditions (upper atmosphere, photochemistry, plasma chemistry, etc.), and from the standpoint of stimulating selective reactions by laser radiation [224]. Information about vibrational relaxation in free radicals in the gas phase is limited because of the considerable experimental difficulties in obtaining it, since their concentrations are so low. Recent progress in this area is associated with highly sensitive laser techniques: laser-induced fluorescence and intracavity laser spectroscopy. Thus, LIF has been used to study vibrational relaxation in $^1C_2$, $^3C_2$, $CF_2(\tilde{X})$, and $CF_2(\tilde{A})$ [4, 225], ICLS has been used to study vibrational relaxation of the HCO [7, 226, 227] and $NH_2$ [227–229] radicals, and IR chemiluminescence has been used to study relaxation of the $CH_3$ radical [230, 231]. All of these results have been obtained by time-resolved methods. It is difficult to obtain information on vibrational relaxation of free radicals in stream systems because, in general, relaxation occurs very rapidly in collisions with the walls, although such studies have been carried out for a few particle species. At present, the use of ICLS and LIF for studying these processes is limited to particles with allowed electronic transitions in the visible and

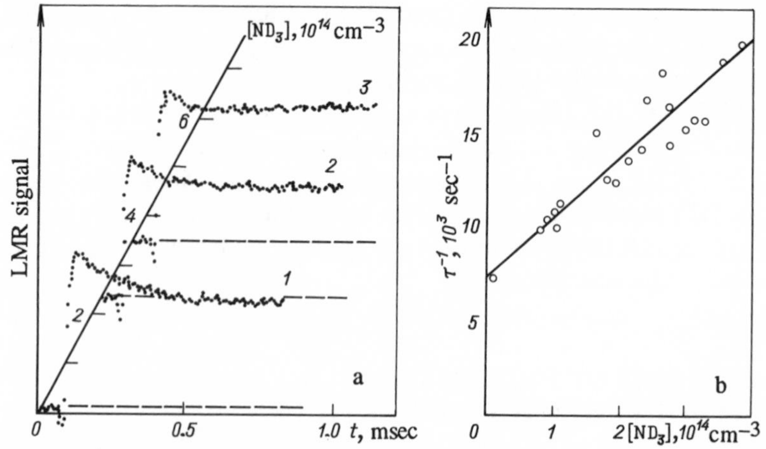


**Fig. 12.** The saturation kinetics of a vibrational transition of the $ND_2$ radical (0, 1, 0 ← 0, 0, 0) during tuning to the transition by means of a rapid jump in the magnetic field with different concentrations of the relaxant $ND_3$ (a) and the dependence of the reciprocal time for establishment of the saturation kinetics on the concentration of the relaxant gas $ND_3$ (b): $[ND_2] = 1.5 \cdot 10^{13}$ cm$^{-3}$; $p_{Ar} = 270$ Pa; and $[ND_3] = 1.4 \cdot 10^{13}$, $23.6 \cdot 10^{13}$, and $35.1 \cdot 10^{13}$ cm$^{-3}$ for curves 1, 2, and 3 [24].

near UV. Although a large number of free radicals have been detected by high-sensitivity LMR (Table 1), this method has not been used recently in studies of relaxation processes, because LMR is mainly used in its steady-state form and it is hard to study relaxation processes in that way. The amplitude-phase method proposed by Gershenzon et al. [45] requires significant degrees of saturation of the free radical transition (fairly easily achieved) and knowledge of the rapidly varying amplitude-phase characteristic of the spectrometer (Fig. 3) (much more difficult to obtain), and has not yet been realized. Time-resolved LMR offers wide-ranging possibilities for research on these processes. Two variants of this method can be used in research on relaxation processes:

1) producing vibrationally excited particles with a suitable pulsed technique (e.g., UV photolysis or IR multiphoton dissociation) and recording the kinetics of the signal by means of time-resolved LMR; and

2) producing radicals through simpler methods in a stream system and studying the saturation kinetics of the LMR signal from a laser by tuning the transition of the free radical to the laser line, which can be done easily with the aid of a rapid jump in the magnetic field strength.

The first variant is standard for time-resolved techniques which are sensitive to certain quantum states of the particle under study and it has wider prospects than the second variant. The second variant, however, may seem more useful for those radicals which cannot be generated in pulses. It is much simpler to produce free radicals in a stream system, but its range of applicability is limited. For simplicity we shall examine a two-level system which can be brought into resonance with the laser light at a given time. Initially, all the particles are in the lower state and the absorption signal is at a maximum. Later, because the transition is saturated, the absorption falls off exponentially and approaches its steady-state saturated value.

The kinetics of the signal evolution obey the following equation [24]:

$$\alpha(t) = \alpha_0 [F + (1 - F) \exp(-t/\tau_{\text{eff}})], \tag{16}$$

where $\alpha_0$ is the unsaturated absorption coefficient, $F = (1 + 2J\sigma\tau)^{-1}$ is the saturation factor, $J$ is the photon flux, and $\sigma$ is the absorption cross section. The time for evolution of the saturation kinetics is given by

$$1/\tau_{\text{eff}} = 1/\tau + 2J\sigma, \tag{17}$$

where the inverse relaxation time ($1/\tau$) and the radiative term ($2J\sigma$) appear additively. Since the inverse lifetime $1/\tau$ is related linearly to the concentration of the relaxant gas M,

$$1/\tau = k_0 + k[\text{M}], \tag{18}$$

where $k_0$ describes relaxation on all other particles, it is possible to determine the rate constant for relaxation on these molecules directly from the slope of the curve obtained by measuring the risetime of the LMR signal and plotting its reciprocal as a function of the concentration of the relaxant gas. On the other hand, only the combination $J\sigma\tau$ can be determined from experiments with steady-state saturation.

The formulas for a two-level system are still valid for saturation of an individual vibrational–rotational transition when the rate of rotational relaxation exceeds the rate of vibrational relaxation and the quantity $\sigma$ is slightly redefined [24]. This method has been used to measure the vibrational relaxation of the first excited mode $\nu_2$ of the $ND_2$ radical [24]. Some difficulties which arose in creating a jump in the magnetic field in the gap of the electromagnet were overcome by using a specially constructed system of coils which made it possible to overcome both the problem of "frozen" magnetic flux in the electromagnet poles (owing to the Foucault effect) and the problem of interactions between the field jump

**TABLE 4.** Vibrational Relaxation of the $\nu_2$ Mode of the $ND_2$ Radical [24]: $\overset{*}{ND_2}(0, 1, 0) + M \xrightarrow{k} ND_2(0, 0, 0) + M$

| M | $k$, $cm^3/sec$ ($\pm 2\sigma$) | $\sigma$, $10^{-16}\,cm^2$ | $Z^*_{10}$ |
|---|---|---|---|
| $ND_3$ | $(2.7\pm1)\cdot10^{-11}$ | 3.3 | 12 |
| $CF_4$ | $(1.9\pm0.4)\cdot10^{-11}$ | 2.9 | 17 |
| Ar | $(2.4\pm0.7)\cdot10^{-14}$ | $3.3\cdot10^{-3}$ | 13 000 |

*The following values have been used for the diameters: $3.5\cdot10^{-8}$ cm for $ND_2$; $3.5\cdot10^{-8}$ cm for $ND_3$; $4.5\cdot10^{-8}$ cm for $CF_4$; and $3.82\cdot10^{-8}$ cm for Ar.

coils and the modulation coils [24]. Figure 12 shows some examples of the saturation kinetics of the LMR signal during rapid tuning to a line of the LMR spectrum of the $ND_2$ radical with the aid of a jump in the magnetic field (8 mT, 10 μsec). Applying Eq. (18) to these curves yields the rate constant for relaxation on $ND_3$. Similar measurements of vibrational relaxation of $ND_2(0, 1, 0)$ in collisions with $CF_4$ and Ar have been made. The results of these measurements are listed in Table 4 [24].

This method is convenient in that it does not require pulsed light sources for production of free radicals. In this regard, determining its range of applicability is of some interest. Studies of relaxation processes by the method of magnetic field jumps in combination with time-resolved LMR require substantial saturation of the vibrational transition, by 10–30%. Estimates show that the power inside a $CO_2$ laser cavity (300–500 $W/cm^2$) is enough for this amount of saturation of the transition under steady-state conditions in a stream at a pressure of several hundred Pascals when the radical under study has a typical diamagnetic moment for a vibrational transition (~0.1 D) and a rotational statistical sum that is not too large ($\leq 10^3$), i.e., for either linear or nonlinear radicals with light peripheral atoms. Thus, the measured dipole moment for the $\nu_2$ vibrational transition of the $ND_2$ radical was equal to $(0.056 \pm 0.02)$ D, while the rotational statistical sum at $T = 300$ K is equal to 134.

Evidently, such radicals as FO, $CH_2$, HCO, and a few others can be studied by this method.

As already mentioned, the number of atoms detected by LMR is small, but under certain conditions, LMR techniques can be used to study processes in which they are involved. It is especially promising for investigating the quenching and reactions of the excited state of the fine structure of the chlorine atom ($^2P_{1/2}$), where until recently the only method for kinetic measurements (from the standpoint of time resolution and sensitivity) was resonance absorption in the vacuum ultraviolet [215]. These data are of great interest, both in terms of the role of the excited fine structure state of the halogens in their chemical reactions, and in terms of improving and developing new lasers employing the fine structure transitions of these atoms.

Observations of the saturation kinetics of an LMR signal with the aid of a jump in the magnetic field have been used [211] to study the quenching of the $^2P_{1/2}$ state by various gases ($O_2$, $Cl_2$, $CF_2Cl_2$, Ar). Although the fine structure transition is magnetic dipole in character, a small number of populated states in this atom leads to large absorption (the absorption cross section is $4 \cdot 10^{-19}$ cm$^2$ for the LMR spectrum of the chlorine atom), which makes it possible to reach substantial levels of saturation. Figure 13 shows an example of saturation kinetics in the LMR signal from chlorine atoms, which can be fit by a double exponential curve, corresponding to relaxation through the hyperfine states of the upper $^2P_{1/2}$ state and to relaxation between the fine structure levels. The problem of mixing in the hyperfine structure of the $^2P_{1/2}$ state was eliminated by adding a paramagnetic gas, molecular oxygen. As a result of this, a single exponential, associated with relaxation of the upper level of the fine structure, was left in the saturation kinetic curve. The values obtained in this way for the rate constants for quenching on several molecules were sharply (by two orders of magnitude) different from data obtained by resonant absorption in the VUV [215].

Processes involving excited chlorine atoms have also been studied in experiments of the first type where they were obtained by pulsed photolysis of ICl [211, 212] and NOCl [214] molecules. When these molecules are photolyzed by the first (530, ICl), third, and fourth (353 and 265 nm, NOCl) harmonics of a neodymium laser, a population inversion develops over the fine structure levels. Observing the kinetics of the LMR signal makes it possible easily to determine the sum of the rate constants for relaxation and for the reaction with the upper state, as well as the rate constant for the reaction with the lower state. The processes that occur in this case are described by the following scheme:

$$Cl^*(^2P_{1/2}) + ICl \xrightarrow{k_q} Cl(^2P_{3/2}) + ICl, \tag{19}$$

$$Cl^*(^2P_{1/2}) + ICl \xrightarrow{k^*} Cl_2 + I, \tag{20}$$

and

$$Cl(^2P_{3/2}) + ICl \xrightarrow{k} Cl_2 + I. \tag{21}$$

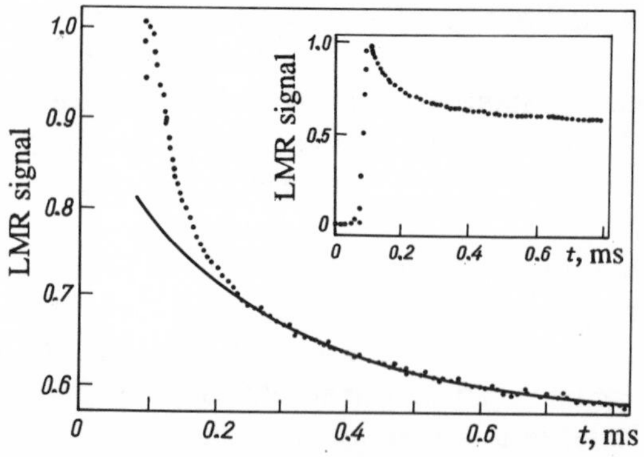


**Fig. 13.** Saturation of the LMR signal from chlorine atoms during rapid tuning to the transition by means of a jump in the magnetic field. The short decay time is caused by mixing in the HFS sublevels within the upper state of the fine structure of $^2P_{1/2}$. The inset at the upper right shows the kinetics of saturation with the full ordinate scale. [Cl] ~ $2 \cdot 10^{13}$ cm$^{-3}$ and $p_{Ar}$ = 1330 Pa (10 Torr) [211, 212].

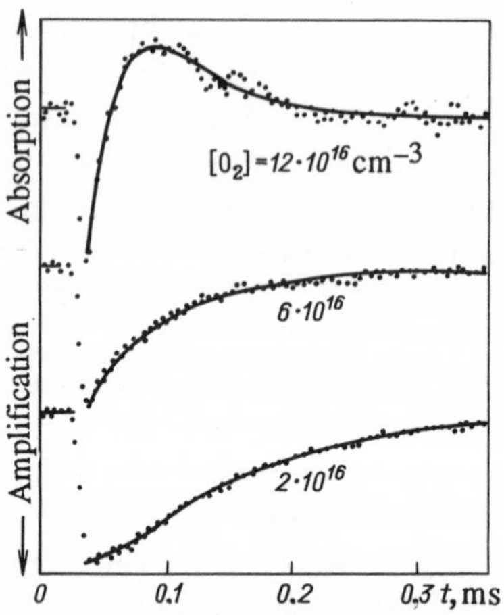


**Fig. 14.** The kinetics of the LMR signal from chlorine atoms during photolysis of ICl at 530 nm for different concentrations of $O_2$ as a relaxant: [ICl] = $3 \cdot 10^{15}$ cm$^{-3}$; $p_{Ar}$ = 1330 Pa (10 Torr) [212].

**TABLE 5.** Quenching of the $^2P_{1/2}$ State of the Chlorine Atom:

$$Cl^* (^2P_{1/2})+M \xrightarrow{k_q} Cl(^2P_{3/2})+M$$

| M | $k_q$, $cm^3/s$ | | |
|---|---|---|---|
| | Resonant absorption in the VUV [215] | Saturation kinetics of LMR signal during rapid tuning to the transition by means of a jump in magnetic field [211] | Photolysis of ICl and NOCl; kinetics of the LMR signal [212, 214] |
| $CF_2Cl_2$ | $(2.1\pm0.4)\cdot10^{-10}$ | $(1.8\pm0.4)\cdot10^{-10}$ | — |
| $O_2$ | $(2.3\pm0.3)\cdot10^{-11}$ | $(1.7\pm0.4)\cdot10^{-13}$ | $(1.7\pm0.4)\cdot10^{-13}$ |
| $Cl_2$ | $(4.5\pm0.4)\cdot10^{-11}$ | $(7.4\pm2.6)\cdot10^{-13}$ | $(7.2\pm2.0)\cdot10^{-13}$ |
| Ar | $(1.1\pm0.3)\cdot10^{-12}$ | $\leqslant1\cdot10^{-14}$ | $<1.5\cdot10^{-14}$ |
| ICl | — | — | $\leqslant(3.3\pm0.4)\cdot10^{-13}$ |
| NOCl | — | — | $\leqslant(1.8\pm0.4)\cdot10^{-11}$ |

An example of the LMR signal kinetics for photolysis of ICl at 530 nm is shown in Fig. 14. These experiments yielded the rate constants for quenching and for reactions of the excited chlorine atom, and complete agreement was obtained with the magnetic field jump technique. The results of these measurements are shown in Table 5. It was found that excitation of chlorine to the $^2P_{1/2}$ state leads to a reduction in the rate of the reaction with ICl by at least a factor of 24. (Similar behavior has been observed in some reactions of excited iodine and bromine atoms [232].) This fact is very important for the development of a laser employing the fine structure transition of the chlorine atom, since it makes it possible to depopulate the lower working state chemically and might be useful for constructing a solar-pumped laser [233]. Thus, when a continuous discharge is produced in ICl, a steady-state amplified signal is observed on the ($^2P_{1/2}$–$^2P_{3/2}$) transition because the $^2P_{3/2}$ state is depopulated more rapidly than the $^2P_{1/2}$ state [212].

Interesting data have also been obtained on the role of the $^2P_{1/2}$ state in the reaction of chlorine atoms with NOCl molecules. It was observed that, as in the case of ICl, the upper state has a lower reactivity. It was also found that the rate constant $1.5\cdot10^{-11}$ $cm^3sec^{-1}$ measured [234] by pulsed photolysis in combination with resonance fluorescence does not correspond to the reaction of the ground state of the chlorine atom $^2P_{3/2}$, but to the quenching of the upper state $^2P_{1/2}$ on this molecule. The rate constant for the reaction of ground state chlorine atoms,

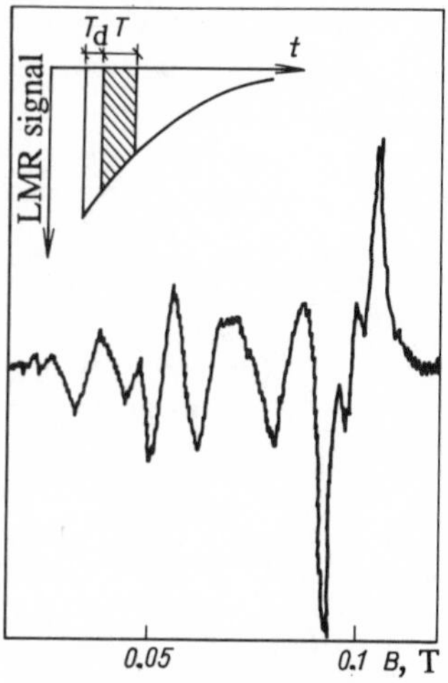


**Fig. 15.** An LMR spectrum of chlorine atoms, inverted as a whole, observed with a time window $T = 20$ μsec and a delay of $T_d = 10$ μsec after photolysis of ICl at 530 nm (lower figure). The upper figure shows the time of the strobing pulse relative to the time of photolysis. The shaded region is sampled and integrated. $p_{ICl} = 90$ Pa (0.7 Torr), $p_{Ar}$ 130 Pa (1 Torr).

$$Cl\,(^2P_{3/2}) + NOCl \rightarrow Cl_2 + NO \qquad (22)$$

on the other hand, was substantially higher than the previously published value ($k = (8 \pm 1)\cdot 10^{-11}$ cm$^3$sec$^{-1}$ [25, 214]).

The appreciable number of ions and electronically excited radicals detected by LMR (Table 1) also opens up the possibility of studying processes involving these particles by using the time-resolved variant of this technique.

Time-resolved LMR is also promising for obtaining LMR spectra of vibrationally excited radicals, including those in quite high vibrational states. Under steady-state conditions, such measurements are difficult because these states relax rapidly, so that the concentrations of highly excited particles may lie below the limit of sensitivity of the method. In order to obtain such spectra, one can use repetitive generation of excited radicals by dissociation of suitable molecules through UV photolysis or multiphoton dissociation and recording the signal that is extracted in some "time window" after the pulse by scanning the magnetic field. This can be done both computationally and by analog strobing of the signal.

This possibility has been tested by recording LMR spectra of chlorine atoms produced by photolysis of ICl at 530 nm. Figure 15 shows the idea of the experiment, as well as an LMR spectrum of chlorine atoms recorded in the time window from 10 to 30 μsec after the photolytic pulse. In this case, analog techniques have been used to strobe the signal. These experiments were undertaken in a search for possible nonstatistical populations in the hyperfine structure levels of the upper and lower fine structure states during photolysis of ICl. It was found, however, that the entire spectrum is inverted as a whole. That may be caused by the rather high total pressures in these experiments. In any case, these experiments demonstrate the feasibility of taking LMR spectra in the highly nonequilibrium situation immediately following a photolytic pulse. It may be hoped that experiments of this type will make it possible, not only to record LMR spectra of vibrationally excited radicals, but also to study relaxation processes and chemical reactions involving them.

## 8. SOME OTHER APPLICATIONS OF THE METHOD

Of the possible areas of application of the laser resonance method, we now mention the following:

The spectral resolution of this method (a resolution of about 1 MHz, determined by the frequency fluctuations of the laser, is easily achieved without difficulty in an IR LMR spectrometer) is considerably smaller than either the Doppler or the collisional linewidths at pressures on the order of several hundred Pascals, so that it is possible to record line shapes and determine the cross sections for line broadening in collisions of radicals with different partners [76]. A study of the shape of a line and its dependence on the pressure could obviously yield information on the interaction potential of free radicals in collisions with molecules.

The high radiant power inside an IR LMR spectrometer (about 400 W/cm$^2$) and the comparatively low pressures at which the experiments are conducted (10–1000 Pa) make it easy to observe the inverse Lamb dip in LMR spectra [10, 13]. The Lamb dip is often used in spectroscopy to resolve hyperfine structure. An investigation of the amplitude and form of the dip can also yield information on the characteristics of molecular collisions.

The high radiant power also makes it possible to attain significant degrees of saturation of an entire transition as a whole [24], which may be used to stimulate selective chemical reactions involving free radicals.

Many schemes have been proposed for controlling selective chemical processes with lasers [220, 224]. One of these is the selective acceleration of chemical reactions by single-photon vibrational excitation of a stable reagent containing a given isotope. It has been demonstrated that fairly high separation coefficients (≈10) can be obtained in this way, but only at very low pressures (on the order of a fraction of a Pascal) because of rapid quasiresonant vibrational energy exchange between molecules containing the different isotopes. In this connection, more promise is offered by selective excitation of active intermediate particles, whose concentration is much lower than that of the stable particles, so that the total pressure could be much greater without loss of selectivity (naturally, at the required laser power).

One attempt (not very successful) to carry out this type of process was by Krasnoperov and Panfilov [235], who studied the effect of laser radiation resonant with the fine structure transition of the chlorine atom on the chlorination of $CH_3F$. The observed acceleration of the reaction lay within the limits of experimental error and could not be established with certainty, perhaps because of reduced reactivity of the excited chlorine atoms.

A number of experiments of this type have been done by Chasovnikov and others at the Institute of Chemical Kinetics and Combustion, Siberian Branch, Academy of Sciences of the USSR in attempts to accelerate several reactions of the $ND_2$, FO, and FCO radicals by vibrationally exciting them in the cavity of an LMR spectrometer. Up to now, these attempts have not yielded positive results, primarily because of insufficient information on chemical processes involving these radicals and a lack of information on the absorption cross sections and relaxation rates.

## 9. CONCLUSION

To summarize, laser magnetic resonance is a powerful method for research on processes involving free radicals in the gaseous phase. The prospects for this method are constantly expanding, both in terms of an increased number of particle species detected, and in terms of the appearance of new technical improvements, such as time-resolved LMR, combined LMR–EPR spectrometers, LMR spectrometers with superconducting solenoids [30, 197], etc. It can be said with certainty that the possibilities for this method are far from exhausted. Undoubtedly, the development of tunable lasers will lead to a further extension of the prospects for this technique and it may be hoped that the time will come when this method can be used to detect an arbitrary free radical in the gaseous phase (naturally, one with resolved vibrational or rotational transitions). Likewise, further prospects exist for raising both the sensitivity and the time resolution of this method and its various combinations with traditional methods (mass spectrometry, resonance fluorescence, etc.).

## REFERENCES


1. A. A. Westenberg, *Prog. React. Kinet.* **7**, 23–82 (1973).
2. A. Carrington, *Microwave Spectroscopy of Free Radicals*, Academic Press, London–New York (1974).
3. V. N. Panfilov, *Free Radical States in Chemistry* [in Russian], Nauka, Novosibirsk (1972), pp. 171–179.
4. H. Reissler, M. Mangir, and K. Wittig, in: *Applications of Lasers in Spectroscopy and Photochemistry* [Russian translation], Mir, Moscow (1983), pp. 140–172.
5. L. A. Pakhomycheva, É. A. Sviridenkov, A. F. Suchkov, et al., *Pis'ma Zh. Éksp. Teor. Fiz.* **12**, 60–63 (1970).
6. O. M. Sarkisov and S. G. Cheskis, *Usp. Khim.* **54**, 396–417 (1985).
7. J. P. Reilly, J. H. Clark, C. B. Moore, et al., *J. Chem. Phys.* **69**, 4381–4394 (1978).
8. K. M. Evenson, R. J. Saykally, D. A. Jennings, et al., in: *Applications of Lasers in Spectroscopy and Photochemistry* [Russian translation], Mir, Moscow (1983), pp. 99–139.
9. K. M. Evenson, *Faraday Discuss. Chem. Soc.* **71**, 7–14 (1981).
10. A. R. W. McKellar, *Faraday Discuss. Chem. Soc.* **71**, 63–76 (1981).
11. B. A. Trache, in: *Problems of Chemical Kinetics* [Russian translation], Mir, Moscow (1979), pp. 29–37.
12. P. B. Davies, *J. Phys. Chem.* **85**, 2599–2607 (1981).
13. V. N. Panfilov and L. N. Krasnoperov, *Khim. Fiz.*, No. 4, 468–477 (1983).
14. Yu. M. Gershenzon, S. D. Il'in, O. P. Kishkovich, et al., *Khim. Fiz.*, No. 4, 478–484 (1983).
15. D. K. Russell, *Electron Spin Resonance, Vol. 8, A Specialist Periodical Report*, The Royal Society of Chemistry, Burlington House, London (1983), pp. 1–30.
16. E. Hirota, in: *Applications of Lasers in Spectroscopy and Photochemistry* [Russian translation], Mir, Moscow (1983), pp. 45–98.
17. I. C. Bowater, J. M. Brown, and A. Carrington, *Proc. R. Soc. London* **A333**, 265–288 (1973).
18. J. T. Hougen, *J. Mol. Spectrosc.* **54**, 447–471 (1975).
19. H. Uehara and K. Hakuta, *Chem. Phys. Lett.* **58**, 287–290 (1978).
20. K. Hakuta and H. Uehara, *Chem. Phys. Lett.* **63**, 496–499 (1979).
21. K. Hakuta and H. Uehara, *Chem. Phys. Lett.* **63**, 500–502 (1979).
22. H. Uehara and K. Hakuta, *J. Chem. Phys.* **74**, 969–978 (1981).
23. L. N. Krasnoperov, E. N. Chesnokov, and V. N. Panfilov, *Chem. Phys.* **89**, 297–305 (1984).

24. A. I. Chichinin and L. N. Krasnoperov, *Chem. Phys. Lett.* **115**, 343–348 (1985).
25. S. A. Chasovnikov and L. N. Krasnoperov, *Chem. Phys. Lett.* **124**, 248–255 (1986).
26. K. M. Evenson, H. P. Broida, J. S. Wells, et al., *Phys. Rev. Lett.* **21**, 1038–1040 (1968).
27. K. M. Evenson, J. S. Wells, and H. E. Radford, *Phys. Rev. Lett.* **25**, 199–202 (1970).
28. S. V. Broude, Yu. M. Gershenzon, S. D. Il'in, et al., *Dokl. Akad. Nauk SSSR* **223**, 366–368 (1975).
29. A. Kaldor, W. B. Olson, and A. G. Maki, *Science* **176**, 508–510 (1972).
30. H. J. Zeiger, F. A. Blum, and K. Nill, *J. Chem. Phys.* **59**, 3968–3970 (1973).
31. K. Hakuta and H. Uehara, *J. Mol. Spectrosc.* **58**, 316–322 (1975).
32. J. Pfeiffer, D. Kirsten, P. Kalkert, et al., *Appl. Phys.* **B26**, 173–177 (1981).
33. W. Urban and W. Herrmann, *Appl. Phys.* **17**, 325–330 (1978).
34. T. J. Bridges and E. G. Burkhardt, *Opt. Commun.* **22**, 248–250 (1977).
35. A. Hinz, J. Pfeiffer, W. Bohle, et al.., *Mol. Phys.* **45**, 1131–1139 (1982).
36. M. Rosenbluh, T. A. Miller, D. M. Larsen, et al., *Phys. Rev. Lett.* **39**, 874–877 (1977).
37. S. D. Il'in, "Construction of magnetic resonance spectrometers for gaseous phase chemical kinetics," Author's Abstract of Candidate's Dissertation, Physicomathematical Sciences, Moscow (1982).
38. H. E. Radford, K. M. Evenson, and C. J. Howard, *J. Chem. Phys.* **60**, 3178–3183 (1974).
39. L. N. Krasnoperov, "The role of electronic and vibrational energy in reactions of halogen atoms with C–H bonds," Author's Abstract of Candidate's Dissertation, Physicomathematical Sciences, Novosibirsk (1982).
40. Yu. M. Gershenzon and B. L. Lifshits, *Kvantovaya Élektron.* **6**, 933–942 (1979).
41. V. R. Braun, L. N. Krasnoperov, and V. N. Panfilov, *Kvantovaya Élektron.* **7**, 1895–1904 (1980).
42. S. F. Luk'yanenko, M. M. Makogon, and L. N. Sinitsa, *Intracavity Laser Spectroscopy* [in Russian], Nauka, Novosibirsk (1985).
43. N. M. Galaktionova, A. A. Mak, O. A. Orlov, et al., *Pis'ma Zh. Éksp. Teor. Fiz.* **18**, 507–510 (1973).
44. Yu. M. Gershenzon, "Vibrational relaxation in collisions of molecules with surfaces and chemically active atoms," Author's Abstract of Doctoral Dissertation, Physicomathematical Sciences, Moscow (1978).
45. Yu. M. Gershenzon, S. D. Il'in, O. P. Kishkovich, et al., *Kvantovaya Élektron.* **8**, 631–634 (1981).
46. E. Arimondo and P. Glorieux, *Appl. Phys. Lett.* **33**, 49–52 (1978).
47. L. N. Krasnoperov and V. N. Panfilov, *Kinet. Katal.* **20**, 540–541 (1979).
48. L. N. Krasnoperov, V. R. Braun, and V. N. Panfilov, *Kinet. Katal.* **19**, 1610–1611 (1978).
49. J. S. Geiger, D. R. Smith, and J. D. Bonnet, *Chem. Phys.* **74**, 239–246 (1983).
50. J. S. Wells and K. M. Evenson, *Rev. Sci. Instrum.* **41**, 226–227 (1970).
51. M. Mizushima, J. S. Wells, K. M. Evenson, et al., *Phys. Rev. Lett.* **29**, 831–833 (1972).
52. K. M. Evenson and M. Mizushima, *Phys. Rev. A* **6**, 2197–2204 (1972).
53. L. Tomuta, M. Mizushima, C. J. Howard, et al., *Phys. Rev. A* **12**, 974–979 (1975).

54. M. Mizushima, K. M. Evenson, J. A. Mucha, et al., *J. Mol. Spectrosc.* **100**, 303–315 (1983).
55. M. Mizushima, L. R. Zink, and K. M. Evenson, *J. Mol. Spectrosc.* **107**, 395–404 (1984).
56. M. Mizushima, K. M. Evenson, and J. S. Wells, *Phys. Rev. A* **5**, 2276–2287 (1972).
57. P. A. Bonczyk, *Chem. Phys. Lett.* **18**, 147–149 (1973).
58. P. A. Bonczyk, *Rev. Sci. Instrum.* **46**, 456–458 (1975).
59. R. M. Dale, J. W. C. Johns, A. R. W. McKellar et al., *J. Mol. Spectrosc.* **67**, 440–458 (1977).
60. R. S. Lowe and A. R. W. McKellar, in: *Laser Spectroscopy, Vol. V*, A. R. W. McKellar, T. Oka, and B. P. Stoicheff (eds.), Springer Verlag (1981).
61. R. F. Curl Jr., K. M. Evenson, and J. S. Wells, *J. Chem. Phys.* **56**, 5143–5151 (1972).
62. S. M. Freund, J. T. Hougen, and W. J. Lafferty, *Can. J. Phys.* **53**, 1929–1938 (1975).
63. K. Hakuta and H. Uehara, *J. Mol. Spectrosc.* **94**, 126–135 (1982).
64. M. Dagenais, J. W. C. Johns, and A. R. W. McKellar, *Can. J. Phys.* **54**, 1438–1441 (1976).
65. V. R. Braun, L. N. Krasnoperov, and V. N. Panfilov, *Opt. Specktrosk.* **52**, 719–723 (1982).
66. V. R. Braun, L. N. Krasnoperov, and V. N. Panfilov, *Khim. Fiz.* **52**, 758–762 (1982).
67. V. R. Braun, E. N. Chesnokov, and V. N. Panfilov, *Dokl. Akad. Nauk SSSR* **277**, 625–630 (1984).
68. V. R. Braun, A. I. Chichinin, L. N. Krasnoperov, et al., *Kinet. Katal.* **25**, 1041–1047 (1984).
69. V. R. Braun, L. N. Krasnoperov, and V. N. Panfilov, *Oxid. Commun.* **5**, 259–271 (1984).
70. V. R. Braun, A. I. Chichinin, and L. N. Krasnoperov, *Kinet. Katal.* **26**, 449–452 (1985).
71. P. B. Davies, B. J. Handy, E. K. Murray–Lloyd, et al., *J. Chem. Phys.* **68**, 1135–1137 (1978).
72. R. J. Saykally and K. M. Evenson, *J. Chem. Phys.* **71**, 1564–1566 (1979).
73. F. Temps and H. G. Wagner, *Ber. Bunsenges. Phys. Chem.* **86**, 119–125 (1982).
74. R. J. Saykally and K. M. Evenson, *Astrophys. J.* **238**, 107–111 (1980).
75. T. Kasuya and K. Shimoda, *Jpn. J. Appl. Phys.* **11**, 1571–1572 (1972).
76. J. P. Burrows, D. I. Cliff, P. B. Davies, et al., *Chem. Phys. Lett.* **65**, 197–200 (1979).
77. P. B. Davies, W. Hack, A. W. Preuss, et al., *Chem. Phys. Lett.* **64**, 94–97 (1979).
78. J. M. Brown, C. M. Kerr, F. D. Wayne, et al., *J. Mol. Spectrosc.* **86**, 544–555 (1981).
79. P. B. Davies, W. Hack, and H. G. Wagner, *Faraday Discuss. Chem. Soc.* **71**, 15–21 (1981).
80. J. M. Brown and J. E. Schubert, *J. Mol. Spectrosc.* **95**, 194–212 (1982).
81. C. J. Howard and K. M. Evenson, *J. Chem. Phys.* **61**, 1943–1952 (1974).
82. C. J. Howard, *J. Chem. Phys.* **65**, 4771–4777 (1976).
83. C. J. Howard and K. M. Evenson, *J. Chem. Phys.* **64**, 197–202 (1976).
84. C. J. Howard and K. M. Evenson, *J. Photochem.* **5**, 184–185 (1976).

85. C. J. Howard and K. M. Evenson, *J. Chem. Phys.* **64**, 4303–4306 (1976).
86. C. J. Howard and K. M. Evenson, *Geophys. Res. Lett.* **4**, 437–440 (1977).
87. W. Hack, A. W. Preuss, and H. G. Wagner, *Ber. Bunsenges. Phys. Chem.* **82**, 1167–1171 (1978).
88. W. Hack, A. W. Preuss, and H. G. Wagner, *Ber. Bunsenges. Phys. Chem.* **83**, 1275–1279 (1979).
89. C. J. Howard, *J. Chem. Phys.* **71**, 2352–2359 (1979).
90. W. Hack, A. W. Preuss, F. Temps, et al., *Int. J. Chem. Kinet.* **12**, 851–860 (1980).
91. B. A. Thrush and J. P. T. Wilkinson, *Chem. Phys. Lett.* **81**, 1–3 (1981).
92. Y. P. Lee and C. J. Howard, *J. Chem. Phys.* **77**, 756–763 (1982).
93. J. S. Geiger, D. E. Smith, and J. D. Bonnet, *Chem. Phys. Lett.* **70**, 600–604 (1980).
94. K. M. Evenson, H. E. Radford, and M. M. Moran, Jr., *Appl. Phys. Lett.* **18**, 426–429 (1971).
95. J. T. Hougen, J. A. Mucha, D. A. Jennings, et al., *J. Mol. Spectrosc.* **72**, 463–483 (1978).
96. J. M. Brown and K. M. Evenson, *J. Mol. Spectrosc.* **98**, 58–67 (1983).
97. R. J. Saykally, K. G. Lubic, A. Scalabrin, et al., *J. Chem. Phys.* **77**, 58–67 (1982).
98. M. A. Gondal, W. Rohrbeck, W. Urban, et al., *J. Mol. Spectrosc.* **100**, 290–302 (1983).
99. H. E. Radford and M. M. Litvak, *Chem. Phys. Lett.* **34**, 561–564 (1975).
100. F. D. Wayne and H. E. Radford, *Mol. Phys.* **32**, 1407–1422 (1976).
101. J. R. Anacona and P. B. Davies, *Chem. Phys. Lett.* **108**, 128–131 (1984).
102. P. B. Davies, B. J. Handy, E. K. Murray-Lloyd, et al., *Mol. Phys.* **36**, 1005–1015 (1978).
103. R. S. Lowe, *Mol. Phys.* **41**, 929–931 (1980).
104. W. Rohrbeck, A. Hinz, and W. Urban, *Mol. Phys.* **41**, 925–927 (1980).
105. K. Kawaguchi, C. Yamada, and E. Hirota, *J. Chem. Phys.* **71**, 3338–3345 (1979).
106. P. B. Davies, B. J. Handy, E. K. Murray-Lloyd, et al., *J. Chem. Phys.* **68**, 3377–3379 (1978).
107. J. M. Brown, A. Carrington, and T. J. Sears, *Mol. Phys.* **37**, 1837–1848 (1979).
108. D. I. Cliff, P. B. Davies, B. J. Handy, et al., *Chem. Phys. Lett.* **75**, 9–12 (1980).
109. J. M. Brown, A. Carrington, and A. D. Fackerell, *Chem. Phys. Lett.* **75**, 13–17 (1980).
110. J. M. Brown and A. D. Fackerell, *Phys. Scr.* **25**, 351–359 (1982).
111. K. Hakuta and H. Uehara, *J. Mol. Spectrosc.* **85**, 97–108 (1981).
112. P. B. Davies, D. K. Russell, and B. A. Thrush, *Chem. Phys. Lett.* **36**, 280–282 (1975).
113. P. B. Davies, D. K. Russell, D. R. Smith, et al., Can. J. Phys. **57**, 522–528 (1979).
114. N. Ohashi, K. Kawaguchi, and E. Hirota, *J. Mol. Spectrosc.* **103**, 337–349 (1984).
115. H. Uehara and K. Hakuta, *J. Chem. Phys.* **74**, 4326–4329 (1981).
116. K. Kawaguchi, S. Saito, and E. Hirota, *J. Chem. Phys.* **79**, 629–634 (1983).
117. J. M. Brown, R. F. Curl, and K. M. Evenson, *J. Chem. Phys.* **81**, 2884–2890 (1984).
118. J. M. Brown and D. Robinson, *Mol. Phys.* **51**, 883–886 (1984).

119. A. R. W. McKellar, *Can. J. Phys.* **57**, 2106–2113 (1979).
120. R. M. Stimpfle, R. A. Perry, and C. J. Howard, *J. Chem. Phys.* **71**, 5183–5190 (1979).
121. Y. P. Lee, R. M. Stimpfle, R. A. Perry, et al., *Int. J. Chem. Kinet.* **14**, 711–732 (1982).
122. R. S. Lowe and A. R. W. McKellar, *J. Mol. Spectrosc.* **79**, 424–431 (1980).
123. A. R. W. McKellar, *J. Mol. Spectrosc.* **86**, 43–54 (1981).
124. H. Uehara, *Chem. Phys. Lett.* **84**, 539–540 (1981).
125. J. M. Brown, K. M. Evenson, and T. J. Sears, *J. Chem. Phys.* **83**, 3275–3284 (1985).
126. C. J. Howard, *J. Chem. Phys.* **67**, 5258–5362 (1977).
127. W. Hack, A. W. Preuss, H. G. Wagner, et al., *Ber. Bunsenges. Phys. Chem.* **83**, 212–217 (1979).
128. J. P. Burrows, D. I. Cliff, G. W. Harris, et al., *Proc. R. Soc. London* **A368**, 463–481 (1979).
129. B. A. Thrush and J. P. T. Wilkinson, *Chem. Phys. Lett.* **66**, 441–443 (1979).
130. H. E. Radford, *Chem. Phys. Lett.* **71**, 195–197 (1980).
131. C. J. Howard and B. J. Finlayson-Pitts, *J. Chem. Phys.* **72**, 3842–3843 (1980).
132. M. S. Zahnizer and C. J. Howard, *J. Chem. Phys.* **73**, 1620–1626 (1980).
133. B. A. Thrush and J. P. T. Wilkinson, *Chem. Phys. Lett.* **84**, 17–19 (1981).
134. J. T. Hougen, H. E. Radford, K. M. Evenson, et al., *J. Mol. Spectrosc.* **56**, 210–228 (1975).
135. J. W. C. Johns, A. R. W. McKellar, and M. Riggin, *J. Chem. Phys.* **68**, 3957–3967 (1978).
136. C. E. Barnes, J. M. Brown, A. Carrington, et al., *J. Mol. Spectrosc.* **72**, 86–101 (1978).
137. S. V. Broude, Yu. M. Gershenzon, A. V. Gorelik, et al., *Kinet. Katal.* **19**, 535–536 (1978).
138. V. V. Azatyan, K. I. Gaganidze, S. A. Kolesnikov, et al., *Kinet. Katal.* **23**, 244–245 (1982).
139. V. B. Rosenshtein, Yu. M. Gershenzon, S. D. Il'in, et al., *Chem. Phys. Lett.* **112**, 473–478 (1984).
140. C. E. Barnes, J. M. Brown, and H. E. Radford, *J. Mol. Spectrosc.* **84**, 179–196 (1980).
141. A. R. W. McKellar, *J. Chem. Phys.* **71**, 81–88 (1979).
142. H. Uehara, *J. Chem. Phys.* **77**, 3314–3318 (1982).
143. F. Temps, H. G. Wagner, P. B. Davies, et al., *J. Chem. Phys.* **87**, 5068–5071 (1983).
144. A. Hinz, W. Seebass, and W. Urban, *Bull. Soc. Chim. Belg.* **92**, 501 (1983).
145. H. Uehara, *Chem. Phys. Lett.* **106**, 554–557 (1984).
146. J. A. Mucha, K. M. Evenson, D. A. Jennings, et al., *Chem. Phys. Lett.* **66**, 244–247 (1979).
147. T. J. Sears, P. R. Bunker, and A. R. W. McKellar, *J. Chem. Phys.* **77**, 5348–5369 (1982).
148. T. Böhland and F. Temps, *Ber. Bunsenges. Phys. Chem.* **88**, 459–461 (1984).
149. T. Böhland, F. Temps, and H. G. Wagner, *Ber. Bunsenges. Phys. Chem.* **88**, 455–458 (1984).
150. S. Dobe, T. Böhland, F. Temps, et al., *Ber. Bunsenges. Phys. Chem.* **89**, 432–441 (1985).
151. T. Böhland, F. Temps, and H. G. Wagner, *Ber. Bunsenges. Phys. Chem.* **89**, 1013–1018 (1985).

152. T. J. Sears, P. B. Bunker, and A. R. W. McKellar, *J. Chem. Phys.* **75**, 4731–4732 (1981).
153. T. J. Sears, P. B. Bunker, and A. R. W. McKellar, *J. Chem. Phys.* **77**, 5363–5369 (1982).
154. P. R. Bunker, T. J. Sears, A. R. W. McKellar, et al., *J. Chem. Phys.* **79**, 1211–1219 (1983).
155. J. M. Cook, K. M. Evenson, C. J. Howard, et al., *J. Chem. Phys.* **64**, 1381–1388 (1976).
156. P. B. Davies, D. K. Russell, D. R. Smith, et al., Lasers Chem. Proc. Conf. London 1977, Amsterdam (1977), pp. 97–100.
157. J. M. Brown, J. Buttenshaw, A. Carrington, et al., *Mol. Phys.* **33**, 589–592 (1977).
158. J. W. C. Johns, A. R. W. McKellar, and M. Riggin, *J. Chem. Phys.* **67**, 2427–2435 (1977).
159. J. M. Brown, J. Buttenshaw, A. Carrington, et al., *J. Mol. Spectrosc.* **79**, 47–61 (1980).
160. R. S. Lowe and A. R. W. McKellar, *J. Chem. Phys.* **74**, 2686–2697 (1981).
161. S. A. Chasovnikov, L. N. Krasnoperov, and V. N. Panfilov, *Khim. Fiz.*, No. 5, 570–574 (1982).
162. C. E. Barnes, J. M. Brown, A. D. Fackerell, et al., *J. Mol. Spectrosc.* **92**, 485–496 (1982).
163. P. B. Davies, D. K. Russell, B. A. Thrush, et al., *J. Chem. Phys.* **62**, 3739–3742 (1975).
164. P. B. Davies, D. K. Russell, B. A. Thrush, et al., *Chem. Phys. Lett.* **42**, 35–38 (1976).
165. P. B. Davies, D. K. Russell, B. A. Thrush, et al., *Proc. R. Soc. London* **A353**, 299–318 (1977).
166. G. W. Hills and A. R. W. McKellar, *J. Mol. Spectrosc.* **74**, 224–227 (1979).
167. K. Kawaguchi, C. Yamada, E. Hirota,, et al., *J. Mol. Spectrosc.* **81**, 60–72 (1980).
168. A. Carrington, J. S. Geiger, D. R. Smith, et al., *Chem. Phys. Lett.* **90**, 6–8 (1982).
169. G. W. Hills and A. R. W. McKellar, *J. Chem. Phys.* **71**, 3330–3337 (1979).
170. K. Hakuta and H. Uehara, *J. Chem. Phys.* **74**, 5995–5999 (1974).
171. Yu. M. Gershenzon, S. D. Il'in, S. A. Kolesnikov, et al., *Kinet. Katal.* **19**, 1405–1410 (1978).
172. Yu. M. Gershenzon, S. D. Il'in, O. P. Kishkovich, et al., *Kinet. Katal.* **23**, 534–541 (1982).
173. Yu. M. Gershenzon, S. D. Il'in, O. P. Kishkovich, et al., *Int. J. Chem. Kinet.* **15**, 399–415 (1983).
174. P. B. Davies, D. K. Russell, B. A. Thrush, et al., *Chem. Phys. Lett.* **37**, 43–46 (1976).
175. P. B. Davies, D. K. Russell, B. A. Thrush, et al., *Chem. Phys.* **44**, 421–426 (1979).
176. G. W. Hills and A. R. W. McKellar, *J. Chem. Phys.* **71**, 1141–1149 (1979).
177. K. Kawaguchi, S. Saito, E. Hirota, et al., *J. Chem. Phys.* **82**, 4893–4902 (1985).
178. T. J. Sears and A. R. W. McKellar, *Mol. Phys.* **49**, 25–32 (1983).
179. H. E. Radford, F. D. Wayne, and J. M. Brown, *J. Mol. Spectrosc.* **99**, 209–220 (1983).

180. R. J. Saykally, L. Veseth, and K. M. Evenson, *J. Chem. Phys.* **80**, 2247–2255 (1984).
181. J. A. Mucha, D. A. Jennings, K. M. Evenson, et al., *J. Mol. Spectrosc.* **68**, 122–124 (1977).
182. P. B. Davies, F. Dransfeld, F. Temps, et al., *J. Chem. Phys.* **81**, 3763–3765 (1984).
183. L. N. Krasnoperov, V. R. Braun, V. V. Nosov, et al., *Kinet. Katal.* **12**, 1332–1334 (1981).
184. V. R. Braun, L. N. Krasnoperov, and V. N. Panfilov, *Dokl. Akad. Nauk SSSR* **260**, 901–903 (1981).
185. H. E. Radford and D. K. Russell, *J. Chem. Phys.* **66**, 2222–2224 (1977).
186. D. K. Russell and H. E. Radford, *J. Chem. Phys.* **72**, 2750–2759 (1980).
187. H. E. Radford, K. M. Evenson, and D. A. Jennings, *Chem. Phys. Lett.* **78**, 589–591 (1981).
188. J. W. C. Johns, A. R. W. McKellar, and M. Riggin, *J. Chem. Phys.* **66**, 3962–3963 (1977).
189. T. J. Sears and A. R. W. McKellar, *Can. J. Phys.* **60**, 345–348 (1982).
190. M. Inguscio, K. R. Leopold, J. S. Mussay, et al., *J. Opt. Soc. Am.* **B2**, 1566–1569 (1985).
191. C. Yamada, K. Kawaguchi, and E. Hirota, *J. Chem. Phys.* **69**, 1942–1944 (1978).
192. A. R. W. McKellar, P. R. Bunker, T. J. Sears, et al., *J. Chem. Phys.* **79**, 5251–5264 (1983).
193. J. T. Hougen, H. E. Radford, K. M. Evenson, et al., *J. Mol. Spectrosc.* **56**, 210–228 (1975).
194. A. Scalabrin, R. J. Saykally, K. M. Evenson, et al., *J. Mol. Spectrosc.* **89**, 344–351 (1981).
195. P. B. Davies and F. Temps, *J. Chem. Phys.* **74**, 6556–6559 (1981).
196. P. B. Davies, A. H. Ferguson, D. P. Stern, et al., *J. Mol. Spectrosc.* **113**, 28–38 (1985).
197. R. J. Saykally and K. M. Evenson, *Phys. Rev. Lett.* **43**, 515–578 (1979).
198. D. Ray, K. G. Lubic, and R. J. Saykally, *J. Mol. Spectrosc.* **46**, 217–221 (1982).
199. A. Hinz, W. Bohle, D. Zeitz, et al., *Mol. Phys.* **53**, 1017–1021 (1984).
200. Z. K. Smedarchina and Yu. M. Gershenzon, *Dokl. Akad. Nauk SSSR* **232**, 638–640 (1977).
201. K. S. Krasnov, N. V. Fillippenko, V. A. Bobkova, et al., *Molecular Constants of Inorganic Substances* [in Russian], Khimiya, Leningrad (1979).
202. J. Higuchi, S. Kubota, T. Kumamoto, et al., *Bull. Chem. Soc. Jpn.* **47**, 2775–2780 (1974).
203. G. Herzberg, *The Spectra and Structure of Simple Free Radicals*, Cornell Univ. Press, Ithaca, New York (1971).
204. T. Böhland, F. Temps, and H. G. Wagner, *Int. J. Res. Phys. Chem. Chem. Phys.* **142**, 129–140 (1984).
205. M. Clyne, *The Physical Chemistry of Fast Reactions*, B. P. Levitt (ed.), Plenum Press, New York (1973).
206. C. J. Howard, *J. Phys. Chem.* **83**, 3–9 (1979).
207. R. V. Ambartzumian and V. S. Letokhov, in: *Chemical and Biochemical Applications of Lasers, Vol. 3*, C. B. Moore (ed.), Academic Press, New York (1977).
208. H. von Hartel and M. Polanyi, *Z. Phys. Chem.* **11**, 97–138 (1930).

209. V. L. Talrose, A. F. Dodonov, V. V. Zelenov, et al., *Int. J. Mass Spectrom. Ion Phys.* **46**, 123–126 (1983).
210. M. Braithwaite and S. R. Leone, *J. Chem. Phys.* **69**, 839–845 (1978).
211. A. I. Chichinin and L. N. Krasnoperov, Abstracts of talks at the 12th All-Union Conf. on Coherent and Nonlinear Optics, Moscow (1985), Part II, p. 97.
212. A. I. Chichinin and L. N. Krasnoperov, Abstracts of talks at the 4th All-Union Symp. on Laser Chemistry, Zvenigorod (1985), p. 99.
213. E. N. Chesnokov and V. N. Panfilov, *Khim. Fiz.*, No. 10, 1349–1355 (1982).
214. S. A. Chasovnikov, A. I. Chichinin, and L. N. Krasnoperov, Abstracts of talks at the 2nd All-Union Conf. on Chemiluminescence, Ufa (1986), p. 6.
215. R. H. Clark and D. Husain, *J. Chem. Soc. Faraday Trans.* **80**, 97–113 (1984).
216. N. N. Semenov, *The Development of the Theory of Chain Reactions and Thermal Ignition* [in Russian], Znanie, Moscow (1969).
217. E. E. Nikitin and V. N. Kondrat'ev, *Kinetics and Mechanism of Gaseous Phase Reactions* [in Russian], Nauka, Moscow (1974).
218. N. N. Semenov, *Some Problems in Chemical Kinetics and Reactivity* [in Russian], Izd-vo. Akad. Nauk SSSR, Moscow (1958).
219. S. A. Arutyunyan and É. N. Sarkisyan, *Arm. Khim. Zh.* **35**, No. 1, 3–6 (1982).
220. E. N. Chesnokov and V. N. Panfilov, *Dokl. Akad. Nauk SSSR* **261**, 925–929 (1981).
221. V. B. Rozenshtein, Yu. M. Gershenzon, O. P. Kishkovich, et al., *Dokl. Akad. Nauk SSSR* **280**, 656–657 (1985).
222. V. N. Panfilov and A. K. Petrov, *Plasma Chemistry, Vol. 6* [in Russian], Atomizdat, Moscow (1979), pp. 54–88.
223. R. T. Bayley and F. R. Cruickshank, *Chemical Society Specialist Periodical Report on Molecular Spectroscopy, Vol. 2*, R. F. Barrow (ed.), Chemical Society, London (1974), pp. 262–356.
224. Yu. N. Molin, V. N. Panfilov, and A. K. Petrov, *Infrared Photochemistry* [in Russian], Nauka, Novosibirsk (1985).
225. D. L. Atkins, D. S. King, and J. S. Stephenson, *Chem. Phys. Lett.* **65**, 257–260 (1979).
226. A. P. Baronavsky, A. Cabello, J. H. Clark, et al., *J. Photochem.* **9**, 322–323 (1978).
227. V. A. Nadtochenko, O. M. Sarkisov, M. P. Frolov, et al., *Kinet. Katal.* **22**, 865–870 (1981).
228. O. M. Sarkisov, S. Ya. Umanskii, and S. G. Cheskis, *Dokl. Akad. Nauk SSSR* **246**, 662–665 (1979).
229. V. A. Lozovoi, O. M. Sarkisov, S. Ya. Umanskii, et al., *Khim. Fiz.*, No. 2, 201–210 (1983).
230. S. L. Baughcum and S. R. Leone, *J. Chem. Phys.* **72**, 6531–6545 (1980).
231. H. W. Hermann and S. R. Leone, *J. Chem. Phys.* **76**, 4766–4774 (1982).
232. H. K. Haugen, E. Weitz, and S. R. Leone, *Chem. Phys. Lett.* **119**, 75–80 (1985).
233. E. F. Gordiets, L. I. Gudzenko, and V. Ya. Panchenko, *Izv. Akad. Nauk SSSR, Ser. Fiz.* **43**, 251–254 (1979).
234. H. H. Nelson and H. S. Johnston, *J. Phys. Chem.* **85**, 3891–3896 (1981).
235. L. N. Krasnoperov and V. N. Panfilov, *Teor. Éksp. Khim.* **15**, 348–358 (1979).
236. W. Bohle, J. Werner, D. Zeitz, et al., *Mol. Phys.* **58**, 85–95 (1986).
237. W. Rohrbeck, A. Hinz, P. Nelle, et al., *Appl. Phys.* **B31**, 139–144 (1983).
238. J. Werner, W. Seebass, M. Koch, et al., *Mol. Phys.* **56**, 453–461 (1985).
239. D. Zeitz, W. Bohle, J. Werner, et al., *Mol. Phys.* **54**, 953–958 (1985)

240. W. Hack, *Int. Rev. Phys. Chem.* **4**, 165–200 (1985).
241. W. Seebass, Dissertation, Bonn University (1986).
242. P. B. Davies, P. Dransfeld, F. Temps, et al., *Max-Planck Inst. für Strömungsforschung*, Report No. 6/1984.
243. E. V. Antsupov and G. I. Ksandopulo, *Khim. Fiz.* **4**, 1677–1681 (1985).

# DEGRADATION SPECTRA OF ELECTRONS IN GASES

V. P. Konovalov and É. E. Son

## 1. INTRODUCTION

In many engineering devices, such as lasers with nonself-sustaining discharges or with nuclear pumping, in systems where beams of electrons or other high-energy particles interact with gases, and in the earth's ionosphere, the problem of calculating the degradation spectrum of electrons, i.e., the energy distribution of secondary product electrons, arises. The interaction of electron beams with plasmas at low pressures, when Langmuir waves are strongly pumped, has been examined elsewhere [1, 2]. In this review we discuss the methods for calculating the degradation spectra of electrons in low-temperature plasmas, present the results of some calculations of these spectra in atomic and molecular gases, and discuss the applications of nonequilibrium plasmas excited by external ionization sources. Because of the limited space, we do not discuss plasma chemical reactions in beam plasmas. A review by Bychkov and Eletskii [3] is partially devoted to this question for high-pressure gases.

When an ionization source in the form of x rays, nuclear reaction products, or neutral or charged particle beams acts on a gas, the first interactions involve the transfer of energy to electrons; hence, the action of any ionization source can be reduced to that of a highly energetic electron beam which is monoenergetic or has a specified energy distribution. As the electrons interact with the gas, successive loss of energy by the primary electrons and formation of secondary electrons take place, that is, the energy of the electrons degrades. The distribution of the secondary electrons, which are $E/U_i$ times more than the primaries (where $E$ is the energy of the primary electrons and $U_i$ is the energy cost of producing an electron–ion pair), is substantially non-Maxwellian. The spectrum characteristically has a "tail," i.e., is enriched in energetic electrons. This form of distribution leads to a sharp increase in the rates of threshold elementary processes involving electrons and determines the differences between the kinetics and ion–molecule composition of such discharges and those of the nonequilibrium plasmas in electrical discharges having the same average parameters (electron density and temperature). In some cases, a study of the combined effect of external ionization sources and an electric field is of interest. Thus, for example, an external ionization source is used in molecular lasers with nonself-sustaining discharges to create the required electron density, while an electric field is applied to obtain the optimum average electron energy at which the upper laser levels are most efficiently pumped. This distinction between the functions of the ionization source and field leads to the possibility of independently varying the density and average energy of the electrons and enhances the stability of the discharges.

The mechanisms for plasma formation by ionization sources are of great importance for excimer [4] and ion [5] lasers excited by electron beams, as well as for plasma lasers and reactor lasers [6], whose active media are formed by fission fragments.

In nature, a plasma is produced by the action of an external ionization source in the earth's ionosphere [7], where penetrating fluxes of highly energetic particles create plasma layers which govern radio communication, auroras, and the chemical composition of the ionosphere.

The basic information on the properties of plasmas excited by ionization sources and laser light is contained in the high-energy electron energy distribution, i.e., in the degradation spectrum. The major characteristics of these plasmas are the rate constants for the plasma chemical reactions and the energy costs of producing electron–ion pairs, radicals, and excited molecules, which depend integrally on the electron energy distribution function.

The importance of determining the degradation spectra of electrons in gases was apparently first pointed out by Fermi [8], who noted that the problem would be complicated because of the need to take account of the many processes by which the electrons interact with the gas molecules. The problem of the degradation spectrum was formulated mathematically by Fano and Spencer [9–11], who obtained an equation describing the degradation of electrons in a medium and calculated the penetration of relativistic electrons through aluminum and lead for energies above threshold. Then [12, 13] the method was generalized to subthreshold energies. Another form of the equation for the degradation spectrum of $\alpha$-

particles was obtained by Fowler [14] and subsequently used for calculating the degradation spectrum of electrons in gases, as well. Specific calculations have been carried out for those gases for which the most detailed data on the cross sections for elementary electron–molecule collision processes are available: helium [15, 16], argon [17, 18], and molecular hydrogen [19, 20]. A special issue of the journal *Radiation Research* [**64**, No. 1 (1975)] was devoted to the topic of electron degradation in gases and contained a number of reviews [21–25] of work completed before 1975. The basic results have also been discussed and supplemented in a book by Nikerov and Sholin [26].

The last decade has been characterized, first of all, by a greater demand for data on degradation spectra in connection with engineering applications, second, by a substantial increase in the available information on elementary processes in different gases, third, by an extension in the computer capabilities needed for taking a large number of elementary processes, especially in molecular gases, into account, and, fourth, by the development of analytic methods for calculating degradation spectra. The trends in research over the last decade and the results of this research are discussed in this review.

## 2. METHODS OF CALCULATING DEGRADATION SPECTRA OF ELECTRONS IN GASES

It is convenient to distinguish the following three ranges of electron energy in discussing the slowing down of fast electrons in a gas:

1) the range extending from the initial energy of a primary electron to energies on the order of the ionization potential, in which the electrons lose energy as they are slowed down through ionization and excitation of electronic states of the molecules;

2) the energy range below the ionization potential extending to the minimum threshold for electronic and vibrational excitation, in which an electron loses energy primarily through excitation of electronic and vibrational states of the molecules, but the energy losses by electrons through elastic collisions with molecules and excitation of their rotational levels must also be taken into account; and

3) the energy range below the thresholds for electronic and vibrational excitation, in which the bulk of the free electrons are concentrated and where they are lost efficiently through recombination with ions and attachment to molecules.

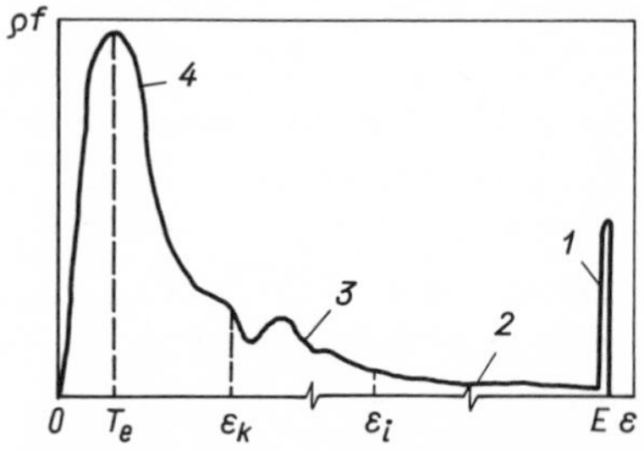


**Fig. 1.** The qualitative form of the degradation spectrum of electrons in a gas: 1) primary source electrons; 2) cascade electrons; 3) electrons in the region of the threshold for electronic excitation; 4) thermal electrons.

Figure 1 shows a qualitative picture of the electron energy distribution [6]. The plasma electrons concentrated in the low-energy region can be referred to as thermal, while the electrons in the threshold-energy region are suprathermal [7].

The energy distribution of an electron as it is slowed down through various interaction channels with the molecules can be calculated using the continuous slowing-down model, in which it is assumed that a highly energetic electron loses energy continuously in the process of slowing down. Because of its simplicity, this model has come to be widely used in ionospheric physics and in calculations for a number of atmospheric gases [27–32].

The energy loss by a fast electron per unit path length is characterized by an effective slowing down [33], in which the energy lost by the electron in all elementary interaction processes with particles in the medium are summed:

$$L = \sum_n (\Delta\varepsilon_n)\,\sigma_n, \tag{1}$$

where $\sigma_n$ is the cross section for a given process and $\Delta\varepsilon_n$ is the energy lost by an electron in that process. In the first approximation of the continuous slowing-down model, one examines the degradation of a primary electron with a large initial energy $E$ without considering the redistribution of the energy by the secondary electrons. The fraction of the energy of a primary electron expended in the $j$th inelastic process in the course of coming fully to a stop is given by

$$p_j = \frac{1}{E} \int_{\varepsilon_j}^{E} \frac{\varepsilon_j \sigma_j}{L} d\varepsilon, \tag{2}$$

where $\varepsilon_j$ is the excitation potential of the $j$th level of a gas molecule.

The next approximation of the continuous slowing-down model takes into account the energy expended in the $j$th process by the secondary electrons generated by the primaries:

$$p_j = \frac{\varepsilon_j}{E} \left[ \int_{\varepsilon_j}^{E} \frac{\sigma_j(\varepsilon)\, d\varepsilon}{L(\varepsilon)} + \int_{\varepsilon_j}^{(E-\varepsilon_i)/2} d\varepsilon_s \int_{\varepsilon_i+2\varepsilon_s}^{E} \frac{\sigma_{ii}(\varepsilon_p, \varepsilon_s)\, d\varepsilon_p}{L(\varepsilon_p)} \int_{\varepsilon_j}^{\varepsilon_s} \frac{\sigma_j(\varepsilon)\, d\varepsilon}{L(\varepsilon)} \right]. \tag{3}$$

Here $\sigma_{ii}(\varepsilon_p, \varepsilon_s)$ is the differential cross section for ionization; $\varepsilon_p$ is the energy of a primary electron; $\varepsilon_s$ is the energy of a newly-created secondary electron; and $\varepsilon_i$ is the ionization potential. In an ionization event, the electron with the lower energy is regarded as the secondary; hence, $\varepsilon_p \geq \varepsilon_i + 2\varepsilon_s$ and the range of energies of the secondary electrons is given by $0 \leq \varepsilon_s \leq (E - \varepsilon_i)/2$. Similarly, in the following approximations we can take into account the corrections associated with the expenditure in the $j$th process of energy from the third, fourth, and all subsequent generations of electrons. In the literature this multiplication of electrons is referred to as a cascade. The infinite series obtained in this fashion can be interpreted as the solution of an integral equation [29]. This series converges rapidly, so that in the calculations we can limit ourselves to the first few terms, usually one to three. The main shortcoming of the continuous slowing-down model is the notion that the energy loss by an electron as it slows down is continuous, which is true only at high energies [30]. Nevertheless, Eq. (2) is extremely convenient for estimating the energy expended in processes with relatively high thresholds, i.e., in ionization and electronic excitation of the molecules.

The most complete representation of the spectrum of the electrons over the entire energy range is given by the Boltzmann kinetic equation for the electron energy distribution function (EEDF). Under quasistationary, uniform conditions, the electron energy distribution can be treated independently of the spatial and time distributions. Then the kinetic equation for the EEDF can be conveniently interpreted as describing the change in the fluxes of electrons along the energy axis owing to various processes. This makes it possible to draw an analogy among various physical problems and may be regarded as a general approach. The kinetic equation for the EEDF in a weakly ionized plasma can be written in the form [34]

$$\rho\, \partial f/\partial t = -\partial j/\partial \varepsilon + \sum_k S_k + S_i + (q - p)/n, \tag{4}$$

where $f$ is the EEDF normalized by the condition

$$\int_0^\infty f\rho\, d\varepsilon = 1, \quad \rho = \varepsilon^{1/2}. \tag{5}$$

Here $n$ is the electron density; $q$ is the rate of production of electrons by the external source; $p$ is the rate of loss of free electrons in elementary acts of recombination and attachment; $S_k$ is the collision integral associated with inelastic excitation of a molecule into the $k$th state; $S_i$ is the collision integral describing the ionization of the molecules by electron impact; and $j$ is the flux of electrons along the energy axis. The dependence on the arguments (the energy $\varepsilon$ and time $t$) has been omitted to simplify the notation.

The flux $j$ includes processes that are characterized by a small change in the electron energy in an elementary act, such as elastic collisions of electrons with gas molecules ($T$), rotational excitation of molecules ($r$), and electron–electron collisions ($ee$):

$$j = \sum_\alpha j_\alpha, \quad \alpha = T,\ r,\ ee.$$

Each of these terms represents the difference of diffusive and convective fluxes, i.e.,

$$j_\alpha = -\rho D_\alpha \frac{\partial f}{\partial \varepsilon} - \rho \mu_\alpha f = j_\alpha^D - j_\alpha^C,$$

where $D_\alpha$ and $\mu_\alpha$ are the diffusion coefficient and mobility of the electrons as they move along the energy axis owing to process $\alpha$. The convective flux is formed by collisions of the first kind which lead to energy loss by the electrons. The corresponding diffusive flux is caused by transitions of both the first and second types which lead to changes in the energy of the electrons.

The coefficients $D_\alpha$ and $\mu_\alpha$ are expressed in terms of the averaged characteristics of the collisions [35] and for an estimate we can set

$$\mu_\alpha \sim \delta_\alpha \varepsilon \nu_\alpha; \quad D_\alpha \sim \delta_\alpha \varepsilon T_\alpha \nu_\alpha,$$

where $\delta_\alpha$ is the average fraction of energy lost by an electron in a single collision of type $\alpha$; $\nu_\alpha$ is the collision frequency of electrons with particles in the $\alpha$th process; and $T_\alpha$ is the temperature characterizing the corresponding particle distribution. For example, for elastic collisions of electrons with gas molecules, we have $\delta_T = 2m/M$, where $m$ and $M$ are the masses of an electron and a molecule, $T$ is the translational temperature of the molecules, and $\nu_T$ is the transport collision frequency. Note that the diffusion coefficient and mobility corresponding to electron–electron collisions generally depend nonlinearly on the EEDF and are integral functionals. In this approach one can also include processes associated with the action of external fields on the plasma electrons by introducing the appropriate diffusion coefficients.

Thus, the steady-state kinetic equation for the electrons can be written in the form

$$\left.\begin{aligned} &\frac{d}{d\varepsilon}\left(\rho D\frac{df}{d\varepsilon}+\rho\mu f\right)+\sum_k S_k+S_i+\frac{q-p}{n}=0,\\ &D=\sum_\alpha D_\alpha,\quad \mu=\sum_\alpha \mu_\alpha. \end{aligned}\right\} \tag{6}$$

The collision integrals for inelastic excitation (of electronic and vibrational levels) and ionization of the molecules are given by

$$S_k=(f\rho\nu_k)\Big|_{\varepsilon}^{\varepsilon+\varepsilon_k},$$

$$S_i=\int_0^\infty f\rho\nu_{ii}d\varepsilon_p-f\rho\nu_i.$$

Here $\varepsilon_k$ is the threshold for excitation of the $k$th level; $\nu_k$ and $\nu_i$ are the rates of $k$th excitation and ionization given by $\nu_k = Nv\sigma_k$ and $\nu_i = Nv\sigma_i$; $\sigma_k$ and $\sigma_i$ are the corresponding cross sections; $N$ is the density of molecules; $v$ is the electron velocity; and $\nu_{ii} = Nv\sigma_{ii}$, where $\sigma_{ii}$ is the differential cross section for ionization, with

$$\sigma_i(\varepsilon_p)=\int_0^{(\varepsilon_p-\varepsilon_i)/2}\sigma_{ii}(\varepsilon_p,\min(\varepsilon_s,\ \varepsilon_p-\varepsilon_i-\varepsilon_s))\,d\varepsilon_s$$

$$=\int_0^{(\varepsilon_p-\varepsilon_i)/2}\sigma_{ii}(\varepsilon_p,\ \varepsilon_s)\,d\varepsilon_s=\int_{(\varepsilon_p-\varepsilon_i)/2}^{\varepsilon_p-\varepsilon_i}\sigma_{ii}(\varepsilon_p,\ \varepsilon_p-\varepsilon_i-\varepsilon_s)\,d\varepsilon_s. \tag{7}$$

We now integrate Eq. (6) with respect to the energy over the interval $(\varepsilon, \infty)$, noting the boundary conditions $f(\infty) = 0$ and $df/d\varepsilon(\infty) = 0$ and using Eq. (7). As a result, we obtain the following integrodifferential equation:

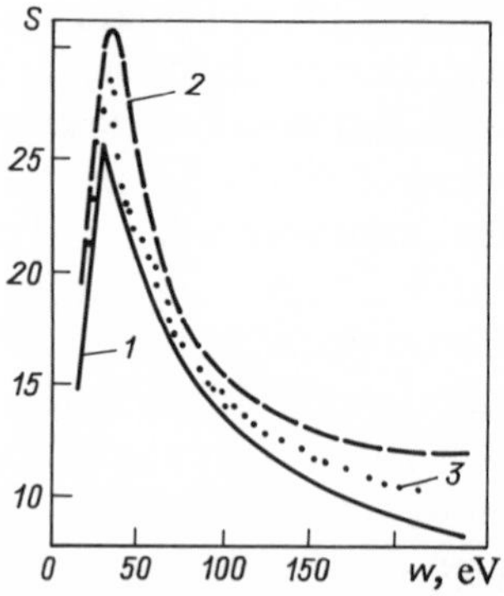


**Fig. 2.** The spectrum of secondary electrons produced during ionization of $N_2$ molecules by electron impact [36]: 1) theory [36]; 2) the approximation of [38]; 3) experiment [37].

$$\rho D \frac{df}{d\varepsilon} + \rho\mu f + \sum_{k,\,i} \int_{\varepsilon}^{\varepsilon+\varepsilon_k} f\rho\nu_k d\varepsilon_p = \int_{\varepsilon+\varepsilon_i}^{\infty} f\rho K d\varepsilon_p + Q. \tag{8}$$

Here $Q = \int_{\varepsilon}^{\infty} (q - p) d\varepsilon'/n$, and the sum of the left-hand side contains all the inelastic processes, including ionization. The kernel of this integral is

$$K(\varepsilon_p, \varepsilon) = \int_{\varepsilon}^{(\varepsilon_p - \varepsilon_i)/2} \nu_{ii}(\varepsilon_p, \min(\varepsilon_s, \varepsilon_p - \varepsilon_i - \varepsilon_s))\, d\varepsilon_s.$$

The integral on the right-hand side of Eq. (8) explicitly characterizes the difference between ionization and the other inelastic processes associated with the production of secondary electrons, and has the form

$$\int_{\varepsilon+\varepsilon_i}^{\infty} f\rho K(\varepsilon_p, \varepsilon)\, d\varepsilon_p = -\int_{\varepsilon_i+\varepsilon}^{\varepsilon_i+2\varepsilon} f\rho d\varepsilon_p \int_{\varepsilon_p-\varepsilon_i-\varepsilon}^{(\varepsilon_p-\varepsilon_i)/2} \nu_{ii}(\varepsilon_p, \varepsilon_s)\, d\varepsilon_s$$

$$+ \int_{\varepsilon_i+2\varepsilon}^{\infty} f\rho d\varepsilon_p \int_{\varepsilon}^{(\varepsilon_p-\varepsilon_i)/2} \nu_{ii}(\varepsilon_p, \varepsilon_s)\, d\varepsilon_s,$$

from which it is evident that the kernel $K(\varepsilon_p, \varepsilon)$ changes sign. Equation (8) can be simplified considerably at high energies $\varepsilon \gg \varepsilon_k, \varepsilon_i$, where an expansion can be taken in the small parameter $\varepsilon_k/\varepsilon \ll 1$. In addition, in the high-energy region it is usually possible to neglect the diffusive fluxes, so that Eq. (8) takes the form of a Fredholm integral equation

$$f\rho\left(\mu + \sum_{k,\,i} \varepsilon_k \nu_k\right) = Q + \int f\rho K d\varepsilon_p.$$

The domain of integration can always be extended from zero to infinity by setting the kernel $K = 0$ in the corresponding energy range. It may also be assumed that all inelastic processes form corresponding convective fluxes of electrons with mobilities given by $\mu_k = \varepsilon_k \nu_k$. Then, the expression in parentheses represents the total mobility of the electrons along the energy axis owing to all processes:

$$\sum \mu = \mu + \sum_{k,\,i} \varepsilon_k \nu_k.$$

The resulting integral equation has a solution in the form of a Neumann series obtained by successive approximations. When $Q = \text{const}$, we have

$$f = \frac{Q}{\rho \Sigma \mu}\left(1 + \int \frac{K d\varepsilon'}{\Sigma \mu} + \int \frac{K d\varepsilon'}{\Sigma \mu} \int \frac{K d\varepsilon''}{\Sigma \mu} + \ldots\right). \tag{9}$$

The character of this solution can be clarified using the example of the simplest ionization model, in which secondary electrons are always formed with a single energy $\varepsilon_*$, i.e., the differential ionization cross section is given by $\sigma_{ii}(\varepsilon_p, \varepsilon_s) = \sigma_i(\varepsilon_p) \cdot \delta(\varepsilon_s - \varepsilon_*)$. In this case, the exact solution has the form

$$f = \frac{(Q/\rho)}{\mu + \sum\limits_k \varepsilon_k \nu_k + (\varepsilon_i + \varepsilon_*)\, \nu_i}\,.$$

The quantity in the denominator is proportional to the energy loss by a slowed-down electron, i.e., to the effective slowing down $L$. Consequently, the solution of the kinetic equation at high energies corresponds to degradation of the electron energy in the continuous slowing-down model. This is also evident from a comparison of Eqs. (3) and (9). A typical dependence of the cross section for differential ionization on the energy of a secondary electron is shown in Fig. 2 [36], where experimental data [37] for a primary electron energy of $\varepsilon_p = 1$ keV in nitrogen are compared with the semiempirical approximation [38] and some more detailed calculations [36]. The abscissa is the energy loss by a primary electron in an ionization event $w = \varepsilon_i + \varepsilon_s$, and the ordinate is the dimensionless quantity $s = \sigma_{ii} w^2 \varepsilon_p / 4\pi a_0^2 \,\mathrm{Ry}^2$, where $a_0$ is the Bohr radius and Ry is the Rydberg constant. The rapid convergence of the series (9) is caused by the resonant character of differential ionization, since secondary electrons with a relatively low energy are created with a high probability. Because of this circumstance, the solution of the kinetic equation at high energies can always be expressed analytically with sufficient accuracy by choosing a specific form for the differential ionization cross section. The Breit–Wigner formula [24, 38]

$$\sigma_{ii}(\varepsilon_p, \varepsilon_s) = \frac{\sigma_i(\varepsilon_p)}{\tan^{-1}\left(\frac{\varepsilon_p - \varepsilon_i - 2\varepsilon_*}{2\Gamma}\right) + \tan^{-1}\left(\frac{\varepsilon_*}{\Gamma}\right)} \left[\frac{\Gamma}{(\varepsilon_s - \varepsilon_*)^2 + \Gamma^2}\right],$$

where $\varepsilon_*$ is the resonance point (the most probable energy for a secondary electron) and $\Gamma$ is its width, is convenient for this purpose. In this case we can recommend the following expression for the EEDF at high energies $\varepsilon \gg \varepsilon_k, \varepsilon_i, \varepsilon_*$:

$$f = \frac{(Q/\rho)}{\mu + \sum_k \varepsilon_k \nu_k + \left\{\varepsilon_i + \varepsilon_* + \Gamma \ln\left(\frac{\varepsilon}{\sqrt{\Gamma^2 + \varepsilon_*^2}}\right) \Big/ \left[\frac{\pi}{2} + \tan^{-1}\left(\frac{\varepsilon_*}{\Gamma}\right)\right]\right\} \nu_i}.$$

It should be emphasized that the character of the differential ionization determines the basic properties of the degradation spectrum of the electrons to a significant extent. When a molecule is ionized by a high-energy electron, two factors are important: first, the resulting secondary electron, as a rule, has a low energy on the order of that for electronic excitation of the molecule and, second, the energy of the secondary electron is practically independent of the energy of the primary electron. Because of this, energy transfer from the electron source to the gas can be represented qualitatively in the following way: the high-energy primary electrons mainly cause ionization and produce low-energy secondary electrons, which themselves mostly expend their energy in exciting molecules and are less able to ionize them. Figuratively speaking, a high-energy electron "has many children but few grandchildren." This is precisely the reason for the rapid convergence of the series (3) and (9), which include the electrons from different generations in sequence.

The energy cost of producing an electron–ion pair can be estimated directly from the Breit–Wigner formula with the aid of some qualitative arguments. It is natural to assume that the main process for energy loss by the high-energy electrons is ionization. From physical considerations, it follows that the cost of an electron–ion pair is the sum of the ionization potential and the average energy of the product secondary electron; thus, it can be estimated as $U_i = \varepsilon_i + \varepsilon_* + \text{const}\,\Gamma$, where const $\approx 1$.

Typical values of the parameters which appear here for gases are $\varepsilon_i \sim$ 15 eV, $\varepsilon_* \sim 5$ eV, and $\Gamma \sim 10$ eV, which yield $U_i \sim 2\varepsilon_i$, a result that is close to exact for most gases.

It is clear from this estimate that the energy cost of ionization (cost of an electron–ion pair) is determined solely by the form of the differential ionization cross section and is practically independent of both the absolute magnitude of the ionization cross section and the energy of the energetic primary electrons. This is, therefore, an important universal characteristic of the degradation spectrum of electrons in a gas. The secondary electron spectrum is independent of the energy of the primaries, and this is what establishes the universality of the electron degradation spectrum in the threshold energy region. We note that there is a case in which this universality fails. If the energy of the primary electrons is low (no more than a few hundred electron volts), then the differential cross section for ionization does not have a sharply distinct resonance, so that the electron distribution in the threshold region depends on the distribution of the primaries.

The exact integral relation (8) can be formally regarded as a first-order linear differential equation for the EEDF:

$$df/d\varepsilon + \beta f = J/\rho D,$$

where

$$\beta = \mu/D \tag{10}$$

and

$$J = -\sum_{k,i} \int_{\varepsilon}^{\varepsilon+\varepsilon_k} f\rho v_k d\varepsilon_p + \int_{\varepsilon+\varepsilon_i}^{\infty} f\rho K d\varepsilon_p + Q. \tag{11}$$

On integrating it, we obtain a general integral relation for the EEDF which is valid over the entire range of energies,

$$f(\varepsilon) = f(0)\exp\left(-\int_0^{\varepsilon} \beta d\varepsilon'\right) + \int_0^{\varepsilon} \frac{J}{\rho D}\exp\left(-\int_{\varepsilon'}^{\varepsilon} \beta d\varepsilon''\right) d\varepsilon'.$$

The value of $f(0)$ should be such that the normalization condition (5) is satisfied. The ratio of the mobility to the diffusion coefficient for the electrons along the energy axis $\beta$ gives a measure of the average electron energy. This is the parameter that represents the fall-off of the EEDF at low energies, where the thermal electrons are concentrated. If these electrons are in equilibrium with the gas molecules, then $\beta = T^{-1}$, where $T$ is the gas temperature.

The quantity $J(\varepsilon)$ characterizes the change in the number of electrons in the energy interval from $\varepsilon$ to infinity. Maintaining the electron balance under quasistationary conditions requires that

$$J(0) = \frac{1}{n}\int_0^\infty (q - p)\, d\varepsilon' + \int_{\varepsilon_i}^\infty f \rho \nu_i d\varepsilon_p = 0.$$

With the aid of this condition, the general integral relationship transforms to

$$f(\varepsilon) = f(0)\, e^{-\int_0^\varepsilon \beta d\varepsilon'} + \frac{J}{\rho\mu} - \int_0^\varepsilon \frac{d}{d\varepsilon'}\left(\frac{J}{\rho\mu}\right) e^{-\int_{\varepsilon'}^{\varepsilon} \beta d\varepsilon''} d\varepsilon'. \qquad (12)$$

Here the first term describes the distribution of the thermal electrons, while the second is predominant at threshold and high energies, where the inequality $\beta\varepsilon \gg 1$ is satisfied. In this energy range we have $\rho\mu f = J$, which corresponds to neglecting the diffusion term in Eq. (8).

There is a similarity between the methods used to determine the degradation spectrum of electrons in a gas and the methods used in other areas of physics. The electron degradation problem is characterized by two main features. The first is related to the formation of secondary high-energy electrons with energies much lower than the primaries, while the second is determined by the disappearance of free electrons at low energies in recombination and attachment reactions.

For an approximate examination of the degradation spectrum of electrons in gases, one can use the simplest model in which the entire range of electron energies is broken up into three regions: subthreshold $\varepsilon < \varepsilon_1$ (where $\varepsilon_1$ is the characteristic energy for electronic excitation), above-threshold $\varepsilon_1 < \varepsilon < \varepsilon_2$, and a region $\varepsilon_2 < \varepsilon < E$ in which electron multiplication takes place ($E$ is the energy of the primary electrons in the beam). In the above-threshold region $\varepsilon_1 < \varepsilon < \varepsilon_2$, the electron source lies on the right-hand boundary and the sink lies on the left, while the average energy loss by the electrons as they are slowed down roughly satisfies the condition $\Delta\varepsilon \ll \varepsilon$. A power-law distribution, determined by the constancy of the flux of electrons over the spectrum, is typical under such conditions.

Power-law spectra are characteristic of problems with the following two features: first, remoteness of the particle (quasiparticle) sources and sinks and, second, small changes in the energy of a particle compared to its total energy. In this case, there is an "inertial" interval, within which the flux of particles along the energy axis is roughly conserved, which also determines its power-law character. This sort of power-law spectrum was first found by Kolmogorov [39] for locally uniform and isotropic turbulence, where the flux of energy in the eddies over the spectrum is determined by the gathering of energy from large eddies and by their breakup and dissipation in small eddies [40]. A power-law spectrum has been obtained by Sagdeev [41] in a study of the mechanism for ion-acoustic turbulence in a weakly turbulent plasma corresponding to two-plasmon decay over short scale lengths. In a nonequilibrium vibrationally excited gas, a power-law spectrum for the vibrational distribution of the molecules develops when the lower levels are pumped and vibrational–translational relaxation to the upper vibrational levels of the molecules takes place. This result was obtained by Brau [42] and Gordiets [43].

In the kinetic theory of high-temperature plasmas, power-law "tails" appear in nonequilibrium ion distributions owing to nuclear reactions or to the action of beams of high-energy particles [44] and have a significant effect on the dispersive properties of the plasma and on the rates of thermonuclear reactions.

The power-law distribution of the electrons in the degradation spectrum is analogous to the intermediate asymptote in the "inertial" interval [45], provided the condition of a constant electron flux along the energy axis is satisfied and the electron distribution is determined by solving the equation corresponding to the constant flux condition. The solution can be refined by including processes which lead to a change in the flux through changes in the number of primary and secondary electrons. The corresponding solution satisfies the general integral relation (12), where the term $J/\rho\mu$ on the right-hand side represents the first term in an expansion with respect to the small parameter $(\beta\varepsilon)^{-1}$ and determines the power-law form of the degradation spectrum at high energies. The following approximations to the EEDF, which use the integral relation (12) together with the definitions (10) and (11), can be used to construct a solution of the kinetic equation over the entire energy range. Equation (12) is convenient for making a qualitative analysis of the solution.

The energy distribution of the electrons as they degrade through excitation of the various molecular levels is described by the electron energy balance equation, which is obtained, as usual, from the initial kinetic equation (6) by multiplying by the electron energy $\varepsilon$ and then integrating over the entire range. It should be noted that electron–electron collisions do not change the electron-energy balance, while the energy expended in molecular ionization is determined solely by the ionization potential. As a result, we obtain

$$\sum_n \int_0^\infty \varepsilon_n f \rho \nu_n \, d\varepsilon = \frac{1}{n} \int_0^\infty (q - p) \, \varepsilon d\varepsilon,$$

where the sum includes all processes involving collisions of electrons with molecules and we must set $\varepsilon_T = \delta_T \varepsilon$ in the case of elastic collisions. Then the fraction of energy contributed by the ionization cascade of the electrons into the $j$th process is

$$p_j = \frac{\int_0^\infty f \rho \varepsilon_j \nu_j d\varepsilon}{\sum_n \int_0^\infty f \rho \varepsilon_n \nu_n d\varepsilon}. \tag{13}$$

Iteration schemes similar to that discussed above for analytically solving the kinetic equation have been used previously in various forms [7, 46–48]. They are all based on constructing a series of successive approximations which satisfy the exact kinetic equation or an approximation. Several convenient schemes [7] exist for constructing an exact solution in those cases where elastic collisions between electrons and molecules can be neglected and the temperature of the thermal electrons can be assumed equal to zero. We note, however, that the need to include electron-impact vibrational excitation of molecules in a rigorous approach does greatly complicate those schemes and leads to a volume of calculations nearly as large as that required for a numerical solution of the kinetic equation.

A modification of the continuous slowing-down model has been proposed by Medvedev and Khokhlov [47]. In this modification, all electron-energy loss processes are divided into two groups. The first includes elastic electron–molecule collisions and the excitation of vibrational and electronic levels of the molecules, while the second includes ionization. The energy losses in the first group were assumed to be continuous, as in the continuous slowing-down model, and the losses in the second group are treated rigorously. This method, which is essentially analogous to a scheme discussed by Krinberg [7], provides a good description of the EEDF for energies above the ionization potential, but its accuracy is limited in the threshold energy range, especially for molecular gases.

Peyraud [49] has proposed a method for solving the kinetic equation based on the assumption that the differential cross section for ionization depends so strongly on the energy of the secondary electron that we can eliminate the differential cross section and express the ionization collision integral directly in terms of the ionization cross section. In this case, the integrated kinetic equation (8) contains only single integrals and can be solved analytically with the aid of the Laplace transform, subject to the additional assumption that the cross sections for the inelastic processes can be expanded in a power series in the electron energy. This method has been used [49] to solve the simpler [compared to Eq. (8)] equation

$$\sum_{k,\,l}\int_{\varepsilon}^{\varepsilon+\varepsilon_k} f\rho\nu_k d\varepsilon_p = Q$$

for energies above the inelastic excitation threshold. The resulting analytic solution has been used to calculate the degradation spectrum in argon and has been compared with numerical calculations [50, 51]. The high-energy part of the degradation spectrum [50] has been used [52] to establish the electron distribution in the subthreshold energy range. Because of the predominance of interelectronic collisions, in this region the degradation spectrum was close to Maxwellian with parameters (electron temperature and density) determined from the continuity condition for the EEDF and its derivative in the threshold region for inelastic electronic excitation. Parametric studies of the resulting analytic solution showed that, for a fixed current density of the primary electron beam, the electron density $n$ rises with increasing gas pressure, while the electron temperature $T_e$ decreases. For a fixed pressure, $n$ and $T_e$ increase with increasing current density in the beam.

As can be seen from Eqs. (13), the electron-energy distribution with vari-ous excitation processes is determined by the ratios, rather than by the absolute values, of the cross sections for the corresponding processes. This makes it possible to simplify the kinetic equation in the threshold and high-energy regions. Suprathermal electrons lose their energy primarily through inelastic excitation and ionization of atoms, while any processes involving a small energy change usually play a secondary role for these electrons and the associated diffusive and convective fluxes are negligibly small. In addition, recombination and attachment of suprathermal electrons can be neglected. Thus, the kinetic equation (6) for the electrons at energies above the threshold can be rewritten in the form

$$f\rho\sum_{k,\,l}\nu_k = \sum_{k}(f\rho\nu_k)\Big|^{\varepsilon+\varepsilon_k} + \int_{\varepsilon+\varepsilon_i}^{\infty} f\rho\nu_{il}d\varepsilon_p + \frac{q}{n}\,. \qquad (14)$$

The probability of realizing an elementary event of the $j$th process for an electron with energy $\varepsilon$ is equal to

$$P_j(\varepsilon) = \frac{\nu_j}{\sum\limits_{k,\,i} \nu_k} = \frac{\sigma_j(\varepsilon)}{\sum\limits_{k,\,i} \sigma_k(\varepsilon)}.$$

A graphic illustration is provided by the case in which a single electron with energy $E$ degrades in a medium. Then one should set $q = \delta(\varepsilon - E)$ and it is convenient to use a new variable, known as the degradation spectrum [6, 53],

$$z = nf\rho \sum_{k,\,i} \nu_k.$$

Here the sum is taken over all inelastic interactions of the electron with molecules, including ionization. With the help of this substitution, Eq. (14) transforms to [53]

$$z(\varepsilon) = \sum_k z(\varepsilon + \varepsilon_k) P_k(\varepsilon + \varepsilon_k) + \int\limits_{\varepsilon+\varepsilon_i}^{E} z(\varepsilon_p) P_{ii}(\varepsilon_p, \varepsilon)\, d\varepsilon_p + \delta(\varepsilon - E). \quad (15)$$

The degradation spectrum $z(\varepsilon)$ essentially represents the density of the energy distribution of electrons formed in the energy interval $(\varepsilon, \varepsilon + d\varepsilon)$ during slowing down of a single fast particle.

This equation for the electron-degradation spectrum was first introduced by Spencer and Fano [10]. It was derived by analogy with the kinetic equation for slowing down of neutrons and heavy charged particles. In the Spencer–Fano theory, the degradation spectrum is introduced as the distance moved by an electron with initial kinetic energy $E$ as its energy drops from $\varepsilon + d\varepsilon$ to $\varepsilon$, so that $dx = y(E, \varepsilon)d\varepsilon$. The yield of particles of type $j$ during complete slowing down to zero energy of an electron with energy $E$ is given by

$$Z_j = \int\limits_0^E N\sigma_j(\varepsilon)\, y(E, \varepsilon)\, d\varepsilon;$$

hence, the Spencer–Fano degradation spectrum $y(E, \varepsilon)$ is proportional to the electron flux $f\rho v$.

It should be emphasized that Eq. (15) for the degradation spectrum is equivalent to the kinetic equation for the EEDF (14) with only inelastic pro-cesses taken into account. Thus, in the equation for the degradation spec-trum, we have neglected elastic collisions of the slowing-down electrons with molecules and thermal electrons. Here the energy balance equation reduces to

$$\int_0^E z \frac{\sum\limits_{k,l} \varepsilon_k \nu_k}{\sum\limits_{k,l} \nu_k} d\varepsilon = E,$$

hence, in accordance with Eq. (13), the fraction of energy expended in the $j$th process is equal to

$$p_j = \frac{\varepsilon_j \int_0^E z P_j d\varepsilon}{E}. \tag{16}$$

The integral in the numerator can be treated as the total number of times the $j$th process occurs during the slowing down of a primary electron,

$$Z_j = \int_0^E z P_j \, d\varepsilon, \tag{17}$$

and the average energy expended in each such event is

$$U_j = \frac{E}{Z_j} = \frac{E}{\int_0^E z P_j d\varepsilon} = \frac{\varepsilon_j}{p_j}, \tag{18}$$

which can be referred to as the energy cost of the $j$th inelastic process.

The convenience of introducing the degradation spectrum lies in the simplicity of Eqs. (16)–(18). It should not be forgotten, however, that the approximations made here are by no means always justified, since energy loss by electrons in elastic collisions with atoms may make a significant contribution to the electron-energy balance. In addition, the neglect of diffusive electron fluxes means that the problem cannot be generalized to the case where external fields and other interactions might have an effect, as can be done for the initial kinetic equation (6). Finally, the degradation-spectrum method does not usually touch upon the question of the distribution of the thermal electrons at subthreshold energies.

We now discuss calculations of the degradation spectrum based on the Fowler equation [14]. Let a fast electron with energy $E$, which is sufficient to create ions and secondary electrons through ionization of molecules, enter a gas. We shall consider the processes that take place after the first collision of this fast electron with a molecule. We denote the number of ions that can be formed during complete slowing down of an electron with energy $E$ by $Z_i(E)$. In the first collision of the fast electron with a molecule, the molecule will be excited with probability $P_k(E)$ into a state with energy $\varepsilon_k$, after which the electron's remaining energy is $E - \varepsilon_k$, and the number of ions which it can produce is equal to $Z_i(E - \varepsilon_k)$. During ionization with probability $P_{ii}(E, \varepsilon_s)$, two electrons are produced with energies $\varepsilon_s$ and $E - \varepsilon_i - \varepsilon_s$, which yield $Z_i(\varepsilon_s)$ and $Z_i(E - \varepsilon_i - \varepsilon_s)$ ions as they are slowed down. In addition, during ionization, which takes place in the first collision with a probability of $P_i(E)$, a single ion is created. Therefore, the total number of ions formed as a fast electron with initial energy $E$ is slowed down obeys the Fowler equation

$$Z_i(E) = P_i(E) + \sum_k P_k(E) Z_i(E - \varepsilon_k) + \int_0^{(E-\varepsilon_i)/2} P_{ii}(E, \varepsilon_s) \times [Z_i(\varepsilon_s) + Z_i(E - \varepsilon_i - \varepsilon_s)]\, d\varepsilon_s.$$

The Fowler equation is to be solved for energies above threshold, beginning with the boundary conditions $Z_i(E) = 0$ for $E < \varepsilon_i$ and $Zi(E) = P_i(E)$ for $\varepsilon_i \leq E \leq \varepsilon_i + \varepsilon_1$, where $\varepsilon_1$ is the energy of the first excited state of the molecule. It describes the cascade multiplication of ions in terms of the probability of processes following the first collision of an electron as it is slowed down, while the Spencer–Fano equation is determined by the probabilities of subsequent collisions by the electron. Thus, these equations are in a sense complementary, and a unified variational principle can be developed for them [54]. The relationship between these equations has been discussed by Inokuti [21] and by Rau et al. [54]

The Fowler equation is convenient in that it yields an explicit result for the energy cost of an electron–ion pair (the cost of ionization) and, thus, is often used in calculations [18–20, 36, 55]. The degradation spectrum has been calculated numerically for helium [15, 16, 25, 48, 53], for mixtures of helium with neon [53], and for molecular hydrogen [48, 56] and molecular nitrogen [57].

The ionization and excitation of a gas by an electron beam can be reduced to a Monte Carlo model of the processes through which the electron energy changes in an ionization cascade. The advantage of this method is that the physical processes involved in individual collisions of electrons with molecules are correlated with a direct statistical analogy, a branching process of random walks by particles in a one-dimensional phase space [58–61]. Then the transition probabilities are determined by the cross sections for the corresponding interactions between electrons and molecules.

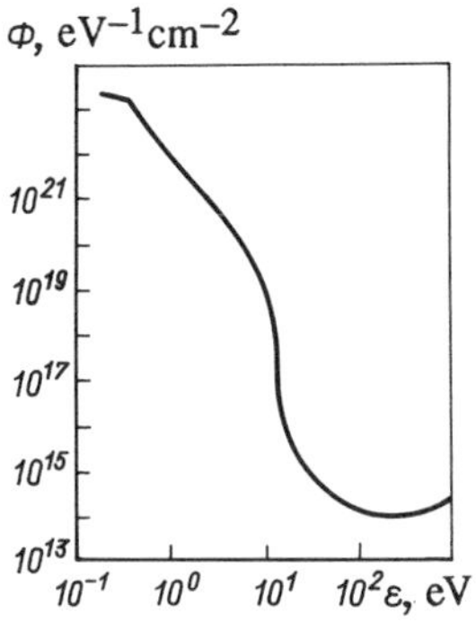


**Fig. 3.** The normalized electron flux in argon [65].

In the model, the electrons in the different generations are examined, while the number of collisions leading to excitation of each level and the energy expended in doing this are added up; thus, the fractions of the energy of a primary electron applied to each process of molecular excitation and ionization are determined. A Monte Carlo approach with "consolidated collisions" in energy space, developed by Lappo et al. [62], requires much less machine time but retains the basic advantages of statistical modelling. In this approach, the electrons evolve in the high-energy region just as in the continuous slowing-down model, while in the more important threshold energy region, the calculations switch to direct modelling by a conventional Monte Carlo method.

Detailed information on the degradation spectrum of electrons in a particular gas can be obtained by numerical solution of the exact kinetic equation (6) over the entire energy range, using the most detailed and reliable data on the cross sections for elementary processes. It can often be assumed that the thermal electrons have a Maxwellian distribution with a temperature equal to the gas temperature and it is possible to limit the calculations to solving the kinetic equation for energies ranging from a few tenths of an electron volt to the energy of the primary electrons. This wide range encompasses the suprathermal electrons in the subthreshold and threshold regions, as well as the high-energy electrons.

In calculating the EEDF, it is convenient to treat the source of primary electrons as monoenergetic, with $q = S\delta(\varepsilon - E)$, and to express the results of the calculation in terms of the normalized electron flux

$$\Phi = \frac{Nnf\rho v}{S(1 + E/U_i)}.$$

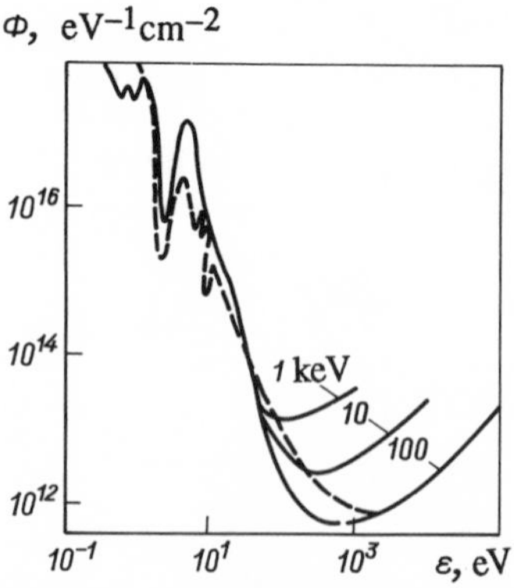


**Fig. 4.** The normalized el ctron flux in air for different primary electron energies [62]. The dashed curve is the solution given in [66].

Various calculations [63–65] have shown that the normalized electron flux in the threshold energy region is practically independent of the initial energy of the primary electrons if the latter is sufficiently high and, therefore, that it represents a universal function for every gas. This extremely important fact is a consequence, first, of the resonant character of the differential ionization cross section and the large contribution of the secondary electrons to the EEDF at low energies, and, second, of the drop in the cross sections for ionization and inelastic excitation of molecules at high energies.

As a typical example of the degradation spectrum of electrons in an atomic gas, Fig. 3 shows the normalized electron flux in argon [65]. The initial energy of the primary electrons was $E = 1$ keV and the calculation covered the energy range down to 0.2 eV. In an atomic gas, the electron spectrum is monotonic over a wide subthreshold energy range and decreases sharply near the thresholds for inelastic excitation and ionization of the atoms.

The picture is more complicated for molecular gases, in which electron-impact vibrational excitation plays an important role. Figure 4 shows the normalized electron flux in air calculated by the Monte Carlo method for different primary electron energies [62], as well as with the aid of a modified continuous slowing-down model [66]. The figure illustrates the fact that the spectrum in the threshold region is independent of the energy

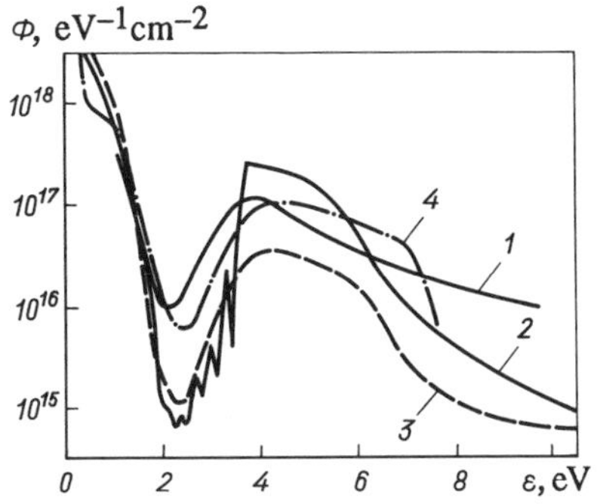


**Fig. 5.** The normalized electron flux in molecular nitrogen [62, 64]: 1) theory [62]; 2) theory [64]; 3) theory [63]; 4) experiment [63].

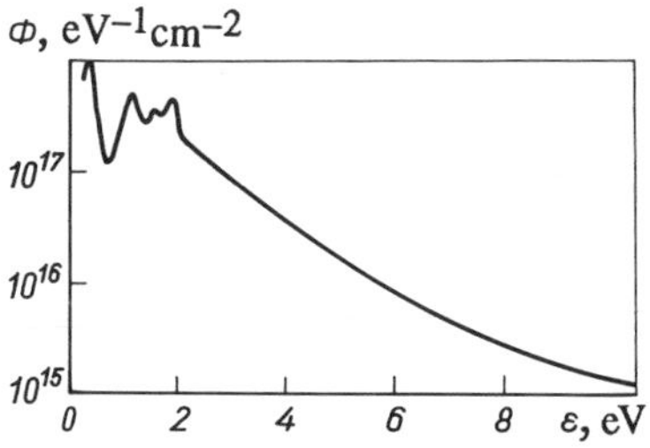


**Fig. 6.** The normalized electron flux in molecular oxygen [64].

of the primary electrons. The electron spectrum is highly non-Maxwellian, varies over several orders of magnitude, is nonmonotonic, and has a spiky structure.

Since the EEDF drops sharply beyond the thresholds for electronic excitation and ionization of molecules, the main contribution to the integral characteristics comes from the EEDF in the threshold region. Figure 5 shows the calculated normalized electron flux in nitrogen in the threshold region [62–64]. The quantitative differences in the numerical results are primarily caused by different choices for the cross sections of the elementary processes, but also by the specific features of the methods employed. The resonance structure of the spectra reported in [64] is a consequence of distinguishing the individual levels of molecular vibrational excitation. A minimum in the electron flux corresponds to a maximum cross section for resonant vibrational excitation of the molecules. The experimental spectrum [63] was obtained from probe measurements using the second derivative method and is in qualitative agreement with the calculations. The electron flux in pure oxygen has similar properties and is shown in Fig. 6 [64].

A comparison of Figs. 3 and 6 reveals the strong difference between degradation spectra in atomic and molecular gases. It follows from this, in particular, that even a small fraction of molecular gas impurity can substantially change the distribution of subthreshold electrons in an atomic gas. This effect has been examined in detail elsewhere [61, 65].

One feature of the numerical solution of the kinetic equation for the EEDF (6) is that the value of the function at each point depends integrally on its values at higher energies up to the energy of the primary electrons. Thus, as the energy of the primary electrons is raised, the volume of calculations should increase substantially. A numerical method has been proposed [67] for reducing this volume which is based on a multigroup method where all the electrons are divided into groups along the energy axis and only transitions among these electron groups are considered. This type of solution for helium [67] is in good agreement with the computational results of [53].

In our opinion, however, there is no need to solve the kinetic equation over the entire energy range by a rigorous numerical method. The initial kinetic equation is valid for initial energies of the primary electrons as high as on the order of 10 MeV, when it is still possible to neglect energy losses by the primaries through bremsstrahlung. At high energies it is always possible to construct an analytic solution with sufficient accuracy by using an expansion in terms of the small parameter given by the ratio of the thresholds for excitation to the electron energies. The electron spectrum in the most important (threshold) region always depends very weakly on the primary electron distribution and, therefore, on the solution in the high-energy region.

The most important point in numerical calculations of the degradation spectrum of electrons in gases, therefore, is to make the correct choice of the cross sections for the elementary interactions of electrons with molecules, since these basically determine the electron spectrum in the threshold energy range. Since the contribution of the high-energy region is taken into account analytically, in almost all cases a numerical calculation of the EEDF can be set to start with electron energies of no more than a few kiloelectron volts.

**TABLE 1.** Energy Costs (eV) of Inelastic Excitation of an He Atom during Degradation of an Electron with an Energy $E = 10$ keV

| State | $2^3S$ | $2^1S$ | $2^3P$ | $2^1P$ | 3 | 4 | 5 | Ionization |
|---|---|---|---|---|---|---|---|---|
| Threshold | 19.8 | 20.6 | 20.9 | 21.2 | 22.7 | — | — | 24.6 |
| Cost | 331 | 844 | 1050 | 133 | 422 | 1050 | 2110 | 46.4 |

**TABLE 2.** Energy Costs (eV) of Exciting an $H_2$ Molecule during Degradation of a High-Energy Electron

| State | $V = 0 \rightarrow 1$ | $b^3\Sigma_u^+$ | $a^3\Sigma_g^+, c^3\Pi_u$ | $B^1\Sigma_u^+$ | $C^1\Pi_u, D^1\Pi_u$ | Ionization |
|---|---|---|---|---|---|---|
| Threshold | 0.55 | 8.8 | 11.8 | 11.55 | 12.3 | 15.6 |
| Cost ($E$ = 1 keV) | 5.32 | 91.7 | 346 | 73.0 | 91.0 | 33.8 |
| Cost ($E$ = 1 MeV) | 5.41 | 87.0 | 327 | 69.9 | 87.0 | 31.5 |

## 3. ENERGY COSTS OF ELEMENTARY PROCESSES IN SPECIFIC GASES

A convenient quantity for characterizing the rate of inelastic molecular excitation during slowing-down of a high-energy electron is the energy cost of the elementary process leading to this excitation. The energy cost $U_j$ of the $j$th process, defined by Eq. (18), is inversely proportional to the fraction of energy $p_j$ (13) delivered by the ionization cascade to the $j$th process. In the model case of a single inelastic process, its cost coincides with the average energy expended by an electron in a single event of this process and is independent of the cross section for the process. The main contribution to $p_j$ is from the part of the EEDF in the neighborhood of the threshold energy. Since the shape of the electron energy spectrum in this region is practically independent of the energy of a primary electron, the energy costs of inelastic processes are also independent of it over a wide range. This result can be seen directly from the estimate (2) if we assume, for example, that the cross sections of all inelastic processes decrease in the same fashion at high energies. Because of this, the energy costs of elementary interactions between electrons and particles can be seen as important universal quantities, which characterize the degradation of a high-energy electron in a given medium.

The largest number of calculations of the degradation spectra of electrons in atomic gases have been devoted to helium. This is because the excitation and ionization cross sections are known more reliably for the He atom than for other gases. The results of various authors are in good agreement. Table 1 shows the energy costs for inelastic excitation of the He atom obtained from the data of Syts'ko and Yakovlenko [53].

The degradation spectrum of electrons in molecular hydrogen has been calculated [48] for different primary electron energies. The energy costs derived from these numerical calculations are listed in Table 2. It shows that, despite a change of three orders of magnitude in the energy of the primary electrons, the energy costs of all the inelastic processes remain practically unchanged. We note that differences in the costs for different initial energies obtained by various authors may be considerably smaller than the uncertainty owing to possible inaccuracies in the cross sections used for the elementary processes, which sometimes reach tens of percent.

**TABLE 3.** Energy Costs (eV) of Inelastic Excitation of an Ar Atom during Degradation of an Electron with an Energy of $E = 1$ keV

| State | $4s_{3/2}$ | $4s_{1/2}$ | — | $3d_{1/2}$ | $5s_{3/2}$ | $3d_{3/2}$ | $5s'_{1/2}$ |
|---|---|---|---|---|---|---|---|
| Threshold | 11.6 | 11.8 | 13.0 | 13.9 | 14.1 | 14.1 | 14.3 |
| Cost [65] | 975 | 261 | 242 | 63 200 | 4130 | 1100 | 12 900 |
| Cost [17] | 773 | 268 | 236 | 63 200 | 4030 | 1180 | 13 000 |

Table 3, continued

| State | $3d'_{3/2}$ | $a$ | $b$ | $c$ | $d$ | $e$ | Ionization |
|---|---|---|---|---|---|---|---|
| Threshold | 14.3 | 14.9 | 15.0 | 15.2 | 15.4 | 15.5 | 15.8 |
| Cost [65] | 894 | 2300 | 3570 | 3660 | 5000 | 2540 | 33.5 |
| Cost [17] | 753 | 1800 | 3140 | 3170 | 4670 | 2280 | 29.0 |

Table 3 lists the energy costs for inelastic excitation of argon derived from the data of Peterson and Allen [17] and Keto [65]. Keto [65] solved the kinetic equation for the EEDF numerically, while Peterson and Allen [17] used the continuous slowing-down model. The good agreement of these results should be noted for most excitation levels and can serve as a validation of the continuous slowing-down model for atomic gases.

Several interesting features show up during the degradation of an electron in a mixture of different gases. The dissipation of the energy of electrons from a beam in mixtures of argon with krypton, xenon, and nitrogen has been modelled using the Monte Carlo method by Ryzhov and Yastremskii [61]. They found that, since the cross sections for excitation of the lowest electron levels, $A^3\Sigma_u{}^+$ and $B^3\Pi_g$, of the $N_2$ molecule are only weakly overlapped by the cross sections for argon, even when the concentration of nitrogen in the mixture is low, the fraction of energy expended in exciting these levels is relatively high. In contrast, the fraction of energy expended by the ionization cascade on ionization and excitation of atoms and molecules is usually proportional to their relative concentrations in the mixture. It should be noted that in atomic gases a noticeable fraction of the electrons' energy (~10–15%) can be spent in elastic collisions with atoms, i.e., in heating the gas. Electron degradation in a mixture of helium with neon has also been examined [53].

In calculations of the degradation spectrum in molecular gases, electron-impact excitation of the vibrational levels of the molecules must be taken into account. Khare and Kumar [36, 55] set themselves the task of calculating the energy cost of an electron–ion pair in molecular nitrogen and oxygen. In doing this, they carefully chose the form of the differential ionization cross section, but examined degradation of the electrons only at energies above the ionization potential, so that electron energy losses through vibrational excitation were neglected. They made a comparison with a continuous slowing-down model calculation and showed that it was sufficient to retain the first three terms in it to obtain a result close to the exact one. From this we may conclude that the continuous slowing-down model is capable of giving an exact estimate of the cost of elementary processes with a high threshold in molecular gases as well.

The EEDF in nitrogen, oxygen, and air excited by an electron beam has been calculated by Konovalov and Son [64] with a fairly complete set of cross sections for the elementary processes. In subsequent calculations using this method, a larger number of levels has been included and the cross sections for the elementary electron collisions with molecules have been chosen more carefully. More refined (than [64]) calculations, fol-

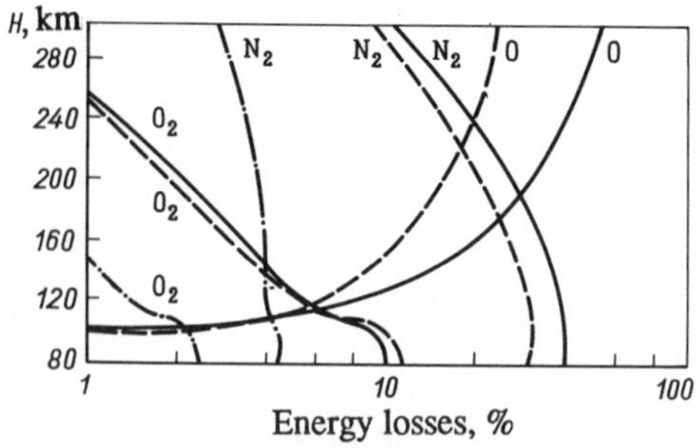


**Fig. 7.** Fractions of the energy lost by high-energy electrons in the ionosphere at different altitudes $H$ [68]: the smooth curves correspond to loss through ionization; the dashed curves, to the loss through excitation of electronic levels; and the dot-dashed curves, to loss through excitation of vibrational levels.

**TABLE 4.** Energy Costs (eV) of Inelastic Excitation of an $N_2$ Molecule during Degradation of an Electron with an Energy $E = 1$ keV in Pure Nitrogen and Air

Electronic excitation of $N_2$

| State | $A^3\Sigma_u^+$ | $B^3\Pi_g$ | $W^3\Delta_u$ | $B'^3\Sigma_u^-$ | $a'^2\Sigma_u^+$ | $a^1\Pi_g$ | $w^1\Delta_u$ | $C^3\Pi_u$ | — | $\Sigma R$* | Dissociation† |
|---|---|---|---|---|---|---|---|---|---|---|---|
| Threshold | 6.17 | 7.35 | 7.36 | 8.16 | 8.40 | 8.55 | 8.89 | 11.03 | 12.25 | 3.75 | 9.76 |
| In nitrogen | 152 | 191 | 250 | 958 | 1170 | 326 | 1040 | 520 | 3080 | 67.5 | 43.5 |
| In air | 258 | 249 | 308 | 1120 | 1370 | 382 | 1200 | 579 | 3610 | 86.9 | 54.8 |

Vibrational excitation of $N_2$

| Transition | 0→1 | 0→2 | 0→3 | 0→4 | 0→5 | 0→6 | 0→7 | 0→8 |
|---|---|---|---|---|---|---|---|---|
| Threshold | 0.292 | 0.585 | 0.877 | 1.17 | 1.46 | 1.75 | 2.05 | 2.34 |
| In oxygen | 9.19 | 75.6 | 195 | 280 | 259 | 378 | 405 | 464 |
| In air | 18.6 | 108 | 212 | 338 | 306 | 460 | 523 | 716 |

Ionization of $N_2$

| Ion state | $X^2\Sigma_g^+$ | $A^2\Pi_u$ | $B^3\Sigma_u^+$ | $D^2\Pi_g$ | $C^2\Sigma_u^+$ | Dissociation† | All |
|---|---|---|---|---|---|---|---|
| Threshold | 15.6 | 16.8 | 18.7 | 22.0 | 23.6 | 25.0 | 17.0‡ |
| In oxygen | 92.3 | 185 | 359 | 616 | 780 | 182 | 33.0 |
| In air | 120 | 240 | 470 | 814 | 1030 | 240 | 43.1 |

*Sum of the Rydberg levels.

†Proceeds through electronic excitation.

‡Averaged over the states of the ion.

**TABLE 5.** Energy Costs (eV) of Inelastic Excitation of an $O_2$ Molecule during Degradation of an Electron with an Energy of $E = 1$ keV in Pure Oxygen and Nitrogen

Electronic excitation of $O_2$

| State | $a^1\Delta_g$ | $b^1\Sigma_g^+$ | $A^3\Sigma_u^+$ | $B^3\Sigma_u^-$ | — | $\Sigma R$* | Dissociation† |
|---|---|---|---|---|---|---|---|
| Threshold | 0.98 | 1.64 | 4.50 | 8.40 | 9.90 | 13.5 | 5.10 |
| In oxygen | 23.1 | 90.7 | 56.2 | 52.4 | 1020 | 66.0 | 20.2 |
| In air | 93.3 | 386 | 170 | 279 | 4990 | 286 | 107 |

Vibrational excitation of $O_2$

| Transition | 0→1 | 0→2 | 0→3 |
|---|---|---|---|
| Threshold | 0.196 | 0.392 | 0.516 |
| In oxygen | 8.06 | 25.7 | 112 |
| In air | 14.5 | 47.4 | 241 |

Ionization of $O_2$

| Ion state | $X^2\Pi_g$ | $a^4\Pi_u$ | $A^2\Pi_u$ | $b^4\Sigma_g^-$ | Dissociation† | All |
|---|---|---|---|---|---|---|
| Threshold | 12.1 | 16.1 | 16.8 | 18.2 | 20.0 | 15.0 ‡ |
| In oxygen | 209 | 120 | 534 | 342 | 125 | 34.4 |
| In air | 977 | 499 | 2160 | 1380 | 487 | 141 |

*Sum of the Rydberg levels.
†Proceeds through electronic excitation.
‡Averaged over the states of the ion.

lowing this method, of the energy costs of inelastic excitation and ionization for $N_2$ and $O_2$ molecules in pure gases and in air are shown in Tables 4 and 5.

Calculations have also been carried out for a three-component mixture consisting of molecular nitrogen, molecular oxygen, and atomic oxygen, as a simulation of the composition in the lower ionosphere of the earth. The overall distribution of the energy losses in the ionosphere by highly energetic electrons through the different channels is shown in Fig. 7 [68].

Other integral characteristics of the EEDF have the same sort of universality, i.e., the energy spectrum being independent of the primary electrons, as the energy costs. For example, it is possible to estimate the emissivity of the plasma excited by the ionization cascade of the electrons. If we neglect collisional deactivation of optically allowed transitions and assume that all the energy expended in exciting a given energy level is lost as radiation, then the radiation can be characterized by an emission coefficient which provides an upper bound on the conversion of power from the primary electrons into radiated power at a given frequency. The emissivity of an air plasma excited by an electron beam has been calculated in this way by Lappo et al. [62]

# 4. DEGRADATION SPECTRUM OF ELECTRONS IN LOW-TEMPERATURE PLASMAS

When calculating the degradation spectrum of electrons in a plasma with a fairly high degree of ionization, where interelectronic collisions must be taken into account, one must, in general, solve the self-consistent problem arising from the action of an ionization source and external electromagnetic fields on the kinetics of electron creation and loss.

Here the time dependence of the primary electron source, which determines the temporal evolution of the electron distribution function, may be important. Numerical solutions of the time-dependent kinetic equation have been obtained for the degradation spectrum in argon [50, 51, 69–71] and xenon [69]. The difficulty in solving the electron kinetic equation with the secondary electrons included is that the appearance of two electrons in different energy groups must be taken into account. This feature of the integral equation leads to a substantial increase in the machine time used in calculating the time-dependent electron distribution. Thus, tens of hours were required on the BÉSM-6 computer [70, 71]. These calculations, which neglected electron–electron collisions, were done for primary electron energies ranging from $10^3$ to $10^6$ eV and yielded the important result that the ratios of the rates of ionization and excitation of the molecules remain constant with high accuracy, changing very slightly with time. This confirms the universality of the values of the energy costs for elementary processes. It was also found that as the energy of the primary electrons increases, there is an increase in the time required for the distribution to reach a steady state, which approaches tens of nanoseconds in gases under standard conditions.

Bretagne et al. [50, 51] have calculated the EEDF numerically for the same range of primary electron energies, but separately for energies above and below threshold. They noticed a rapid evolution of the steady state. Interelectronic collisions were neglected in the region above threshold, but were taken into account below the threshold, along with electron–ion recombination and collisions of electrons with excited molecules. The difference between the energy cost for creation of an electron–ion pair in argon $U_i = 25.4$ eV obtained by Bretagne et al. [50] and the data of Peterson and Allen [17] and Keto [65] is noteworthy and is apparently related to a different form for the free term in the kinetic equation.

A substantial deviation of the electron distribution from Maxwellian at energies above threshold has been found to exist throughout the time over which the EEDF evolves [50, 69].

As the power of the primary electron source rises, the degree of ionization of the plasma increases; thus, it becomes necessary to include interelectronic collisions in a calculation of the electron-degradation spectrum. This sort of problem is of interest in connection with various practical applications, especially in excimer lasers excited by electron beams, where the degree of ionization may be quite high and electron–electron collisions have to be taken into account. If the ionization source is spatially isotropic, then the EEDF is also isotropic and Coulomb interelectronic collisions control the transition to a Maxwellian electron distribution as the degree of ionization is raised.

A complete solution of the problem must include an analysis of the kinetics of the electrons and molecules together with a solution of the kinetic equation for the EEDF. This problem is very complicated, so at first it is best to examine the effect of interelectronic collisions on the EEDF at a fixed electron density.

In a weakly ionized low-temperature plasma, the change in the energy of an electron as it collides with other electrons is comparable to the energy loss in elastic collisions with atoms at low degrees of ionization: $\alpha = n/N \sim \delta_T\sigma_m/\sigma_{ee}$, where $\sigma_m$ and $\sigma_{ee}$ are the transport cross sections for collisions of an electron with an atom and an electron, respectively, which for an average electron energy of $\langle\varepsilon\rangle \sim 1$ eV gives $\alpha \sim 10^{-7}$. At higher degrees of ionization, interelectronic collisions can have a substantial effect on the EEDF and, therefore, on the rate of excitation and ionization of the molecules. Thus, the electron–electron collision integral must be included when solving the kinetic equation. The basic qualitative features can be understood from the general integral equation (12).

Let us consider the excitation of an atomic gas by a high-energy electron source, limiting ourselves to the first two terms on the right-hand side of Eq. (12) and taking $\beta = T_e^{-1}$ = const, which corresponds to a Maxwellian distribution for the thermal electrons at temperature $T_e$. With the normalization (5), we obtain the following expression for the EEDF:

$$f = \left(1 - \int_0^\infty \frac{J}{\mu}\, d\varepsilon\right) \frac{2}{\sqrt{\pi T^3}}\, e^{-\varepsilon/T_e} + \frac{J}{\rho\mu}, \tag{19}$$

where $J$, as before, is defined by Eq. (11) and gives the power-law tail of the distribution. Depending on the degree of ionization, $\alpha$, three characteristic cases can be distinguished. These are described below.

For low degrees of ionization $\alpha \ll \delta_T\sigma_m/\sigma_{ee}$, interelectronic collisions have no effect on the electron distribution. In this case, the EEDF of the thermal electrons is Maxwellian with a temperature equal to that of the gas $T$, and at energies above the threshold, the distribution evolves through inelastic collisions of the electrons with atoms, including ionization. The first term in Eq. (19) is negligibly small over almost the entire energy range, when the overall rate of excitation of the $j$th level of the atom is determined by the power-law tail of the distribution and is given by

$$\langle \nu_j \rangle = \int_{\varepsilon_j}^{E} f \rho \nu_j \, d\varepsilon = QE/U_j.$$

Thus, at low degrees of ionization, the rates of inelastic atomic excitation by electron impact are determined by the energy cost of the corresponding processes.

For degrees of ionization $\delta_T \sigma_m / \sigma_{ee} \ll \alpha \ll \Sigma \sigma_k / \sigma_{ee}$, interelectronic collisions are dominant at energies up to the threshold for electronic excitation of the atoms, but they do not compete with inelastic collisions above the threshold region. In this case, the EEDF is close to Maxwellian almost everywhere up to the threshold for inelastic excitation, and falls off very rapidly above the threshold because inelastic electron–atom collisions predominate. Then the rate of inelastic excitation is determined by the value of the EEDF immediately adjacent to the threshold, so that the overall integral rate of inelastic collisions,

$$\langle \Sigma \nu \rangle = \sum_{k,\,l} \int_{\varepsilon_k}^{E} f \rho \nu_k d\varepsilon_p \simeq \sum_{k,\,l} \int_{\varepsilon_k}^{\varepsilon_k + \varepsilon} f \rho \nu_k \, d\varepsilon_p,$$

depends weakly on the upper limit of integration. In this case, however, Eq. (11) implies that the flux $J = -\langle \Sigma \nu \rangle$ can be regarded as roughly constant in most of the subthreshold energy region. If we neglect the contribution of the region above the threshold to the normalization of the EEDF, then Eq. (19) transforms to

$$f = \frac{2}{\sqrt{\pi T_e^3}} e^{-\varepsilon/T_e} - \frac{\langle \Sigma \nu \rangle}{\rho \mu} = f_M - \frac{\langle \Sigma \nu \rangle}{\rho \mu}, \quad \varepsilon < \varepsilon_k,$$

where $f_M$ is the Maxwellian distribution function.

In this case, the overall rate of inelastic excitation is determined by the rate at which the EEDF evolves in the subthreshold region through electron–electron collisions. These conditions correspond to the general solution of the kinetic equation in the diffusion approximation by the "infinite sink" method [72], which leads to an EEDF that equals zero at the lowest

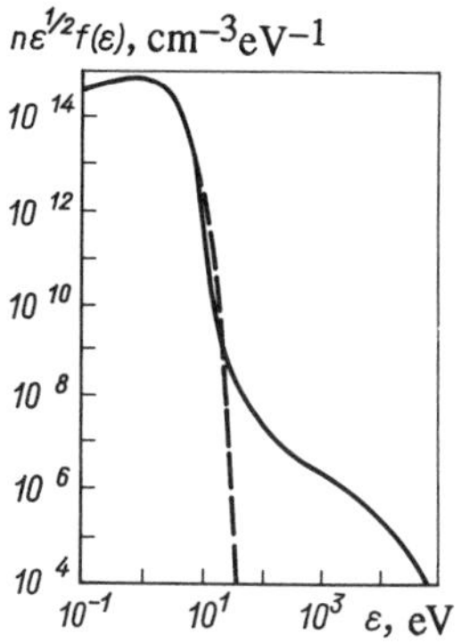


**Fig. 8.** The electron energy distribution function in argon excited by an electron beam [51]. (The dashed curve represents the corresponding Maxwellian distribution.)

threshold for inelastic excitation. In this case, the equality $f(\varepsilon_1) = 0$ yields

$$\langle \Sigma \nu \rangle = \frac{2\rho\mu}{\sqrt{\pi T_e^3}} \exp\left(-\frac{\varepsilon_1}{T_e}\right) = \rho\mu f_M(\varepsilon_1), \quad \varepsilon_1 = \min\{\varepsilon_k\};$$

i.e., the overall rate of inelastic excitation is determined by the value of the Maxwellian EEDF at the excitation threshold and is independent of the excitation cross section. This result was first obtained by Brau [73] especially for evaluating the rates of excitation of atoms in the plasmas typical of excimer lasers.

Figure 8 shows the steady-state EEDF in argon at a pressure of 0.3 MPa with an electron density of $n = 1.94 \cdot 10^{15}$ cm$^{-3}$ and an electron temperature of $T_e = 1.54$ eV obtained from the data of Bretagne et al. [51] These parameters correspond to the degree of ionization in the case examined above. The figure illustrates the basic features of the EEDF in an atomic gas excited by a source of high-energy electrons: a Maxwellian form at energies below threshold, a sharp exponential drop beyond the threshold for inelastic excitation, and a gently sloping power-law tail at high energies. Note that the above estimate for the overall rate is roughly a factor of 2 higher than that obtained numerically.

Finally, if the degree of ionization of the plasma is so high that $\Sigma\sigma_k/\sigma_{ee} \ll \alpha < 1$, and the electron–electron collision frequency is substantially higher than the rate of inelastic excitation above the threshold, then the EEDF is Maxwellian, not only below the threshold, but also far beyond the excitation threshold. In this case, the overall electron–impact excitation rate for the $j$th level of the atom is equal to

$$\langle v_j \rangle = \int_{\varepsilon_j}^{E} f_M \rho v_j d\varepsilon,$$

and the contribution of the power-law distribution in the high-energy region to the integral characteristics is small for all processes, except possibly for ionization, where the "tail" of the power-law distribution may have an effect.

It is interesting to compare an electron-degradation spectrum with the electron distribution in a gas discharge. Research on the EEDF in discharges has been reviewed by Aleksandrov and Son [74]. The disequilibrium of the plasma in a gas discharge is determined by the applied electric field, so it differs from the recombination disequilibrium of the plasma formed when a gas is excited by an external source of high-energy electrons.

This difference shows up in the corresponding electron distributions: in a gas discharge the EEDF has a sharp exponential drop above the thresholds for inelastic molecular excitation, while the electron-degradation spectrum is characterized by a power-law distribution over a wide range of energies above the threshold. The electron–impact molecular excitation rates in a gas discharge depend strongly on the applied electric field, while the rates of molecular excitation during degradation of high-energy electrons are determined by the constant values of the energy costs for the corresponding excitation processes. In a non-self-sustaining discharge controlled by an electron beam, the plasma is excited by both these mechanisms together and it is necessary to solve for the EEDF in this case. This difference in the electron distributions in gas discharge and beam plasmas makes it possible to ascertain the form of the EEDF in a non-self-sustaining discharge over wide ranges of variation in the external electric field and in the power of the primary electron beam.

The motion of the electrons along the energy axis under gas discharge conditions is characterized by the diffusion coefficient of the electrons in the electric field:

$$D_E = \frac{2}{3} \frac{(eE)^2}{m} \frac{\varepsilon}{\Sigma \nu},$$

where $e$ is the electronic charge, $E$ is the external electric field, and $\Sigma\nu$ is the overall rate of collisions of electrons with gas particles. Energy gain by the electrons in an external electric field through a diffusive process is fundamental in the subthreshold energy region. Typical values of the reduced field in gas discharges are usually below $E/N \sim 10^{-15}$ V·cm$^2$, so that beyond the thresholds for electronic excitation of atoms, the condition $\varepsilon\Sigma\varepsilon_k\nu_k/D_E \gg 1$, which determines the amount by which the convective flux owing to inelastic excitation exceeds the diffusive flux, is satisfied.

Under these conditions, an external electric field has little effect on the electron-degradation spectrum at energies beyond the threshold, and excitation of the gas reduces approximately to the independent action of two independent ionization sources: an external electric field without an electron beam and a beam of primary electrons without an electric field [34]. The rate of inelastic excitation is then given roughly by the sum of these two terms.

## 5. CONCLUSION

We now state the practical results of importance for determining the degradation spectra of electrons in gases. For estimates of the electron energy losses, we can use Eq. (2), which was obtained from an elementary theory that only includes energy losses by the primary electrons through molecular excitation and ionization.

Cascade multiplication of electrons with formation of the first, second, and subsequent generations of electrons can be taken into approximate account in the continuous slowing-down model.

Calculating the degradation spectrum of the electrons in all generations requires solution of a kinetic equation for the electrons that includes all of the many elementary processes through which the electrons interact with the molecules. The contribution of individual processes within the different energy intervals is determined by the cross sections for the corresponding processes and can be interpreted in terms of fluxes of electrons along the energy axis. This approach, with an expansion in the small parameter given by the ratio of the energy lost by an electron in a collision with a molecule to the energy of the electron, can be used to obtain an analytic solution which accurately fits the electron-degradation spectrum at high energies. An accurate determination of the degradation spectrum in the threshold and thermal energy range, as a rule, requires a numerical solution of the kinetic equation.

Approximate calculations of the molecular excitation rates based on assuming a Maxwellian electron distribution with a temperature determined by the electron energy balance can lead to large errors, since they do not take into account the formation of a power-law tail in the electron energy distribution. This tail develops because of the remoteness (on the energy axis) of the primary electron source from the major energy sink (at low energies). The power-law distribution for the electrons with energies above the threshold is the reason for the much higher rates of inelastic processes compared to those for a Maxwellian distribution with the same average energy.

The rate of each inelastic excitation process is characterized by a universal quantity: the energy cost of this process. Note that, thus far, only a few specific calculations of the energy costs of inelastic processes in different gases have been carried out, mainly because of limited information on the cross sections for the elementary processes.

The degradation spectrum of the electrons determines many kinetic properties of a beam plasma. Knowledge of this spectrum is required in research on lasers pumped by electron beams [75, 76] and by non-self-sustained discharges [77], in which the degree of excitation of the gas resulting from the passage of an electron beam is important.

Very few experimental studies of beam plasmas have been made. The accuracy of electrical measurements is limited by the fact that the electron density in a beam plasma is low, while the perturbations induced by the primary electron source are large. Probe diagnostics are limited by the need to take the anisotropy of the electron fluxes into account, especially at high energies.

A plasma excited by an external ionization source can be used to determine kinetic constants such as the rate of three-body attachment of electrons to oxygen molecules [78]. In general, that rate constant is related to the degradation spectrum of the electrons [79] and plays a significant role in the kinetics of nonequilibrium air plasmas [80].

One important source of information on the parameters of a beam plasma is its radiation. The basis of optical diagnostic techniques is a model according to which the electrons excite molecules into one of their electronic–vibrational–rotational states, after which the molecule undergoes a radiative transition to an intermediate state which is observed optically. Spectroscopic measurements of the emission from a mixture of oxygen and argon excited by an electron beam have been carried out in conjunction with calculations of the electron-degradation spectrum and of the kinetics of the molecular particles [65, 81, 82].

Rebrov et al. [83, 84] have discussed electron-beam diagnostics in nitrogen. They can be used to study the excitation of radiation in the first negative band system of nitrogen and to determine the rotational populations of the molecules. Piotrovskii and Tolmachev [85] have made an experimental study of the relationship between the intensities of spectrum lines and the electron distribution, and have demonstrated the feasibility of creating large populations of long-lived states in gases excited by an electron beam.

The possible practical applications of electron-degradation spectra in a wide range of problems associated with modern technological processes should be pointed out. These include the cutting and processing of materials with electron beams where the material boils and is evaporated, the breakup of rocks by electron beams in the atmosphere, industrial radiation technology, electron-beam technology outside a vacuum [86, 87], and so on.

## REFERENCES


1. A. A. Ivanov and T. K. Soboleva, "Electron beams in plasmas," in: *Plasma Chemistry*, Vol. 3 [in Russian], Atomizdat, Moscow (1976), pp. 240–301.
2. A. A. Ivanov and V. G. Leiman, "Electron beams in plasma chemistry," in: *Plasma Chemistry*, Vol. 5 [in Russian], Atomizdat, Moscow (1978), pp. 176–221.
3. V. L. Bychkov and A. V. Eletskii, "High-pressure beam plasmas," in: *Plasma Chemistry*, Vol. 12 [in Russian], Énergoatomizdat, Moscow (1985), pp. 119–158.
4. A. V. Eletskii, "Excimer lasers," *Usp. Fiz. Nauk* **125**, 279 (1978).
5. S. K. Searles, "Superfluorescent laser emission from electron-beam pumped Ar–$N_2$ mixtures," *Appl. Phys. Lett.* **25**, 735 (1974).
6. L. I. Gudzenko and S. I. Yakovlenko, *Plasma Lasers* [in Russian], Atomizdat, Moscow (1978).
7. I. A. Krinberg, *Electron Kinetics in the Earth's Atmosphere* [in Russian], Nauka, Moscow (1978).
8. E. Fermi, *Nuclear Physics* [Russian translation], IL (1951).
9. U. Fano, "Degradation and range straggling of high-energy radiation," *Phys. Rev.* **92**, 328 (1953).
10. L. V. Spencer and U. Fano, "Energy spectrum resulting from electron slowing down," *Phys. Rev.* **93**, 1172 (1954).
11. U. Fano and L. V. Spencer, "Quasiscaling of electron-degradation spectra," *Int. J. Radiat. Phys. Chem.* **7**, 63 (1975).
12. R. L. Platzman, "Subexcitation electrons," *Radiat. Res.* **2**, No. 1, 1–7 (1955).
13. R. L. Platzman, "Total ionization in gases by high energy particles: An appraisal in our understanding," *Int. J. Appl. Radiat. Isotop.* **10**, 116 (1961).
14. R. H. Fowler, "Contributions on the theory of the motion of α-particles through matter. Part II. Ionizations," *Proc. Cambridge Philos. Soc.* **21**, 531 (1923).
15. G. D. Alkhazov, "Ionization cascade of nonrelativistic electrons in helium," *Zh. Tekh. Fiz.* **41**, 2513 (1971).
16. D. A. Douthat, "Electron-degradation spectra in helium," *Radiat. Res.* **61**, No. 1, 1–20 (1975).
17. L. R. Peterson and J. E. Allen, "Electron impact cross sections for argon," *J. Chem. Phys.* **56**, 6068 (1972).

18. E. Eggarter, "Comprehensive optical and collision data for radiation action. II. Ar," *J. Chem. Phys.* **62**, 833 (1975).
19. W. M. Jones, "Some calculated quantities in the radiation chemistry of molecular hydrogen: Average energy per ion pair and numbers of singlet and triplet excitations per ion pair," *J. Chem. Phys.* **59**, 5688 (1973).
20. D. E. Gergart, "Comprehensive optical and collision data for radiation action. I. $H_2$," *J. Chem. Phys.* **62**, 821 (1975).
21. M. Inokuti, "Ionization yields in gases under electron irradiation," *Radiat. Res.* **64**, No. 1, 6 (1964).
22. Y.-K. Kim, "Energy distribution of secondary electrons," *Radiat. Res.* **64**, No. 1, 96 (1964).
23. S. P. Khare, "Ionizing collisions with atoms and molecules," *Radiat. Res.* **64**, No. 1, 106 (1964).
24. A. E. S. Green, "The role of secondary electrons in charged particle degradation," *Radiat. Res.* **64**, No. 1, 119 (1964).
25. D. A. Douthat, "Energy deposition by electrons and degradation spectra," *Radiat. Res.* **64**, No. 1, 141 (1964).
26. V. A. Nikerov and G. V. Sholin, *The Kinetics of Degradation Processes* [in Russian], Énergoatomizdat, Moscow (1985).
27. R. S. Stolarski and A. E. S. Green, "Calculations of auroral intensities from electron impact," *J. Geophys. Res.* **72**, 3967 (1967).
28. A. E. S. Green and C. A. Barth, "Calculations of the photoelectron excitation of the dayglow," *J. Geophys. Res.* **72**, 3975 (1967).
29. L. R. Peterson and A. E. S. Green, "The relation between ionization yields, cross sections and loss functions," *J. Phys. A* **1**, 1131 (1968).
30. L. R. Peterson, "Discrete deposition of energy by electrons in gases," *Phys. Rev. A* **187**, 105 (1969).
31. S. P. Khare, "Energy spectrum of secondary electrons and the fluorescent efficiency of electrons in the 3914 Å band," *Planet. Space Sci.* **17**, 1257 (1969).
32. S. P. Khare, "Mean energy expended per ion pair by electrons in atmospheric gases," *J. Phys. B* **3**, 971 (1970).
33. L. D. Landau and E. M. Lifshits, *Quantum Mechanics* [in Russian], Nauka, Moscow (1974).
34. V. P. Konovalov and É. E. Son, "The electron energy distribution function in discharges controlled by an electron beam," *Zh. Tekh. Fiz.* **51**, 547 (1981).
35. E. M. Lifshits and L. P. Pitaevskii, *Physical Kinetics* [in Russian], Nauka, Moscow (1979).
36. S. P. Khare and Ashok Kumar, Jr., "Mean energy expended per ion pair by electrons in molecular nitrogen," *J. Phys. B* **10**, 2239 (1977).
37. C. B. Opal, E. C. Beaty, and W. K. Peterson, "Table of energy and angular distributions of electrons ejected from simple gases by electron impact," *At. Data* **4**, 209 (1972).
38. A. E. S. Green and T. Sawada, "Ionization cross sections and secondary electron distributions," *J. Atmos. Terr. Phys.* **34**, 1719 (1972).
39. A. N. Kolmogorov, "Local structure of turbulence in an incompressible fluid at very high Reynolds numbers," *Dokl. Akad. Nauk SSSR* **30**, 299 (1941).
40. A. S. Monin and A. M. Yaglom, *Statistical Fluid Mechanics* [in Russian], Nauka, Moscow (1972), Part II.
41. L. A. Artsimovich and R. Z. Sagdeev, *Plasma Physics for Physicists* [in Russian], Atomizdat, Moscow (1979).

42. C. A. Brau, "Classical theory of vibrational relaxation of anharmonic oscillators," *Physica* **58**, 533 (1972).
43. B. F. Gordiets, A. I. Osipov, and L. A. Shelepin, *Kinetic Processes in Gases and Molecular Lasers* [in Russian], Nauka, Moscow (1980).
44. V. N. Mel'nik, A. V. Katz, and V. M. Kontorovich, "Steady-state nonequilibrium distributions of ions formed through an interaction with an electron thermostat," *Zh. Éksp. Teor. Fiz.* **78**, 966 (1980).
45. G. I. Barenblatt, *Similarity, Self-Similarity, and Intermediate Asymptotics*, Consultants Bureau, New York (1980).
46. I. A. Krinberg, L. A. Garifullina, and L. A. Akatova, "Photoelectron fluxes and electron gas heating at the lower ionosphere," *J. Atmos. Terr. Phys.* **36**, 1727 (1974).
47. Yu. A. Medvedev and V. D. Khokhlov, "A modified model for slowing down of electrons and its application for determining the distribution function of secondary electrons in a weakly ionized gas," *Zh. Tekh. Fiz.* **49**, 309 (1979).
48. V. A. Nikerov and G. V. Sholin, "Slowing down of fast electrons in helium and molecular hydrogen in the approximation of a degradation cascade," *Fiz. Plazmy* **4**, 1256 (1978).
49. N. Peyraud, "High-energy kinetic theory of particle-beam-generated plasma," *J. Plasma Phys.* **32**, 35 (1984).
50. J. Bretagne, G. Delouya, J. Godart, and V. Puech, "High-energy electron distribution in an electron-beam-generated argon plasma," *J. Phys. D* **14**, 1225 (1981).
51. J. Bretagne, J. Godart, and V. Puech, "Low-energy electron distribution in an electron-beam-generated argon plasma," *J. Phys. D* **15**, 2205 (1982).
52. N. Peyraud, "Energy transfer theory in particle-beam-generated plasma," *Phys. Lett.* **106A**, 37 (1984).
53. Yu. I. Syts'ko and S. I. Yakovlenko, "Ionization and excitation kinetics of a gas by a hard source," *Fiz. Plazmy* **2**, 63 (1976).
54. A. R. P. Rau, M. Inokuti, and D. A. Douthat, "Variational treatment of electron degradation in molecular oxygen," *Phys. Rev. A* **18**, 971 (1978).
55. S. P. Khare and Ashok Kumar, Jr., "Mean energy expended per ion pair by electrons in molecular oxygen," *J. Phys. B* **11**, 2403 (1978).
56. D. A. Douthat, "Energy deposition and electron energy degradation in molecular hydrogen," *J. Phys. B* **12**, 663 (1979).
57. E. A. Oks, V. D. Rusanov, and G. V. Sholin, "Computation of the plasma chemical effect of a relativistic electron beam in molecular nitrogen by the degradation spectrum method," *Fiz. Plazmy* **5**, 211 (1979).
58. M. G. Heaps and A. E. S. Green, "A Monte Carlo approach to the spatial deposition of energy by electrons in molecular hydrogen," *J. Appl. Phys.* **45**, 3183 (1974).
59. V. V. Ryzhkov and A. G. Yastremskii, "Computation of the characteristics of the interaction of high-energy electrons with a gas by a Monte Carlo method," *Izv. Vyssh. Uchebn. Zaved., Fiz.* No. 12, 150 (1975).
60. V. V. Ryzhkov and A. G. Yastremskii, "Energy distribution of an electron beam in a nitrogen plasma," *Fiz. Plazmy* **4**, 1262 (1978).
61. V. V. Ryzhkov and A. G. Yastremskii, "Generation of low-temperature plasmas through ionization of a gas mixture by an electron beam," *Zh. Tekh. Fiz.* **49**, 2141 (1979).
62. G. B. Lappo, M. M. Prudnikov, and V. G. Chicherin, "Electron distribution function of a beam plasma in air," *Teplofiz. Vys. Temp.* **18**, 677 (1980).

63. D. R. Suhre and J. T. Verdeen, "Energy distributions of electrons in electron-beam-produced nitrogen plasma," *J. Appl. Phys.* **47**, 4484 (1976).
64. V. P. Konovalov and É. E. Son, "Electron distribution function and composition of a molecular plasma excited by an electron beam," *Zh. Tekh. Fiz.* **50**, 300 (1980).
65. J. W. Keto, "Electron-beam-excited mixtures of $O_2$ in argon. II. Electron distributions and excitation rates," *J. Chem. Phys.* **74**, 4445 (1981).
66. Yu. A. Medvedev and V. D. Khokhlov, "Distribution function of secondary electrons in weakly ionized air," *Zh. Tekh. Fiz.* **49**, 317 (1979).
67. B. S. Punkevich, N. L. Stal', B. M. Stepanov, and V. D. Khokhlov, "Computation of the energy distribution of a fast electron in a beam plasma by numerical methods based on a multigroup approximation," *Izv. Vyssh. Uchebn. Zaved., Radiofiz.* **27**, 174 (1984).
68. V. P. Konovalov and É. E. Son, "Formation of ions and excited molecules by high energy electrons in lower ionosphere," in: Abstracts of talks at 13th Int. Symp. Rarefied Gas Dynamics, Vol. 2, Novosibirsk, USSR (1982), pp. 551–552.
69. C. J. Elliott and A. E. Green, "Electron energy distributions in e-beam-generated Xe and Ar plasmas," *J. Appl. Phys.* **47**, 2946 (1976).
70. Yu. P. Vysotskii and V. N. Soshnikov, "Extension of an iteration method to numerical solution of the equation for the electron energy distribution during slowing down of high-energy electrons in a gas. I," *Zh. Tekh. Fiz.* **50**, 1682 (1980).
71. Yu. P. Vysotskii and V. N. Soshnikov, "Extension of an iteration method to the degradation equation for slowing down of electrons in a gas for primary electron energies of up to $10^5$–$10^6$ eV. II," *Zh. Tekh. Fiz.* **51**, 996 (1981).
72. L. M. Biberman, V. S. Vorob'ev, and I. T. Yakubov, *Kinetics of Nonequilibrium Low-Temperature Plasmas*, Consultants Bureau, New York (1987).
73. C. A. Brau, "Electron distribution function in electron-beam-excited plasmas," *Appl. Phys. Lett.* **29**, 7 (1976).
74. N. L. Aleksandrov and É. E. Son, "Energy distribution and kinetic coefficients of electrons in gases in an electric field," in: *Plasma Chemistry* , Vol. 7 [in Russian], Atomizdat, Moscow (1980), pp. 35–75.
75. G. I. Bashmakova and I. I. Magda, "Model of a nitrogen laser pumped by a relativistic electron beam," *Kvantovaya Élektron.* **4**, 76 (1977).
76. B. L. Borovich, V. V. Buchanov, and É. I. Molodykh, "Numerical modelling of an electron-beam-pumped copper vapor laser. I. Electron distribution function in the laser plasma," *Kvantovaya Élektron.* **11**, 1007 (1984).
77. J. Bretagne, G. Delouya, C. Gorse, M. Capitelli, and M. Bacal, "Electron energy distribution functions in electron-beam-sustained discharges: application to magnetic multicusp hydrogen discharges," *J. Phys. D* **18**, 811 (1985).
78. H. Shimamori and Y. Hatano, "Thermal electron attachment to $O_2$ in the presence of various compounds as studied by microwave technique combined with pulse radiolysis," *Chem. Phys.* **21**, 187 (1977).
79. É. E. Son, "The electron energy distribution function and rate of three-body attachment to oxygen when an ionization source acts on a gas," *Teplofiz. Vys. Temp.* **19**, 16 (1981).
80. S. M. Antipov, V. P. Konovalov, and A. S. Koroteev, "Experimental study of a weakly ionized air plasma excited by a stationary electron beam," *Teplofiz. Vys. Temp.* **23**, 170 (1985).
81. J. W. Keto, C. F. Hart, and Chien-Yu Kuo, "Electron-beam-excited mixtures of $O_2$ in argon. I. Spectroscopy," *J. Chem. Phys.* **74**, 4433, 4444 (1981).

82. J. W. Keto, C. F. Hart, and Chien-Yu Kuo, "Electron-beam-excited mixtures of $O_2$ in Argon. III. Energy transfer to $O_2$ and $O_3$," *J. Chem. Phys.* **74**, 4450 (1981).
83. A. K. Rebrov, G. I. Sukhinin, and R. G. Sharafutdinov, "Mechanism for excitation of the $B_2\Sigma_u{}^+$ state of the nitrogen ion in an electron beam with energies of 10–50 keV," in: *Plasma Chemistry-79*, Vol. 1 [in Russian], Nauka, Moscow (1979), pp. 128–132.
84. A. K. Rebrov, G. I. Sukhinin, R. G. Sharafutdinov, and Zh.-K Lengran, "Electron beam diagnostics in nitrogen. Secondary processes," *Zh. Tekh. Fiz.* **51**, 1832 (1981).
85. Yu. A. Piotrovskii and Yu. A. Tolmachev, "Spectroscopic studies of plasmas created by high-power electron beams in inert gases," *Zh. Prikl. Spektrosk.* **32**, 974 (1980).
86. A. N. Didenko, V. P. Grigor'ev, and Yu. P. Usov, *High-Power Electron Beams and Their Applications* [in Russian], Atomizdat, Moscow (1977).
87. E. A. Abramyan, B. A. Al'terkop, and G. D. Kuleshov, *Intense Electron Beams* [in Russian], Énergoatomizdat, Moscow (1984).

# CHARGED PARTICLE PRODUCTION AND LOSS PROCESSES IN NITROGEN–OXYGEN PLASMAS

A. Kh. Mnatsakanyan and G. V. Naidis

## 1. INTRODUCTION

The kinetics of plasmas in nitrogen, oxygen, and mixtures of the two, especially air, is of considerable interest for a number of applications in electronics, the physics of gas lasers, plasma chemistry, aeronomy, etc. The processes which control the charged particle balance in such plasmas vary greatly and, despite the existence of extensive experimental and theoretical information on both the properties of these plasmas as a whole and on individual elementary reactions, many questions remain unanswered.

Thus, detailed studies have been made of the kinetics of plasmas in high electric fields with low specific input powers (low current discharges, the initial stages of breakdown) and with high temperatures, $T \geq$ 3000–5000 K, under conditions close to local thermodynamic equilibrium (strong shock waves, high-pressure arcs). Plasmas in the intermediate temperature range and those in which the internal and translational degrees of freedom of the molecules are far from equilibrium have been studied much less.

The goal of the present review is to present the available information on charged particle formation and loss processes under different conditions in nitrogen–oxygen plasmas (electron densities $n_e \sim 10^8$–$10^{18}$ cm$^{-3}$ and gas temperatures $T \sim 200$–10000 K). We examine direct and stepwise ionization of molecules and atoms by electron impact, associative and Penning ionization, photoionization, electron–ion recombination, and processes leading to the creation and loss of negative ions. Most attention is paid to questions which have not been discussed in sufficient detail in the many existing monographs and reviews.

## 2. DIRECT IONIZATION OF MOLECULES BY ELECTRON IMPACT

Direct ionization is taken to mean ionization from the ground electronic state of a molecule with all its vibrational–rotational levels. The rate constant for direct ionization, $K_i$, is usually found experimentally under conditions such that the lower vibrational level is primarily populated. The dependence of $K_i^{(0)}$ (the value of $K_i$ in the absence of vibrational excitation) on the ratio of the electric field strength $E$ to the density of molecules $n$ for nitrogen and air [1, 2] for $E/n \leq 3 \cdot 10^{-15}$ V·cm² can be approximated by the formulas

$$\log K_{i\,(N_2)}^{(0)} = -8.25 - 37.5\ n/E$$

and

$$\log K_{i\,(\mathrm{air})}^{(0)} = -8.25 - 34.0\ n/E$$

(with $E/n$ in units of $10^{-16}$ V·cm² and $K_i^{(0)}$, in cm³/sec). The range over which $K_i^{(0)}$ has been measured is bounded below (for $N_2$ by $E/n \simeq 8 \cdot 10^{-16}$ V·cm² and for air by $1 \cdot 10^{-15}$ V·cm²). Values of $K_i$ for lower $E/n$ and when the gas is vibrationally excited can be obtained by calculation.

The rate constant for direct ionization in a one-component molecular gas is given by

$$K_i = \left(\frac{2}{m}\right)^{1/2} \int \varepsilon f(\varepsilon)\, d\varepsilon \sum_{v'',\, v'} x_{v''} \sigma_{v''v'}(\varepsilon), \tag{1}$$

where $x_{v''} = n_{v''}/n$; $n_{v''}$ is the population of the vibrational level $v''$ of the electronic ground state; and, $\sigma_{v''v'}$ is the cross section for ionization of a molecule in the vibrational level $v''$ by an electron with energy $\varepsilon$ leading to formation of a molecular ion in the vibrational state $v'$ (the generalization to a mixture of gases is obvious). The electron-energy distribution function $f(\varepsilon)$ in Eq. (1) is normalized according to the condition $\int \varepsilon^{1/2} f(\varepsilon) d\varepsilon = 1$. The dependence of $K_i$ on the degree of vibrational excitation (or on the vibrational temperature $T_V$ when the vibrational levels obey a Boltzmann distribution) shows up both through the effective ionization cross section $\sigma_i = \Sigma_{v''v'} x_{v''} \sigma_{v''v'}$, and through $f(\varepsilon)$.

The cross sections for ionization of the $N_2$ and $O_2$ molecules from the lower vibrational levels are known from experiment [3]. Integrating them with the electron energy distribution function $f^{(0)}(\varepsilon)$ corresponding to $T_V =$

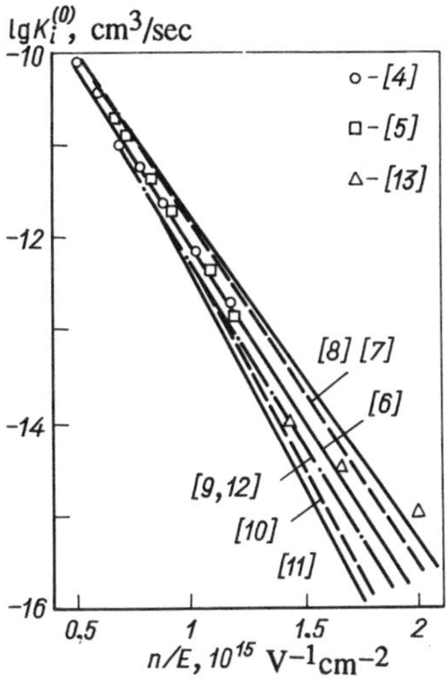


**Fig. 1.** The rate constant for direct ionization in nitrogen.

0 yields the value of $K_i^{(0)}$. Figure 1 shows the dependence of $K_i^{(0)}$ on $E/n$ in nitrogen obtained from drift experiments [4, 5] and theoretically [6–12]. It is clear that the slope of the theoretical $\log K_i^{(0)}$ plotted as a function of $E/n$ is constant over a wide range of $E/n$, including where there are no experimental data (the differences among the calculated values are a consequence of choosing different sets of cross sections for elastic and inelastic scattering of electrons on $N_2$ molecules). This fact allows us to extrapolate the experimental values of $K_i^{(0)}$ to weaker fields. An attempt to determine $K_i^{(0)}$ in nitrogen for $E/n \simeq 5 \cdot 10^{-16}$ V·cm$^2$ from the ionization balance in the positive column of a discharge in the limit of low currents [13] (Fig. 1) was unconvincing, since under these conditions secondary processes such as Penning ionization involving easily ionized impurities cannot be excluded (see below).

Ionization rate constants $K_i^{(0)}$ for nitrogen–oxygen mixtures with different concentration ratios of $N_2$ and $O_2$ have been obtained both experimentally [14] and theoretically [15].

The effect of vibrational excitation of the molecules on the rate of direct ionization can be taken into account in the following way: it is known that in nitrogen [7, 16] and in mixtures of nitrogen with oxygen, the electron energy distribution function in the range of electron energies responsible for ionization has the form

$$f(\varepsilon) = f(\tilde{\varepsilon})\, e^{-(\varepsilon-\tilde{\varepsilon})/T_H}, \qquad (2)$$

where $\tilde{\varepsilon} \sim 10$ eV, and that the effect of the degree of vibrational excitation (the vibrational temperature $T_V$) shows up only through $f(\tilde{\varepsilon})$, while the "temperature" of the electrons in the high-energy region $T_H$ is almost independent of $T_V$. The ionization cross section $\sigma_{v''v'}(\varepsilon)$ in the Born–Oppenheimer approximation is given by [17]

$$\sigma_{v'v'}(\varepsilon) = q_{v''v'}\varphi\left(\frac{\varepsilon - \varepsilon_{\text{thr}}}{\varepsilon_{\text{thr}}}\right), \quad \text{with} \quad \varepsilon_{\text{thr}} = I + \omega' v' - \omega'' v'', \qquad (3)$$

where $q_{v''v'}$ is the Franck–Condon factor, $\omega''$ and $\omega'$ are the vibrational quanta in the electronic ground states of the molecule and the molecular ion, $I$ is the ionization potential of the molecule, and $\varphi$ is a universal function which is nonzero only when its argument is positive. Substituting Eqs. (2) and (3) in Eq. (1) and assuming the existence of a Boltzmann vibrational distribution with temperature $T_V$, for $T_v \ll I$ we obtain

$$K_i = \zeta(T_H) f(\tilde{\varepsilon}) \left(1 - e^{-\omega''/T_V}\right) \sum_{v',\, v''} q_{v''v'} e^{-\frac{\omega' v'}{T_H} - \omega'' v'' \left(\frac{1}{T_V} - \frac{1}{T_H}\right)}, \qquad (4)$$

where

$$\zeta(T_H) = \left(\frac{2}{m}\right)^{1/2} I e^{-(I-\tilde{\varepsilon})/T_H} \int_0^\infty d\varepsilon\varepsilon\, e^{-\varepsilon/T_H} \varphi(\varepsilon/I).$$

The sum in Eq. (4) can be taken analytically using the expression for the Franck–Condon factors in the simple harmonic oscillator approximation [18]. The resulting formula can be simplified in the case where the $q_{v''v'}$ are calculated using the wave functions for harmonic oscillators with different equilibrium internuclear distances $r_e'$ and $r_e''$ and the same frequency $\omega = (\omega' + \omega'')/2$ (i.e., the relative shift of the molecular terms plays a controlling role). Then

$$\frac{K_i}{K_i^{(0)}} = \frac{f(\tilde{\varepsilon})}{f^{(0)}(\tilde{\varepsilon})} \exp\left\{\Lambda \frac{z(1-t)^2}{t(1-z)}\right\}, \qquad (5)$$

where $\Lambda = \mu\omega(r_e' - r_e'')^2/2\hbar$, $z = \exp(-\omega/T_V)$, $t = e^{-\omega/T_H}$, and $\mu$ is the reduced mass of the molecule.

The influence of the degree of vibrational excitation on the electron energy distribution function [i.e., the first factor on the right-hand side of

Eq. (5)] has been examined by many authors [7–10, 12, 19–22], primarily for molecular nitrogen. Thus, Son [19, 23] has obtained the following dependence of $f(\tilde{\varepsilon})/f^{(0)}(\tilde{\varepsilon})$ on $T_V$ and the reduced electric field strength $E/n$ in $N_2$:

$$\log\{f(\tilde{\varepsilon})/f^{(0)}(\tilde{\varepsilon})\} = Cz/\gamma^2, \quad \gamma = 10^{16}\, E/n\ (\mathrm{V \cdot cm^2}), \tag{6}$$

where the factor $C$ is proportional to the integral with respect to the energy of the product of the transport cross section $\sigma_m$ and the effective vibrational excitation cross section $\sigma_v$. A value of $C = 43.5$ was obtained [19] by comparing Eq. (6) with the results of numerical calculations [8]. Elsewhere the function $f(\tilde{\varepsilon})/f^{(0)}(\tilde{\varepsilon})$ has a form similar to Eq. (6), but the coefficient $C$, obtained by normalizing the vibrational excitation cross sections, is somewhat different (Fig. 2). The currently available experimental data do not allow a reliable choice for the cross section normalization to be made [24]. A comparison between theoretical and experimental electron energy distribution functions for high $T_V$ in [7, 9] corresponds to high values of $E/n$, where the electron energy distribution function is relatively insensitive to variations in $T_V$. A comparison of theoretical and experimental rates of excitation for various electronic states [9, 17] also fails to provide an unambiguous answer because of the possible contribution to populating these levels from other processes that were not included in the calculations. In this regard, a measurement of the electron energy distribution function in a non-self-sustaining discharge with a low $E/n$ and a high specific energy input (high $T_V$) would be of some interest.

In molecular oxygen, as well as in nitrogen–oxygen mixtures, energy loss by electrons through excitation of molecular vibrations is significant only in weak fields, with $E/n \simeq 1 \cdot 10^{-16}$ V·cm² [25, 26], when the rate of direct ionization is negligible. Thus, in the range of $E/n$ of practical interest, the effect of the degree of vibrational excitation of $O_2$ on the electron energy distribution function can be neglected (this has been confirmed computationally [22]). The dependence of the ratio $f(\tilde{\varepsilon})/f^{(0)}(\tilde{\varepsilon})$ on the vibrational temperature of $N_2$ in nitrogen–oxygen mixtures can be calculated as in nitrogen using Eq. (6), with $C = C_{N_2} x_{N_2} (\sigma_m^{N_2} x_{N_2} + \sigma_m^{O_2} x_{O_2})/\sigma_m^{N_2} \simeq C_{N_2} x_{N_2} (x_{N_2} + 0.25 x_{O_2})$, since in the region where nitrogen is excited (2–3 eV), the ratio of the transport cross sections is $C_{N_2}$ [27] [here $x_{N_2}$ and $x_{O_2} = 1 - x_{N_2}$ are the molar fractions of $N_2$ and $O_2$, and $\sigma_m^{N_2}/\sigma_m^{O_2} \simeq 4$ is the value of the factor $C$ in Eq. (6) corresponding to pure $N_2$]. Thus, for air we obtain $C \simeq 0.65 C_{N_2}$.

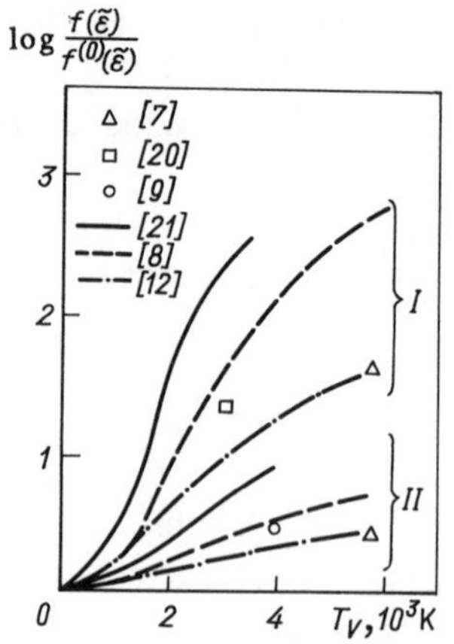


**Fig. 2.** The dependence of the ratio $f(\tilde{\varepsilon})/f^{(0)}(\tilde{\varepsilon})$ on the vibrational temperature in nitrogen: $E/n = 3 \cdot 10^{-16}$ (I), $6 \cdot 10^{-16}$ V·cm$^2$ (II).

Let us now examine the contribution of the second factor in Eq. (5). Using data on the molecular constants of $N_2$, $O_2$, $N_2^+$, and $O_2^+$ [28], we obtain $\Lambda_{N_2} \simeq 0.09$, and $\Lambda_{O_2} = 1.7$. Since the level shift in nitrogen is small, the change in the effective cross section for ionization of $N_2$ can usually be neglected. The influence of the vibrational temperature of $O_2$ on the ionization cross section for oxygen, however, can be substantial. Thus, taking $T_V{}^{O_2} \simeq 0.5$ eV and $T_H \simeq 1$ eV (which corresponds to $E/n \approx 6 \cdot 10^{-16}$ V·cm$^2$ for mixtures with a low oxygen content), we obtain $\exp\{\Lambda(z/t)(1-t)^2/(1-z)\} \simeq 1.2$.

In the absence of a Boltzmann vibrational distribution the contribution of the second factor may be greater. The quantity $f(\tilde{\varepsilon})$ is determined primarily by the population of the lower vibrational levels. If the upper vibrational levels are overpopulated relative to a Boltzmann distribution with temperature $T_V$ (for example, when the distribution contains a plateau [29, 30]), then their contribution to the effective ionization cross section can substantially increase the second factor in Eq. (5) while leaving the first unchanged. In general the contribution of each vibrational level of $K_i$ must be taken into account separately (this kind of calculation has been done for $N_2$ [31]).

In order to determine the rate of ionization in an rf field, the analog with a dc field is customarily employed [32, 33]: an effective field $E_e = E/(1 + \omega^2/\nu_m{}^2)^{1/2}$ is introduced, where $E$ and $\omega$ are the root mean square amplitude and the frequency of the rf field and $\nu_m$ is the effective collision frequency for momentum transfer, and it is assumed that the dependence of the rate constants for excitation and ionization of molecules on $E_e/n$ in an rf field is the same as that on $E/n$ in a dc field. This approach is not rigorous, since the momentum transfer frequency depends on the electron energy, so that the electron energy distribution functions in dc and rf fields have different forms [34]. Nevertheless, for the appropriate choice of $\nu_m$ ($\nu_m = 1.6 \cdot 10^{-7} n$ sec$^{-1}$ for nitrogen and air, with $n$ in cm$^{-3}$) this analogy yields good results. Thus, the dependence of $K_i$ on $E_e/n$ in nitrogen calculated for the conditions $\omega \gg \nu_m$ [10] and shown in Fig. 1 is in good agreement with data for a dc field.

The analogy with a dc field is valid when the frequency of the rf field satisfies the condition $\omega \gg \nu_u$, where $\nu_u$ is the energy transfer collision rate (for nitrogen and oxygen molecules this means that the inequality $\omega \geq 0.1\nu_m$ must be satisfied); in the opposite limiting case of $\omega \ll \nu_u$, the electron energy distribution function is able to track the periodic change in the field and this leads to a periodic change in the ionization rate constant. In this case the value of $K_i$ averaged over the period may be much greater than the value calculated under the assumption that the electron energy distribution function is constant in time. Thus, taking a dependence of the form $K_i(E) = A \exp(-b/E(t))$ for $K_i$, on averaging we obtain $\overline{K}_i \simeq (2\sqrt{2}\, E/\pi b)^{1/3} A \exp -b/E\sqrt{2})$, where $E$ is the root mean square value of the field. For $b/E \gg 1$, this is considerably greater than the value $K_i = A \exp (-b/E)$ corresponding to a constant field (for nitrogen, when $E/n = 7 \cdot 10^{-16}$ V·cm$^2$ we have $\overline{K}_i/K_i \simeq 10$). The ionization rate for nitrogen in the transition region where $\omega \sim \nu_u$ has been calculated by Dyatko et al. [35] The rate constants for ionization of oxygen in rf fields (with $\omega \gg \nu_u$) have been calculated by Masek and Rohlena [36].

The $T_V$ dependence of the rate constant for direct ionization in an rf field is given, as before, by Eqs. (5) and (6) with $E$ replaced by the effective field $E_e$ and with a slight change in the coefficient $C$ in Eq. (6), since the momentum transfer collision rate in the vibrational excitation region exceeds the average value of $\nu_m$. Thus, generalizing the method of [19] to the case of an rf field, for nitrogen in the limit $\omega \gg \nu_m$ we obtain a value of $C$ which is roughly half that in a dc field. The conclusion that the dependence of the electron energy distribution function on $T_V$ is weaker in an rf field has been confirmed computationally [10].

This analogy with a dc field has been used for calculating the ionization rate constants in laser light, as long as the energy of a laser photon is small compared to the vibrational quanta of the molecules (e.g., during breakdown of nitrogen–oxygen mixtures by $CO_2$ laser light) [37].

The rate constant for ionization of air and nitrogen heated to temperatures of 3000–4500 K in an rf field has been measured [38–41]. Figure 3 shows the measured values of $K_i$ for air taken from a review by Taylor et al. [32]. Also plotted there for comparison is a curve corresponding to 100% ionization efficiency (obtained under the assumption that all the energy absorbed in the gas is expended in ionization of the molecules). The ionization rate constant has been determined behind the front of a shock wave at molecular densities of $n = 10^{17}$–$10^{18}$ cm$^{-3}$. Measurements made at different points along the relaxation zone behind the shock front showed that the growth in $K_i$ for fixed $E_e/n$ occurs gradually. Here the distance from the front at which the values of $K_i$ shown in the figure are reached correlates with the characteristic relaxation length for the vibrational temperature of nitrogen molecules. The following model has been proposed [41] on this basis. Suppose that the ionization rate constant $K_i$ is uniquely determined by the average energy $\varepsilon_{av}$ of the electrons determined from the balance equation

$$\frac{e^2 E_e^2}{m \nu_m} = \eta \nu_m \varepsilon_{av},$$

where the inelastic loss factor $\eta$ is proportional to $1 - z_{N_2}$. Here the value of $E_e/n$ corresponding to fixed $\varepsilon_{av}$ and $K_i$ is related to the vibrational temperature of the nitrogen molecules by $E_e/n \propto (1 - z_{N_2})^{1/2}$. Calculations of $K_i$ using this model are in good agreement with measurements. On the other hand, calculations of the electron energy distribution function in vibrationally excited nitrogen (see above) show that the contribution of vibrational excitation to the ionization rate is small in the range of $E_e/n$ characteristic of the experiments [38–41]. The main contribution to the energy balance of the electrons is from losses through excitation of electronic states. Thus, there is still no satisfactory explanation for the results of [38–41]. In this regard, it would be interesting to study experimentally the breakdown of hot nitrogen and air by $CO_2$ laser light.

## 3. STEPWISE IONIZATION

It is known that in an atomic plasma, the rate of stepwise ionization usually is considerably greater than that of direct ionization [42]. In molecular plasmas the relative contribution of stepwise ionization is usually somewhat smaller for a variety of reasons. These reasons include the rapid quenching of excited electronic states of molecules by heavy particles [17], the presence of significant predissociation from highly excited levels [43], and the characteristic shape of the electron energy distribution function, which has a rapid drop at low energies owing to the excitation of molecular vibrations. The complexity and variety of transitions among excited molecular states, inadequate experimental information, and the absence of universal models for such transitions mean that it has not been possible to create a sufficiently general theory of stepwise ionization similar to that developed for atomic plasmas [42]. In practice, stepwise ionization is usually taken into account by selecting a group of levels (in nitrogen, as a rule, these are the triplet states $A^3\Sigma$, $B^3\Pi$, and $C^3\Pi$) for which the mechanisms of excitation and deexcitation are comparatively well known and considering direct ionization from these levels [44–46].

The rate constants for electron–impact ionization of nitrogen molecules in the $A^3\Sigma$ state have been calculated by several authors [8, 12, 20] and rate constants for ionization of nitrogen molecules in the $B^3\Pi$ state have also been obtained [8]. The cross section for ionization of $N_2(A^3\Sigma)$ was calculated in the semiclassical approximation [8] or assumed equal to the cross section for ionization of the ground state (with a corresponding energy shift) [20]. The experimentally determined [47] ionization cross section for $N_2(A^3\Sigma)$ used in the calculations of $K_{Ai}$ in [12] are somewhat smaller than the values used in [8, 20]. The values of $K_{Ai}$ from [8, 12] are shown in Fig. 4. Also shown there are the rate constants for excitation of the $N_2(A^3\Sigma)$ level from the ground state ($K_A$) calculated in [12]. Note that these values of $K_{Ai}$ and $K_A$ correspond to low vibrational temperatures in the ground ($T_V$) and excited ($T_A$) electronic states of $N_2$. The effect of vibrational excitation on $K_A$ and $K_{Ai}$ can be taken into account by a method analogous to that used above for the ionization rate constants $K_i$. Both $K_A$ and $K_{Ai}$ should be multiplied by the coefficient (5), the first factor of which is given, as before, by Eq. (6). Since the potential curve for the $N_2(A^3\Sigma)$ state is strongly shifted relative to both $N_2(X^1\Sigma)$ and $N_2(X^2\Sigma)$ [28], the second factor in Eq. (5) may be important. Using the data of Huber and Herzberg [28], we obtain $\Lambda_A = 7.1$ (the $X \rightarrow A$ transition) and $\Lambda_{Ai} = 5.5$ (ionization from the $A$ state). Note that the quantity $z$ in Eq. (5) is equal to $\exp(-\omega/T_A)$ for ionization from the $A$ state. The influence of vibrational excitation in the ground electronic state of $N_2$ on the effective cross section for populating the $A^3\Sigma$ level when the vibrational distribution has a plateau has been examined by Aleksandrov et al. [48].

We now consider the conditions under which stepwise ionization becomes important in nitrogen. We shall limit ourselves to ionization through the $A^3\Sigma$ level alone and assume that the vibrational temperature $T_V$ is low. The balance equation for the $A^3\Sigma$ state then has the form [17]

$$\frac{dn_A}{dt} = n_e n\,(K_A + K_B) - n_A^2 K_{TA} - n_A n_a K_{Ta}, \tag{7}$$

where $K_B$ is the rate constant for excitation of the $B^3\Pi$ state and the rate of radiative and collisional transitions from this state into state $A$ is fairly high: $K_{TA} = 1.5 \cdot 10^{-9}$ cm$^3$/sec [49] is the rate constant for deactivation in $A + A \rightarrow Y + X$ processes followed by a transition from higher states $Y$ into the $A$ state. Here $n_a$ is the concentration of nitrogen atoms and $K_{Ta} = 5 \cdot 10^{-11}$ cm$^3$/sec [50] is the rate constant for quenching of state $A$ by N atoms. For the quasistationary case (relative to $n_A$), the condition that stepwise ionization be important $n_A K_{Ai} \geq nK_i$ can be rewritten with the aid of Eq. (7) in the form

$$n_e/n > \begin{cases} \varphi, & \text{for} \quad n_a/n < \psi, \\ \varphi/\psi, & \text{for} \quad n_a/n > \psi; \end{cases}$$

$$\varphi \equiv \left(\frac{K_i}{K_{Ai}}\right)^2 \frac{K_{TA}}{K_A + K_B}, \qquad \psi \equiv \frac{K_i K_{Ta}}{K_{Ai} K_{TA}}. \tag{8}$$

Note that electron-impact processes leading to the depopulation of $N_2(A^3\Sigma)$ molecules have been neglected in Eq. (7). When the degree of ionization is high, these processes limit the density of $N_2(A^3\Sigma)$ to a maximum attainable level. Then the rate of stepwise ionization for $E/n = (4–15) \cdot 10^{-16}$ V·cm$^2$ is one or two orders of magnitude greater than the rate of direct ionization.

Under pulsed conditions, when the electron density varies as $n_e = n_{e0} \exp\{K_i nt\}$ [46] (losses owing to attachment and diffusion are neglected), the density of $N_2(A^3\Sigma)$ is given by $n_A(t) = n_e(t)\,(K_A + K_B)/K_i$ with only the first term in Eq. (7) taken into account. Stepwise ionization is included here when

$$n_e(t)/n > \eta, \quad \eta \equiv K_i^2/[K_{Ai}(K_A + K_B)]. \tag{9}$$

The dependences of $\varphi$, $\psi$, and $\eta$ on $E/n$ obtained using the values of $K_A$, $K_B$, $K_{Ai}$, and $K_i$ from [12] are shown in Fig. 4.

As the vibrational temperature $T_V$ increases, stepwise ionization becomes important for lower values of $n_e/n$ and $n_e/n_a$. As noted above, with increasing $T_V$ the rate constant $K_A$ increases more rapidly than $K_i$ (owing to the difference in the corresponding values of $\Lambda$). In addition, as $T_V$

increases, the effective rate constant for quenching of $N_2(A^3\Sigma)$ by N atoms decreases because of the reverse process [17]. Stepwise ionization in $N_2$ can also proceed through the singlet states $a^1\Pi$ and $a'^1\Sigma$, for which quenching by molecules is relatively weak [51, 52].

Additives of oxygen reduce the populations in excited electronic states of the $N_2$ molecule [17]. Thus, the $N_2(A^3\Sigma)$ state is quenched by $O_2$ molecules with a rate constant of $K_{TO2} \simeq (3–5)\cdot 10^{-12}$ cm$^3$/sec [53]. The effect of $O_2$ on the rate of stepwise ionization through $N_2(A^3\Sigma)$ can be evaluated using Eq. (8) with $n_a$ replaced by $n_{O2}$ and $K_{Ta}$, by $K_{TO2}$. Stepwise ionization in oxygen can only proceed through the low-lying electronic states $a^1\Delta$, $b^1\Sigma$, and $A^3\Sigma$, since excitation of higher levels leads to dissociation.

When the concentrations of N and O atoms are high, stepwise ionization of the atoms through electron impact may be important. The methods of calculating the rate constants for stepwise ionization of atoms $\alpha$ have been discussed in detail by Biberman et al. [42] The dependence of $\alpha$ in nitrogen on the electron temperature $T_e$ for different values of $n_e$ has been obtained by Potapov et al. [54] For $n_e \geq 10^{16}$ cm$^{-3}$, when radiative transitions become unimportant in the kinetics of the higher atomic states, the dependence of $\alpha_N$ on $T_e$ calculated in the modified diffusion approximation [42] can be written roughly as

$$\alpha_N \simeq 3\cdot 10^{-8} T_e^{-3} e^{-14.53/T_e} \text{ cm}^3/\text{s}, \quad 0.6 < T_e < 2.5 \text{ eV}.$$

This formula is in good agreement with the numerical calculations of Potapov et al. [54].

## 4. ASSOCIATIVE IONIZATION

Besides electron-impact ionization, associative ionization processes, $A + B \rightarrow AB^+ + e$, where $A$ and $B$ represent an atom or molecule in the ground or an excited state, may play an important role. These processes cause primary ionization behind shock fronts in molecular gases [55, 56] and control the electron balance in arc discharge plasmas [57]. There are also many indications that, under certain conditions, associative ionization involving excited atoms and molecules can serve as the main ionization source in glow discharges at low gas temperatures [17, 45, 58–61].

Of the associative ionization processes which take place in nitrogen–oxygen mixtures, the ones that have been studied in most detail are those involving two atoms:

$$N + N \rightarrow N_2^+ + e, \quad (10)$$

$$N + O \rightarrow NO^+ + e, \quad (11)$$

and

$$O + O \rightarrow O_2^+ + e. \quad (12)$$

The rate constants for these reactions have been determined mainly by analyzing data on the electron density profiles behind shock fronts [62]. This, however, leads to a significant uncertainty associated with the fact that atoms in a variety of electronic states can participate in each of these reactions: $N(^4S, ^2D, ^1P)$ and $O(^3P, ^1D, ^1S)$. The populations of excited atomic states in the ionization zone behind a shock front may differ from the equilibrium values.

In this regard, a beam experiment [63] in which the cross section for the reaction $N(^2D) + O(^3P) \rightarrow NO^+ + e$ (with a threshold of 0.38 eV) was determined is of some interest. The rate constant for the reaction was obtained by averaging the observed cross section, which depends linearly on the energy of relative motion of the colliding particles near threshold, over a Maxwellian distribution [63]. In addition, assuming that the densities of $N(^2D)$ and $N(^4S)$ were in equilibrium and taking only one of the above-mentioned channels for associative ionization of N and O molecules into account, the total rate constant for associative ionization, $\alpha_a^{NO^+}$, was determined [63]. An analysis of the dissociative recombination reaction opposite to reaction (11), showed that 70–80% of the nitrogen atoms are formed in the $^2D$ state and that the fraction of $N(^2D)$ atoms is practically independent of the initial vibrational distribution of the $NO^+$ ion [64]. With a correction for the relative contribution of the channel involving $N(^2D)$ atoms, the expression for $\alpha_a$ takes the form

$$\alpha_a^{NO^+} = 1.0 \cdot 10^{-11} T^{0.5} (T + 0.19) \exp(-2.76/T)$$

($T$ in eV), which is 4–5 times greater than the rate constant for this process obtained indirectly by Lin and Teare [55] for $T \simeq 0.4$–$0.6$ eV.

The only data on the rate of reaction (10) have been obtained using shock wave methods [65]:

$$\alpha_a^{N_2} = 2.7 \cdot 10^{-11} \exp(-5.80/T)$$

(for $T = 0.6$–$0.9$ eV), which is roughly an order of magnitude greater than the value obtained [55] by calculating back from the rate constant for the reverse reaction using the principle of detailed balance. This discrepancy originates from the fact that it was assumed [55] that the rate of dissociative recombination, the reverse of reaction (10), also depended on the temperature as $T^{-1.5}$, while the later measurements revealed a weaker temperature dependence (see below).

In a nonequilibrium gas discharge plasma or afterglow, associative ionization should proceed primarily along channels for which the activation energy is low. In particular, the process

$$N(^2P) + N(^2D) \rightarrow N_2^+ + e \tag{13}$$

has been examined [59, 66] as a possible channel for ionization in nitrogen. An estimate of the maximum possible rate constant for this process using the principle of detailed balance yields $\sim 1 \cdot 10^{-12}$ cm$^3$/sec for $T \leq$ 3000 K (an order of magnitude smaller than the estimate used in [59]).

In order to explain the ionization fluxes observed in nitrogen discharges and afterglows, various associative ionization processes have been proposed, including some involving an excited atom or molecule [67]

$$N_2^* + N^* \rightarrow N_3^+ + e, \tag{14}$$

two vibrationally excited molecules [17, 58]

$$N_2(X,\ v \geqslant 32) + N_2(X,\ v \geqslant 32) \rightarrow N_4^+ + e \tag{15}$$

with a rate constant $\alpha_a \leq 1.9 \cdot 10^{-15} \exp(-1160/T)$ cm$^3$/sec, vibrationally and electronically excited molecules [17, 58]

$$N_2(X,\ v) + N_2(Y) \rightarrow N_4^+ + e \tag{16}$$

with a rate constant $\alpha_a \leq 3.9 \cdot 10^{-12} \exp(-640/T)$ cm$^3$/sec [if all the molecules involved are $N_2(X, v \geq 13)$] where $Y = (a''^1\Sigma)$, or two electronically excited molecules [61, 68]

$$N_2(Y) + N_2(Z) \rightarrow N_4^+ + e\,. \tag{17}$$

(in the last reaction one of the states $Y$ is the $a'^1\Sigma$ level and the other is $A^3\Sigma$ or $a'^1\Sigma$, while the rate constant $\alpha_a$ was taken to be $10^{-10}$ cm$^3$/sec). There are no direct experimental data on the rate constants for associative ionization involving molecules. The channels for dissociative recombination of the $N_3^+$ and $N_4^+$ ions have not been investigated. Estimates using the principle of detailed balance show that for associative ionization lead-

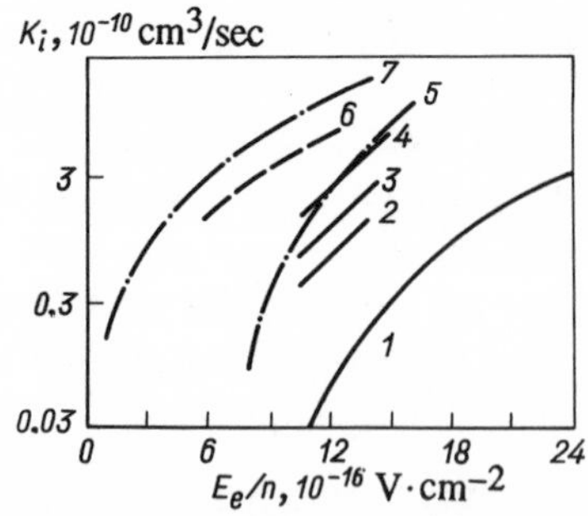


**Fig. 3.** The rate constant for ionization during microwave breakdown of hot air [32]: 1) $K_i^{(0)}$, $T$ = 300 K; 2) [39], $T$ = 3100 K; 3) [39], $T$ = 3300 K; 4) [39], $T$ = 3700 K; 5) [38], $T$ = 3350 K; 6) [40], $T$ = 4500 K; 7) 100% ionization efficiency.

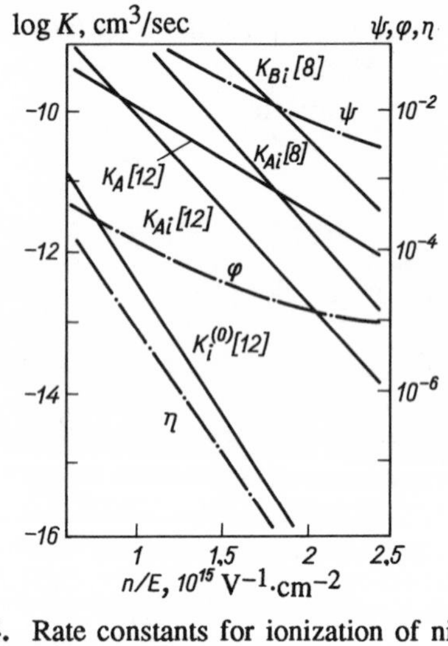


**Fig. 4.** Rate constants for ionization of nitrogen molecules from excited states.

ing to formation of the $N_4^+$ ion, rate constants on the order of $10^{-10}$ cm$^3$/sec are attainable. The correlation found [45, 58, 60] between the ionization flux in an $N_2$ discharge and the calculated degree of excitation in the upper vibrational levels indicates that mechanisms (15) and (16) predominate. At the same time, data [59, 61, 69, 70] exist which do not fit into the framework of these mechanisms and require that additional channels for ionization of the types (13) and (17) be included. The effect of small amounts of $O_2$ on the ionization flux is also not entirely clear [71].

Reaction (15) has also been invoked to explain experimental data on the electron density in an adiabatically expanding supersonic flow of nitrogen [72] and on the behavior of the electron density in the afterglow of a nitrogen discharge [73]. In an afterglow the high vibrational levels can be populated by the recombination of atoms [74]. (The influence of atomic recombination on the vibrational energy balance in a nitrogen afterglow has also been examined by Antonov et al. [75].)

On the whole, the role of different ionization mechanisms in a highly nonequilibrium nitrogen plasma is not completely clear. In mixtures of nitrogen and oxygen with fairly large concentrations of the latter, especially in air, the contribution of processes similar to (14)–(17) should be considerably lower than in pure nitrogen [17]. Oxygen efficiently quenches the electronically excited states of $N_2$. The existence of rapid vibrational exchange between the lower vibrational levels of $O_2$ and levels of $N_2$ with $v = 25$–$30$ causes a sharp dip in the vibrational distribution in this range of $v$ [30, 76] and, thereby, a reduction in the flux owing to associative ionization (15). On the other hand, adding oxygen to a discharge opens up an additional channel for associative ionization (11) involving excited N and O atoms [71, 77].

## 5. PHOTOIONIZATION

Unlike the processes examined above, where ionization occurs through particle collisions, which have a localized character, photoionization becomes important under inhomogeneous conditions such that emission and absorption of the ionizing photons occur in different spatial regions. Photoionization plays an important role in the initial stage of electrical breakdown in gases [78, 79] and in the propagation of discharges in dc [80] and rf [81, 82] fields. It can affect the plasma parameters in the cathode region of glow discharges and in discharges with a transverse flow of gas [83]. The ionization of air ahead of a shock front is determined by this process [84, 85]. By using pulsed discharges that slide over the surface of a dielectric as a source of ionizing radiation, it is possible to obtain an electron density of $n_e \sim 10^{13}$ cm$^{-3}$ in atmospheric pressure nitrogen at distances of 2–3 cm from the radiation source [86, 87] and thereby organize a high-power non-self-sustaining volume discharge.

The number of ionization events per unit volume per unit time owing to the absorption of photons emitted per unit volume at a distance $r$ in an optically homogeneous medium is given by the expression

$$Q = \frac{\varphi}{4\pi r^2}, \quad \varphi(r) = \int d\nu \varepsilon_\nu \xi_\nu k_\nu e^{-k_\nu r}, \tag{18}$$

where $\varepsilon_\nu$ is the emissivity of the gas [$\varepsilon_\nu d\nu$ is the number of photons emitted per unit volume per unit time in the frequency interval ($\nu$, $\nu + d\nu$)]; $k_\nu$ is the absorption coefficient; and $\xi_\nu$ is the probability of ionization through absorption of a photon with a frequency of $\nu$. The integral with respect to the frequency can be approximated as a sum over frequency regions $\Delta\nu$ characterized by different values of the absorption coefficient (which makes it possible to distinguish these regions experimentally):

$$\varphi(r) = \sum_i \frac{\xi_i q_i}{\Delta\nu_i} \int_{\Delta\nu_i} k_\nu d\nu \exp(-k_\nu r), \tag{19}$$

where $\xi_i$ is the average value of $\xi_\nu$ in the interval $\Delta\nu_i$, and $q_i = \int_{\Delta\nu_i} \varepsilon_\nu d_\nu$. The transition from Eq. (18) to Eq. (19) is justified by the fact that in each characteristic interval $\Delta\nu_i$, a large number of molecular bands and atomic lines whose peaks are not correlated to the maxima and minima of $k_\nu$ contribute to $q_i$.

In analyzing the ionization of gases by the radiation from discharges, by analogy with the first Townsend coefficient $\alpha$, one introduces a parameter $\omega$, which gives the number of ionizing photons in a given frequency interval that will produce an electron over a path length of 1 cm. The relationship between $q_i$ and $\omega_i$ is given by

$$q_i = n_e \omega_i v_g / (1 + n/n_{Ti}), \tag{20}$$

where $v_g$ is the drift speed of the electrons and the factor in parentheses, where $n_{Ti}$ is the parameter, takes quenching of the radiating states by heavy particles into account. Experimental studies of the ionizing properties of low-current discharges in nitrogen, oxygen, and mixtures of the two have been conducted by several authors [88–92]. Corona [88–91] and spark [90] discharges, as well as a non-self-sustained discharge in a uniform field [92], have been used as radiation sources. The effective values of the absorption coefficient $k_i$ and the corresponding quantities $\xi_i\omega_i/\alpha$ and $n_{Ti}$ have been measured.

The available data for mixtures of $N_2$ and $O_2$ with an oxygen content of 1–50% have been analyzed in detail [93]. In these mixtures, photoionization proceeds through the absorption by oxygen of the molecular band emission from nitrogen in the spectral range $98 < \lambda < 102.5$ nm. In this region the absorption coefficient of $O_2$ is a sharp function of frequency of the form [84] $k_\nu = k_1(k_2/k_1)^{(\nu - \nu_1)/(\nu_2 - \nu_1)}$, where $\varkappa_1 = k_1/p_{O_2} = 0.035$ $Torr^{-1}cm^{-1}$ and $\varkappa_2 = k_2/p_{O_2} = 2$ $Torr^{-1}cm^{-1}$ (here $p_{O_2}$ is the pressure of the oxygen reduced to $T = 300$ K). Evaluating the integral with respect to the frequency in Eq. (19) yields

$$\varphi(r) = \xi q f(r), \quad f(r) = \frac{\exp(-k_1 r) - \exp(-k_2 r)}{r \ln(k_2/k_1)}. \tag{21}$$

Figure 5 [93] shows a comparison of the values of $\varkappa = -d[\ln(\varphi/p_{O_2})]/d(p_{O_2}r)$, calculated from Eq. (21) with the experimental data of [88, 89, 91, 92, 94]. The points corresponding to different percent amounts of $O_2$ (from 1–50%) are grouped quite densely, which indicates that $O_2$ plays a dominant role in the absorption of the ionizing radiation. Note that a similar expression for $f(r)$ has been used to calculate $n_e$ ahead of a shock wave in air [84] with the source $q$ given by the integral of the Planck black-body radiation function over this spectral range.

The value of $\xi\omega/\alpha$ for air obtained in [93] by analyzing the experimental data of [92] varies from 0.12 to 0.06 for $E/n = (1.7\text{–}7)\cdot 10^{-15}$ $V\cdot cm^2$, while for $E/n = 10^{-13}$ $V\cdot cm^2$ it is about 0.03 according to the data of [91]. The value of $n_{Ti}$ in Eq. (20) for air is $1\cdot 10^{18}$ $cm^{-3}$ [92].

In pure oxygen in the range $pr = 2\text{–}100$ Torr·cm, the dependence of $\varkappa$ on $pr$ is the same as in $N_2$–$O_2$ mixtures [91] (Fig. 5), i.e., the main contribution to photoionization is from frequencies near the ionization threshold, where the radiation source is lines of atomic oxygen O [95]. $\xi\omega/\alpha$ varies over the range $(4\text{–}10)\cdot 10^{-3}$ when $E/n = (2\text{–}13)\cdot 10^{-15}$ $V\cdot cm^2$ and $n_{Ti} \simeq 8\cdot 10^{16}$ $cm^{-3}$ [92].

When $p_{O_2}r \geq 100$ Torr·cm the coefficient $\varkappa_{O_2}$ decreases sharply in both $O_2$ and $N_2$–$O_2$ mixtures [90, 92], becoming significantly smaller than $\varkappa_1$ (the minimum absorption coefficient in the $O_2$ photoionization region). In this case, the ionization processes appear to be determined by the presence of organic impurities with low ionization potentials [94, 96].

Measurements in pure nitrogen, as well as in nitrogen with small, controlled amounts of oxygen, have been made only with a corona discharge [88, 89]. In pure $N_2$, ionization is caused by radiation at wavelengths $\lambda < 80$ nm, where the absorption cross section depends weakly on $\lambda$ [97]. In this spectral region, with $\varkappa = k/p_{N_2} \simeq 1$ $Torr^{-1}\cdot cm^{-1}$ and $\xi\omega/\alpha \simeq (1\text{–}3)\cdot 10^{-3}$ [88, 89], the lines of the $N^+$ ion serve as the radiation source

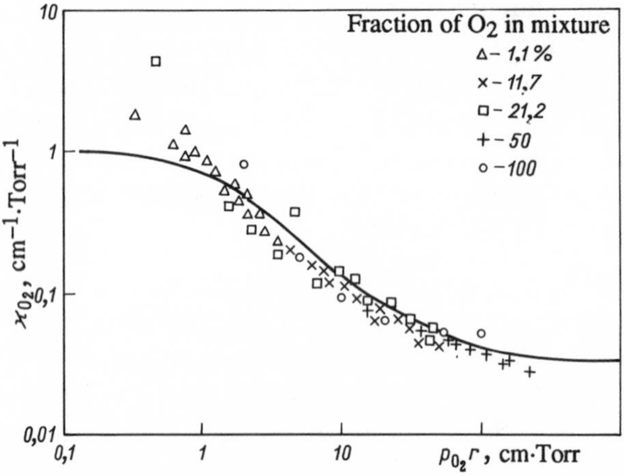


**Fig. 5.** The effective absorption coefficient of oxygen in an $N_2$–$O_2$ mixture: points denote experimental data [88, 89, 91, 92, 94], and the curve is a calculation using Eq. (21).

[95]. In nitrogen with small amounts of oxygen (≤1%), the contribution from two terms must be included in the sum (19). One of them is the same as in pure $N_2$, while the other is given by Eq. (21).

The photoionization observed in nitrogen in [86, 87] was apparently caused by the presence of oxygen impurities. Thus, when the fraction of oxygen was ~$10^{-3}$ in nitrogen at atmospheric pressure, the absorption coefficient $k = \varkappa_{O_2} p_{O_2}$ at a distance of about 1 cm from the source was ~0.6 $cm^{-1}$ (Fig. 5), which corresponds roughly to the measurements reported in [87].

The ionizing radiation from hot mixtures of nitrogen and oxygen, where the degree of dissociation of $O_2$ is fairly high, may be predominantly from atomic O lines [98], even when the fraction of oxygen in the mixture is relatively low (e.g., in air).

## 6. PENNING IONIZATION

Penning ionization $A^* + M \rightarrow A + M^+ + e$, where $A^*$ is an excited molecule or atom of nitrogen or oxygen and $M$ is an easily ionized impurity molecule or atom, may make a significant contribution to the electron balance in a discharge with low $E/n$. Molecules of organic compounds, with ionization potentials of 7–9 eV that form about $10^{-6}$ of gas mixtures under typical conditions, may serve as impurities in this case [96]. For a quasistationary density of excited particles $n^* = n_e n K_{exc} \tau$ [where $n_{im}$ is the density of impurity molecules, $K_P$ is the rate coefficient for the Penning reaction, $K_{exc}$ is the rate coefficient for excitation of the $A^*$ level, and $\tau = (nK_T)$ is the quenching time for this level in collisions with molecules of the primary gas], the Penning ionization flux $j_P = n^* n_{im} K_P$ can be written in the form

$$j_P = nn_e K_{exc} \frac{n_{im} K_P}{nK_T} . \tag{22}$$

The flux $j_P$ has the same form as the direct ionization flux $j_P = nn_e K_i^{eff}$, where the effective rate constant for ionization is $K_i^{eff} = K_{exc} n_{im} K_P / nK_T$.

Measurements [99] show that the ratio $K_P/K_T$ for an admixture of tripropylamine in nitrogen is ~$10^{-3}$. The excited $N_2$ levels $a^1\Pi$ and $a'^1\Sigma$ were considered to participate in the Penning reaction [99, 100]. In nitrogen with $E/n = 5 \cdot 10^{-16}$ V·cm$^2$, the calculated rate constant for excitation of these levels [12] exceeds the rate constant for direct ionization by four orders of magnitude. Thus, for $n_{im}/n \sim 10^{-6}$ we have $K_i^{eff} \sim 10 K_i$. The resulting value of $K_i^{eff}$ is in agreement with [13].

## 7. ELECTRON–ION RECOMBINATION

The main process for recombination of electrons and positive ions in a molecular plasma is usually dissociative recombination, the reverse of the associative ionization process discussed above. For atmospheric ions, extensive information exists both on the rate constants for dissociative recombination and their temperature dependences and on the cross sections for this process (see the reviews by Biondi [101], Eletskii and Smirnov [102], and Mitchell and McGowan [103]). Table 1 lists values of the rate constants for dissociative recombination $\beta_{DR}$ and the yields of atoms in state $i$, $\beta_i$ [the number of events in which $A(i)$ atoms are formed per unit volume per unit time in the reaction $AB^+ + e \rightarrow A(i) + B(j)$ is equal to $\beta_i \beta_{DR} n_{AB^+} n_e$]. The constants $\beta_{DR}$ are shown in the form $\beta_{DR}(T_e) = \beta_{DR}(300\ \mathrm{K})\ (T_e/300)^{-\gamma}$.

Strictly speaking, $\beta_{DR}$ depends both on the electron temperature (in general on the electron energy distribution) and on the vibrational temperature of the ions (on the vibrational distribution). Experiments [106, 121] show, however, that for the $N_2^+$, $O_2^+$, and $NO^+$ ions the dependence of $\beta_{DR}$ on the vibrational distribution is extremely weak and cannot explain

**TABLE 1.** Rate Constants and Channels for Dissociative Recombination

| Ion | $\beta_{DR}$ (300 K), $10^{-7}$ cm$^3$/sec | $T$, K | $T_e$, K | $\gamma$ | References |
|---|---|---|---|---|---|
| $N_2^+$ | 2.7 | 200—480 | $T_e$=$T$ | 0.02 | [104] |
| | 1.8 | 300 | 300—5000 | 0.39 | [105] |
| | 3.5 | 300 | — | 0.5 | [106]* |
| | $\beta_{^2D}$=0.5—1.0 | — | — | — | [107] |
| | $\beta_{^3P}$<0.1 | — | — | — | [108] |
| $O_2^+$ | 2.2 | 200—700 | $T_e$=$T$ | 1.0 | [109] |
| | 2.0 | 300 | 300—5000 | 0.56—0.7 | [105] |
| | 2.0 | 200—600 | $T_e$=$T$ | 0.7 | [110] |
| | 1.9 | 300 | — | 0.5 | [106]* |
| | $\beta_{^3P}$=1.0 | — | — | — | [111] |
| | $\beta_{^1D}$=0.9 | — | — | — | [111] |
| | $\beta_{^1D}$=1.3 | — | — | — | [112] |
| | $\beta_{^1S}$<0.1 | — | — | — | [113] |
| NO+ | 4.1 | 200—450 | $T_e$=$T$ | 1.0 | [114] |
| | 4.3 | 300 | 380—5500 | 0.37 | [115] |
| | 4.3 | 300 | 600—3300 | 1.0 | [116] |
| | 4.0 | 200—600 | $T_e$=$T$ | 0.9 | [110] |
| | 2.3 | 300 | — | 0.5 | [106]* |
| | 1.0 | — | — | 0.4 | † |
| | $\beta_{N(^2D)}$=0.76 | — | — | — | [64] |
| $N_4^+$ | 30 | 300—480 | $T_e$=$T$ | 0.5 | [117] |
| | 14 | 300 | 300—5600 | 0.41 | [118] |
| | 23 | 300 | 2000—11 000 | 0.53 | [119] |
| $O_4^+$ | 20 (200 K) | 200 | 200 | — | [109] |
| $N_2O_2^+$ | 17 | 300 | 300 | — | [114] |

*Calculated from the measured cross section for dissociative recombination.
†Calculated from the rate constant for the reverse process [63] using the equilibrium constant [120].

the differences between the data obtained in various experiments. According to calculations for the atmospheric ions $\gamma = 0.5$ [102].

Data on the cross sections for dissociative recombination $\sigma_{DR}(\varepsilon)$ [106] can be used to estimate $\beta_{DR}$ for discharges in which the electron energy distribution function differs substantially from Maxwellian. Estimates of $\beta_{DR}$ in nitrogen and air for $E/n = (1\text{–}10)\cdot 10^{-16}$ V·cm$^2$ yield $\beta_{DR}^{N_2^+} =$

(5–7)$\cdot 10^{-8}$ cm$^3$/sec, $\beta_{DR}{}^{O2^+}$ = (3–4)$\cdot 10^{-8}$ cm$^3$/sec, and $\beta_{DR}{}^{N4^+}$ = (2–4)$\cdot 10^{-7}$ cm$^3$/sec (for $\sigma_{DR}{}^{N4^+} \propto \varepsilon^{-1}$ [102]).

At high molecular densities, three-body recombination

$$AB^+ + e + M \rightarrow AB + M \searrow A + B + M \tag{23}$$

may compete with dissociative recombination. Typical rate constants for three-body recombination of atomic ions in a medium consisting of nitrogen molecules at $T$ = 300–500 K are ~$10^{-27}$ cm$^6$/sec [122]. The rate of three-body recombination of molecular ions may be higher owing to the possibility of dissociation of highly excited molecules $AB^*$ formed during capture of electrons by $AB^+$ ions [123]. Based on the data of [117], we can place an upper limit on the rate constants for three-body recombination (23) for $N_4^+$ ions in a medium consisting of $N_2$ molecules: $\beta_{3B} < 3\cdot 10^{-26}$ cm$^6$/sec at $T$ = 300 K.

For high densities and low temperatures of the electrons (e.g., in a beam plasma [124]), three-body recombination involving an electron as the third particle, with a rate constant that is independent of the type of ion when $T_e \leq 2000$ K and is equal to $7\cdot 10^{-20}(300/T_e)^{9/2}$ cm$^6$/sec ($T_e$ in K) [42], may become important.

The atomic ions $N^+$ and $O^+$ recombine with electrons primarily through three-body collisions involving electrons or molecules (in rarefied plasmas, radiative recombination with a rate constant of $10^{-12}$ cm$^3$/sec [125] must be taken into account). The rate constant for three-body recombination $N^+ + 2e \rightarrow N + e$, calculated in a modified diffusion approximation [42], can be approximated by $\beta_{3B} \simeq 2\cdot 10^{-28}\, T^{-9/3}$ cm$^3$/sec ($T_e$ in eV) for $n_e \geq 10^{16}$ cm$^{-3}$ and $0.6 < T_e < 2.5$ eV. These rate constants have been calculated numerically for different $n_e$ and $T_e$ [54]. The recombination of atomic ions in a nonisothermal plasma (with $T_e \neq T$) when collisions occur with both electrons and molecules has been studied [122, 42].

## 8. PRODUCTION AND LOSS OF NEGATIVE IONS

Negative ions are formed in nitrogen–oxygen mixtures through dissociative and three-body attachment of electrons to oxygen molecules,

$$O_2 + e \rightarrow O + O^- \tag{24}$$

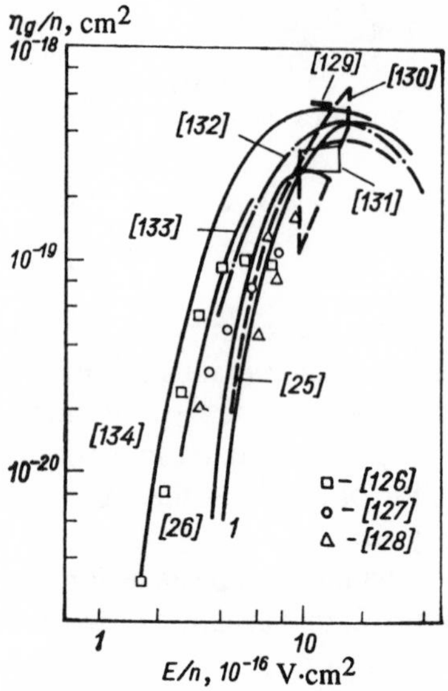


**Fig. 6.** Dissociative attachment coefficient in air: curve 1 is the result of Phelps (given in [135]).

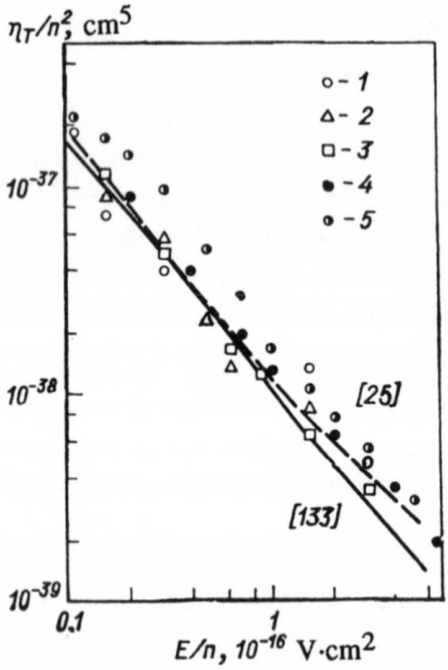


**Fig. 7.** The coefficient of three-body attachment in air: experiment [128] for $n = 1.65{\cdot}10^{18}$ cm$^{-3}$ (1); $3.3{\cdot}10^{18}$ (2); $6.6{\cdot}10^{18}$ (3); theory: 4) results of Phelps for dry air (given in [135]); 5) same for air containing 1.5% $H_2O$.

and

$$O_2 + e + M \rightarrow O_2^- + M. \quad (25)$$

Figure 6 shows experimental [126–131] and theoretical values of the dissociative attachment coefficient $\eta_D/n = K_{DA}/v_g$ [$K_{DA}$ is the rate constant for reaction (24)] in air. The discrepancy among the results of different authors is extremely large, especially for low values of $E/n$. The difference among the calculated values of $\eta_D/n$ is caused by a difference between the electron energy distributions owing to a different choice for normalizing the cross sections for vibrational excitation of nitrogen molecules. Based on the available data about the electron energy distribution function in pure nitrogen (see above), the results of [25, 26] are to be preferred.

The dependence of the rate constant for dissociative attachment (for fixed $E/n$) on the vibrational temperature of the nitrogen molecules can be taken into account with the aid of the factor $f(\tilde{\varepsilon})/f^{(0)}(\tilde{\varepsilon})$. In addition, since the cross sections for reaction (24) with different vibrational levels of the $O_2$ molecules are different [136, 137], $K_{DA}$ also depends on the vibrational temperature of the oxygen molecules.

In the three-body attachment of $O_2$, reaction (25), oxygen molecules $O_2$ are an order of magnitude more effective as a third particle $M$ than $N_2$ molecules, while the efficiency of $H_2O$ molecules is roughly another order of magnitude higher [138]. Figure 7 shows experimental [138] and theoretical values of the three-body attachment coefficient $\eta_T/n^2 = K_{3BA}/v_g$ in air at $T = 300$ K. The approximation

$$\eta_T/n^2 = 1.6 \cdot 10^{-37} (E/n)^{-1.1} \text{ cm}^5$$

($E/n$ in $10^{-17}$ V·cm$^2$) has been proposed [135] for this coefficient. As the gas temperature increases, $K_{DA}$ varies slowly [138].

The primary ions $O^-$ and $O_2^-$ formed in nitrogen–oxygen mixtures generate a large set of negative ions ($O_3^-$, $O_4^-$, $NO^-$, $NO_2^-$, $NO_3^-$) through ion–molecule association and charge exchange reactions [136, 139]. The presence of small admixtures of $CO_2$ and $H_2O$ greatly extends this set, leading, in particular, to the appearance of cluster ions [140]. Since the channels for loss of different sorts of negative ions are different, in order to determine the effective rate at which electrons are removed from negative ions one must know the ion composition of the plasma. (An exception is those cases where electron detachment processes are unimportant, as when negative ions are lost primarily through ion recombination reactions whose rates depend only weakly on the type of ion, or when negative ions recombine at the walls or are carried out of the discharge zone by a flow.) The negative ion composition of nitrogen–oxygen plasmas depends substantially on the whole range of plasma parameters, such as the temperature, the relative densities of various excited particles [in particular metastable $O_2(a^1\Delta)$ molecules] which efficiently destroy negative ions, the degrees of dissociation of $N_2$ and $O_2$ molecules, and so on. In particular, the formation of nitrogen–containing negative ions requires the presence of such molecules as NO, $N_2O$, $NO_2$, and $NO_3$ in the mixture. In pure oxygen, the ratio of the densities of $O_3$ and $O_4$ depends on the relative density of ozone molecules [141]. Thus, determining the ion composition of a nitrogen–oxygen plasma in general will require solution of more than ten equations for the densities of different charged and neutral species involving more than a hundred reactions [44, 141–144]. Extensive information on the rate constants for these reactions is given in [145–147].

The problem of calculating the ion composition is greatly simplified when the densities of active particles (oxygen atoms, ozone and nitrogen oxide molecules, excited molecules) are low and it is sufficient to include the reactions of negative oxygen ions with unexcited $N_2$ and $O_2$ molecules. This situation occurs in low-current discharges (in particular, corona discharges [148]), in the initial stage of electrical breakdown [79], etc. In the following discussion, we restrict ourselves to an analysis of data on the reactions of negative ions with unexcited molecules.

The rate constants for endothermic reactions involving negative ions depend strongly on $E/n$, which determines the average energy of the ions. In order to calculate the rate constants when the cross section is known or to solve the opposite problem of finding the cross section when the dependence of the rate constant on $E/n$ is known, we must know the ion energy distribution. The simplest approach is the following (see [149], for example): the ion energy distribution is assumed to be Maxwellian with a temperature $T_i$ determined by the Wannier formula [150],

$$T_i = T + \frac{1}{3}(m_i + M)\, V_D^2(1 + \delta), \tag{26}$$

where $m_i$ and $M$ are the masses of the ion and buffer gas molecules, $V_D$ is the drift velocity of an ion for given $E/n$, and the small correction $\delta$ is given by [151]

$$\delta = \frac{m_i M\,(5 - 2A^*)}{5\left(m_i^2 + M^2\right) + 4m_i M A^*}\;\frac{d\ln[V_D/(E/n)]}{d\ln(E/n)},$$

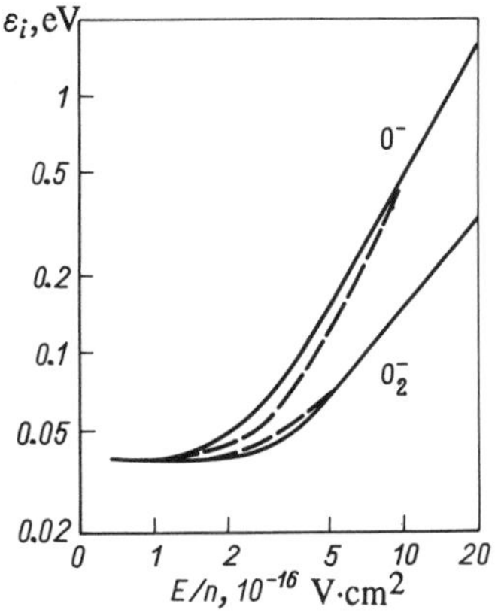


**Fig. 8.** The average energy of $O^-$ and $O_2^-$ ions in oxygen: the smooth curves are from a Monte Carlo calculation [155] and the dashed curves are from a calculation using Eq. (26).

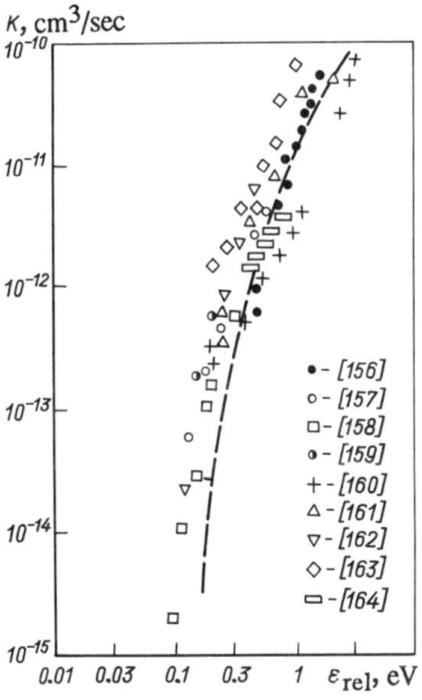


**Fig. 9.** The rate constant for charge exchange (31). The dashed curve is a calculation using Eq. (30).

where $A^* = 1.1–1.2$. In this case, the rate constant for the reaction of an ion with a molecule of mass $M_p$ is given by

$$K_p = \left(\frac{8}{\pi\mu T_{\rm eff}^3}\right)^{1/2} \int \varepsilon\sigma_p(\varepsilon)\exp(-\varepsilon/T_{\rm eff})\, d\varepsilon, \tag{27}$$

with

$$T_{\rm eff} = (m_i T + M_p T_i)/(m_i + M_p), \qquad \mu = \frac{m_i M_p}{m_i + M_p}, \tag{28}$$

where $\sigma_p(\varepsilon)$ is the cross section for the ion–molecule reaction which depends on the energy $\varepsilon$ of relative motion. Taking the widely used form

$$\sigma(\varepsilon) = \begin{cases} 0, & \varepsilon < \varepsilon_p, \\ \sigma_m(1 - \varepsilon_p/\varepsilon), & \varepsilon > \varepsilon_p \end{cases} \tag{29}$$

for the cross section, where $\varepsilon_p$ is the energy threshold for the reaction, we obtain

$$K_p = \sigma_m \left(\frac{8T_{\rm eff}}{\pi\mu}\right)^{1/2} e^{-\varepsilon_p/T_{\rm eff}}. \tag{30}$$

Other approximate methods of calculating $K_p$ have been proposed in which one or another assumption about the form of the ion energy distribution is used [152, 153] or $K_p$ is calculated with the aid of a series [154]. Recently, the ion energy distribution has been calculated by the Monte Carlo method [155, 156]. Figure 8 shows a calculated [155] dependence of the average energy $\varepsilon_i = 3T_i/2$ of the $O^-$ and $O_2^-$ ions in oxygen at $T = 300$ K. Also shown there is the value of $\varepsilon_i$ calculated using Eq. (26) with $V_D$ taken from [157] (the correction $\delta$ is less than 5%). Clearly, the Wannier formula fits $\varepsilon_i$ well, especially for large $E/n$.

Lin et al. [156] have measured the dependence of the rate constant for charge exchange

$$O^- + O_2 \rightarrow O_2^- + O - 1.02 \text{ eV} \tag{31}$$

on $E/n$ in helium and calculated the ion energy distribution for the experimental conditions using a Monte Carlo method [156]. By comparing the measured and calculated rate constants for charge exchange, the energy dependence of the cross section for the process could be made more precise. Figure 9 shows the experimental values of the rate constant for reaction (31) in helium [156] and in oxygen [157–164] as functions of the average energy of relative motion $\varepsilon_{\rm rel} = 3T_{\rm eff}/2$. The latter was calculated in accordance with Eqs. (26) and (28). This figure also shows the rate constant for charge exchange according to Eq. (30) (the cross section obtained in [156] is approximated by Eq. (29) with $\sigma_m = 2 \cdot 10^{-16}$ cm$^2$).

The dependence of the rate constant for detachment,

$$O^- + M \rightarrow O + M + e - 1.46 \text{ eV}, \tag{32}$$

on $E/n$ in oxygen (with $M = O_2$) is given in Fig. 10, which shows experimental values from [131, 164, 165], the calculations of [155], and calculations using Eq. (30) with $\sigma_m = 1.5 \cdot 10^{-16}$ cm$^2$ (an approximation for the cross section from [166]). Figure 11 shows the rate constant for the same reaction in air [130, 131] (determined from the experimental magnitude of the detachment coefficient $\eta_0/n = K_0/V_D$ under the assumption that the drift velocity of the $O^-$ ions in air is the same as in oxygen) and the calculated rate of reaction (32) when the efficiency of $N_2$ molecules in these reactions is assumed to be the same as that of $O_2$. Shown in the same figure is the rate constant for associative detachment

$$O^- + N_2 \rightarrow N_2O + e + 0.21 \text{ eV} \tag{33}$$

in nitrogen [167, 168]. We note that a calculation of the rate constant $K_{a0}$ for this reaction using the experimental value of the cross section for the reverse process (dissociative attachment) [169] and the equilibrium constant from [120] yields an upper bound of $K_{a0} < 10^{-13}$ cm$^3$/sec. The difference between this estimate and the measured value of $K_{a0}$ can be explained by the fact that dissociative attachment proceeds much more efficiently when the $N_2O$ molecules are vibrationally excited. Then excited $N_2O$ molecules are predominantly formed in the direct reaction (33), as well, and this may explain the presence of an effective threshold for reaction (33) of ~0.1–0.2 eV.

At high pressures the conversion process

$$O^- + O_2 + M \rightarrow O_3^- + M \tag{34}$$

competes with the charge exchange and detachment reactions discussed above. The rate constant for this process has been measured only in pure $O_2$ [170] and varies from $10^{-30}$ cm$^6$/sec for low $E/n$ to $10^{-31}$ cm$^6$/sec for $E/n \approx (1\text{–}2) \cdot 10^{-15}$ V·cm$^2$.

Detachment of electrons from $O_2^-$ ions in a hot gas or in a strong electric field proceeds efficiently via the reaction

$$O_2^- + M \rightarrow O_2 + e + M, \tag{35}$$

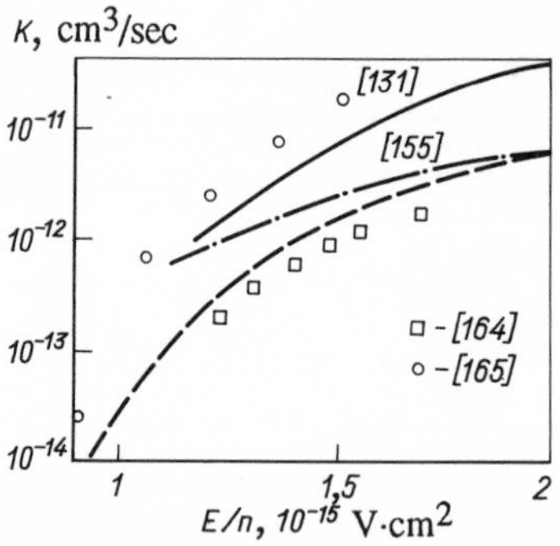


**Fig. 10.** The rate constant for detachment (32) in oxygen. The dashed curve is a calculation using Eq. (30).

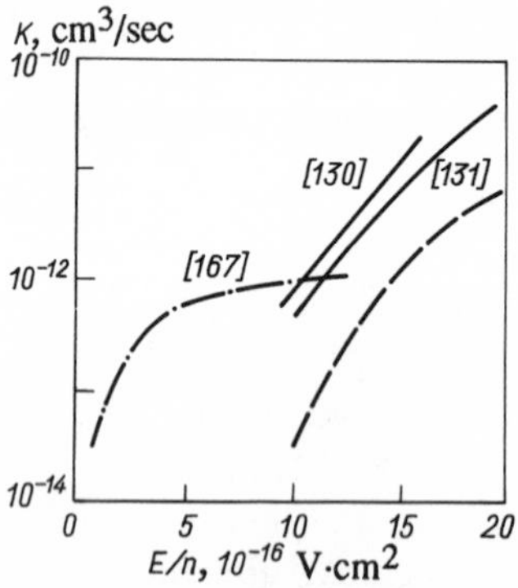


**Fig. 11.** The rate constant for detachment (32) in air and for associative detachment (33) in nitrogen (dot–dashed curve). The dashed curve is a calculation using Eq. (30).

which is the reverse of reaction (25). The rate constant for this reaction can be calculated using a modified Bloch–Bradbury scheme which includes the successive stages of excitation of the $O_2^-$ ion into an autoionizing state followed by its decay [138]. It depends on two parameters: the vibrational temperature of the $O_2^-$ ion (which differs little from $T$ owing to resonance charge exchange in mixtures containing a significant fraction of $O_2$, such as air) and the effective temperature $T_{\text{eff}}$ defined by Eq. (28). Assuming that when the lowest autoionizing level $O_2^-$ ($v = 4$) is quenched, the lower-lying levels $O_2^-$ ($v = 0$–$3$) are populated with equal probabilities, we obtain the following expression for the rate constant of reaction (35) in oxygen:

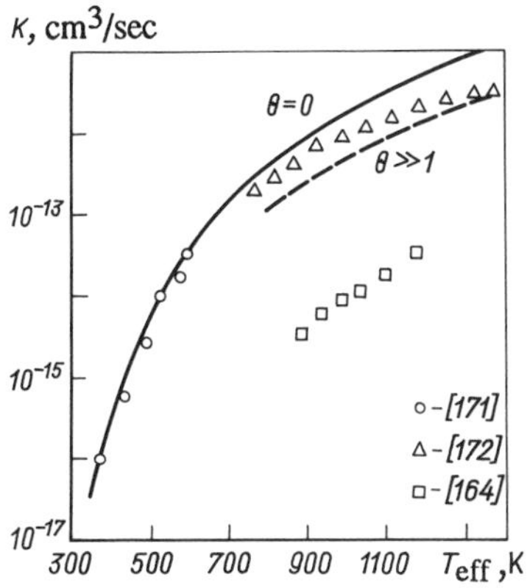


**Fig. 12.** The rate constant for reaction (35) in oxygen.

$$K_0^{(O_2^-)} = 2\cdot 10^{-10} e^{-4\omega_-/T_{\text{eff}}} \frac{1 - e^{-4\theta}}{1 - e^{-\theta}}, \tag{36}$$

with

$$\theta \equiv \omega_- (1/T - 1/T_{\text{eff}}),$$

where $\omega_- = 0.13$ eV is the vibrational quantum of the $O_2^-$ ion (the preexponential numerical factor is determined from measured data [171] with $T_{\text{eff}} = T$). The dependence of the rate constant for reaction (35) on $T_{\text{eff}}$ in the absence of an electric field [171], as well as in an electric field in a cold gas [164, 172], is shown in Fig. 12. Also shown there are the results of a calculation using Eq. (36) when $T_{\text{eff}} = T$ ($\theta = 0$) and when there is a substantial difference between $T_{\text{eff}}$ and $T$ ($\theta \gg 1$). Since the efficiency of $N_2$ molecules in reactions (25) and (35) is roughly one and a half orders of magnitude lower than that of $O_2$ molecules, Eq. (36) can also be used for calculating the rate coefficients of reaction (35) in air (when the relative fraction of $O_2$ molecules is taken into account).

In a cold gas at high pressures, the conversion reaction

$$O_2^- + O_2 + M \rightarrow O_4^- + M \tag{37}$$

may be important. When the gas is heated to temperatures of 400–500 K, $O_4^-$ ions break up rapidly in the inverse of reaction (37) [173]. The $O_3^-$ ion is more stable. At high temperatures or in strong electric fields it can break up in the inverse reaction to reaction (34) or through charge exchange with formation of $O_2^-$. There are no reliable data on the rate constants for these processes [170].

The loss of negative ions through pairwise and three-body ion–ion recombination has been examined in detail by Smirnov [136].

## REFERENCES


1. J. Dutton, *J. Phys. Chem. Data* **4**, 577 (1975).
2. J. W. Gallagher et al., *J. Phys. Chem. Data* **12**, 109 (1983).
3. D. Rapp and P. Englander-Golden, *J. Chem. Phys.* **43**, 1464 (1965).
4. S. C. Haydon and O. M. Williams, *J. Phys. B* **9**, 523 (1976).
5. T. N. Daniel and F. M. Harris, *J. Phys. B* **3**, 363 (1970).
6. L. A. Newman and T. A. DeTemple, *J. Appl. Phys.* **47**, 1912 (1976).
7. A. Kh. Mnatsakanyan and G. V. Naidis, *Fiz. Plazmy* **2**, 152 (1976).
8. N. L. Aleksandrov, A. M. Konchakov, and É. E. Son, *Fiz. Plazmy* **4**, 169–175; 1182–1186 (1978).
9. I. V. Kochetov et al., *Plasma Chemical Processes* [in Russian], Nauka, Moscow (1979), pp. 4–71.
10. S. V. Nikonov, A. P. Osipov, and A. T. Rakhimov, *Kvantovaya Élektron.* **6**, 1258 (1979).
11. K. Tachibana and A. V. Phelps, *J. Chem. Phys.* **71**, 3544 (1979).
12. H. Brunet, P. Vincent, and J. Rocca Serra, *J. Appl. Phys.* **54**, 4951 (1983).
13. A. V. Bondarenko, F. I. Vysikailo, and M. M. Smakotin, *Teplofiz. Vys. Temp.* **22**, 602 (1984).
14. M. C. Siddagangappa, C. S. Laksminarasimha, and M. S. Naidu, Proc. 16th International Conf. on Phenomena in Ionized Gases, Dusseldorf (1983), Vol. 4, pp. 566–567.
15. K. Nishijima, T. Kamikawaji, and M. Hara, *Technol. Rep. Kyushu Univ.* **52**, 609–618 (1979).
16. Yu. A. Ivanov and L. S. Polak, *Plasma Chemistry*, Vol. 2 [in Russian], Atomizdat, Moscow (1975), pp. 161–198.
17. D. I. Slovetskii, *The Mechanism of Chemical Reactions in Nonequilibrium Plasmas* [in Russian], Nauka, Moscow (1980).
18. A. Kh. Mnatsakanyan, *Opt. Spektrosk.* **30**, 1015 (1971).
19. É. E. Son, *Teplofiz. Vys. Temp.* **16**, 1162 (1978).
20. A. P. Osipov and A. T. Rakhimov, *Fiz. Plazmy* **3**, 644 (1977).
21. M. Capitelli, M. Dilonardo, and C. Gorse, *Chem. Phys.* **56**, 29 (1981).
22. M. Capitelli, M. Dilonardo, and C. Gorse, *Beitr. Plasmaphysik* **20**, 83 (1980).
23. N. L. Aleksandrov and É. E. Son, *Plasma Chemistry*, Vol. 7 [in Russian], Atomizdat, Moscow (1975), pp. 35–74.
24. A. V. Phelps and L. C. Pitchford, *Phys. Rev. A* **31**, 2932 (1985).
25. G. Fournier, J. Bonnet, and D. Pigache, *J. Phys. Lett.* **41**, L173–L174 (1980).
26. N. L. Aleksandrov et al., *Teplofiz. Vys. Temp.* **19**, 22–29; 485–490 (1981).
27. A. V. Eletskii, L. A. Palkina, and B. M. Smirnov, *Transport Phenomena in Weakly Ionized Plasmas* [in Russian], Atomizdat, Moscow (1975).
28. K. P. Huber and G. Herzberg, *Constants of Diatomic Molecules* [Russian translation], Mir, Moscow (1984).

29. B. F. Gordiets, A. I. Osipov, and L. A. Shelepin, *Kinetic Processes in Gases and Molecular Lasers*, Nauka, Moscow (1980).
30. A. A. Likal'ter and G. V. Naidis, *Plasma Chemistry* , Vol. 8 [in Russian], Énergoizdat, Moscow (1981), pp. 156–189.
31. M. Cacciatore, M. Capitelli, and C. Gorse, *Chem. Phys.* **66**, 141 (1982).
32. W. C. Taylor, W. E. Scharfman, and T. Morita, *Adv. Microwaves* **7**, 59–130 (1971).
33. A. V. Gurevich, *Usp. Fiz. Nauk* **132**, 685 (1980).
34. Yu. A. Ivanov, Yu. A. Lebedev, and L. S. Polak, *Fiz. Plazmy* **3**, 146 (1977).
35. N. A. Dyatko, I. V. Kochetov, and A. P. Napartovich, *Fiz. Plazmy* **11**, 739 (1985).
36. K. Masek and K. Rohlena, *Czech. J. Phys.* **B34**, 1227 (1984).
37. N. Kroll and K. M. Watson, *Phys. Rev. A* **5**, 1883 (1972).
38. W. C. Taylor, J. B. Chown, and T. Morita, *J. Appl. Phys.* **39**, 191 (1968).
39. G. C. Light and E. C. Taylor, *J. Appl. Phys.* **39**, 1591 (1968).
40. J. T. Mayhan and R. V. DeVore, *J. Appl. Phys.* **39**, 5746 (1968).
41. G. C. Light, *J. Appl. Phys.* **40**, 1715 (1969).
42. L. M. Biberman, V. S. Vorob'ev, and I. T. Yakubov, *Kinetics of Nonequilibrium Low-Temperature Plasmas*, Consultants Bureau, New York (1987).
43. E. C. Zipf and R. W. McLaughlin, *Planet. Space Sci.* **26**, 449 (1978).
44. K. V. Baiadze et al., *Fiz. Plazmy* **11**, 352 (1985).
45. Yu. S. Akishev et al., *Fiz. Plazmy* **11**, 999 (1985).
46. V. A. Pivovar and T. D. Sidorova, *Zh. Tekh. Fiz.* **55**, 519 (1985).
47. P. B. Armendrout et al., *J. Chem. Phys.* **75**, 2786 (1981).
48. N. L. Aleksandrov, I. V. Kochetov, and A. P. Napartovich, *Teplofiz. Vys. Temp.* **23**, 849 (1985).
49. G. N. Hays and H. J. Oskam, *J. Chem. Phys.* **59**, 1507–1513; 6089–6097 (1973).
50. J. W. Dreyer and D. Perner, *J. Chem. Phys.* **58**, 1195 (1973).
51. M. F. Golde and B. A. Thrush, *Rep. Prog. Phys.* **36**, 1285 (1973).
52. J. W. Dreyer and D. Perner, *Chem. Phys. Lett.* **16**, 169 (1972).
53. J. M. Thomas and F. Kaufman, *J. Chem. Phys.* **83**, 2900 (1985).
54. A. V. Potapov, L. E. Tsvetkova, V. I. Antropov, and G. I. Volkova, *Opt. Spektrosk.* **43**, 112 (1977).
55. S. C. Lin and J. D. Teare, *Phys. Fluids* **6**, 355 (1963).
56. L. M. Biberman, A. Kh. Mnatsakanyan, and I. T. Yakubov, *Usp. Fiz. Nauk* **102**, 431 (1970).
57. V. Ya. Aleksandrov, D. B. Gurevich, I. V. Podmoshenskii, and S. F. Khlopina, *Zh. Tekh. Fiz.* **46**, 519 (1976).
58. L. S. Polak, P. A. Sergeev, and D. I. Slovetskii, *Teplofiz. Vys. Temp.* **15**, 15 (1977).
59. M. G. Berdichevskii and V. V. Marusin, *Izv. Sib. Otd. Akad Nauk SSR, Ser. Tekh. Nauk*, No. 8, issue 2, 72–80 (1984).
60. Yu. B. Golubovskii and V. M. Telezhko, *Teplofiz. Vys. Temp.* **22**, 428 (1984).
61. H. Brunet and J. Rocca-Serra, *J. Appl. Phys.* **57**, 1574 (1985).
62. L. S. Polak, A. A. Ovsyannikov, D. I. Slovetskii, and F. B. Vurzel', *Theoretical and Applied Plasma Chemistry* [in Russian], Nauka, Moscow (1975).
63. G. Ringer and W. R. Gentry, *J. Chem. Phys.* **71**, 1902 (1979).
64. D. Kley, G. W. Lawrence, and E. J. Stone, *J. Chem. Phys.* **66**, 4157 (1977).
65. G. D. Smekhov and M. S. Yalovik, *Nauch. Tr. Inst. Mekh. Mosk. Gos. Univ.*, No. 3, 5–67 (1970).

66. J. Tulip and H. Seguin, *Phys. Lett.* **58A**, 173 (1976).
67. H. H. Brömer and J. Hesse, Z. *Phys.* **219**, 269 (1969).
68. R. E. Lund and H. J. Oskam, Z. *Phys.* **219**, 131 (1969).
69. A. G. Basiev, F. I. Vysikailo, V. A. Gurashvili, and E. Yu. Shchekotov, *Fiz. Plazmy* **9**, 1076 (1983).
70. S. M. Kurkin and V. M. Shashkov, *Teplofiz. Vys. Temp.* **22**, 999 (1984).
71. Yu. B. Golubovskii and V. M. Telezhko, *Zh. Tekh. Fiz.* **54**, 1262 (1984).
72. O. V. Achasov, S. A. Zhdanok, D. S. Ragozin, et al., *Zh. Éksp. Teor. Fiz.* **81**, 550 (1981).
73. L. S. Polak, P. A. Sergeev, and D. I. Solovetsky, Proc. 13th. Int. Conf. on Phenomena in Ionized Gases, Berlin (1977), pp. 55, 56.
74. A. A. Likal'ter and G. V. Naidis, *Teplofiz. Vys. Temp.* **20**, 180 (1982).
75. G. G. Antonov, F. G. Baksht, V. S. Borodin, et al., *Zh. Tekh. Fiz.* **55**, 1053 (1985).
76. A. Kh. Mnatsakanyan and G. V. Naidis, *Teplofiz. Vys. Temp.* **23**, 640 (1985).
77. S. A. Zhdanok and A. V. Krauklis, *Heat- and Mass-Transfer* [in Russian], ITMO AN BSSR, Minsk (1985), pp. 94–98.
78. J. Koppitz and K. Stühm, *J. Appl. Phys.* **12**, 23 (1977).
79. M. B. Zheleznyak, A. Kh. Mnatsakanyan, and S. V. Sizykh, Institute of High Temperatures, Preprint No. 80–146, Moscow (1984).
80. E. D. Lozanskii and O. B. Firsov, *The Theory of Sparks* [in Russian], Atomizdat, Moscow (1975).
81. Yu. Ya. Brodskii, S. V. Golubev, V. G. Zorin, et al., *Zh. Éksp. Teor. Fiz.* **84**, 1695 (1983).
82. A. Kh. Mnatsakanyan, G. V. Naidis, and S. V. Sizikh, *Kratk. Soobshch. Fiz.*, No. 4, pp. 15–18 (1985).
83. G. V. Naidis, *Zh. Tekh. Fiz.* **52**, 868 (1982).
84. M. B. Zheleznyak and A. Kh. Mnatsakanyan, *Zh. Tekh. Fiz.* **47**, 2497 (1977).
85. A. G. Kuz'min, *Zh. Tekh. Fiz.* **50**, 503 (1980); **51**, 215 (1981).
86. Yu. I. Bychkov, D. Yu. Zaroslov, N. V. Karlov, et al., *Kvantovaya Élektron.* **9**, 1718 (1982).
87. N. K. Berezhetskaya, E. F. Bol'shakov, A. A. Dorofeyuk, et al., FIAN Preprint No. 224, Moscow (1983).
88. A. Przybylski, Proc. 4th. Int. Conf. on Phenomena in Ionized Gases, Uppsala (1959), pp. 215–218.
89. A. Przybylski, Z. *Naturforsch.* **16a**, 1232 (1961).
90. A. Przybylski, Z. *Phys.* **168**, 304 (1962).
91. G. W. Penney and G. T. Hummert, *J. Appl. Phys.* **41**, 572 (1970).
92. T. H. Teich, Z. *Phys.* **199**, 378–394, 395–410 (1967).
93. M. B. Zheleznyak, A. Kh. Mnatsakanyan, and S. V. Sizykh, *Teplofiz. Vys. Temp.* **20**, 423 (1982).
94. H. J. Seguin, J. Tulip, and D. C. McKen, *IEEE J. Quantum Electron.* **QE–10**, 311 (1974).
95. H. D. Morgan and J. E. Mentall, *J. Chem. Phys.* **78**, 1747 (1983).
96. R. V. Babcock, I. Liberman, and W. D. Partlow, *IEEE J. Quantum Electron.* **QE–12**, 29 (1976).
97. V. L. Carter, *J. Chem. Phys.* **56**, 4195 (1972).
98. N. A. Bogatov, S. V. Golubev, and V. G. Zorin, *Pis'ma Zh. Tekh. Fiz.* **9**, 888 (1983).
99. V. V. Apollonov, N. Akhunov, S. I. Derzhavin, et al., *Pis'ma Zh. Tekh. Fiz.* **6**, 1047 (1980).

100. R. T. Reits, *J. Appl. Phys.* **48**, 3697 (1977).
101. M. A. Biondi, *Plasmas in Lasers* [Russian translation], Énergoatomizdat, Moscow (1982), pp. 145–187.
102. A. V. Eletskii and B. M. Smirnov, *Usp. Fiz. Nauk* **136**, 25 (1982).
103. J. B. Mitchell and J. W. McGowan, in: *Physics of Ion–Ion and Electron–Ion Collisions*, Plenum, New York (1983), pp. 279–324.
104. W. H. Kasner, *Phys. Rev.* **164**, 194 (1967).
105. F. J. Mehr and M. A. Biondi, *Phys. Rev.* **181**, 264 (1969).
106. P. M. Mul and J. W. McGowan, *J. Phys. B* **12**, 1591 (1979).
107. J. E. Frederick and D. W. Rusch, *J. Geophys. Res.* **82**, 3509 (1977).
108. E. C. Zipf, P. J. Espy, and C. F. Boyle, *J. Geophys. Res.* **85**, 687 (1980).
109. W. H. Kasner and M. A. Biondi, *Phys. Rev.* **174**, 139 (1968).
110. E. Alge, N. G. Adams, and D. Smith, *J. Phys. B* **16**, 1433 (1983).
111. E. C. Zipf, *Bull. Am. Phys. Soc.* **15**, 418 (1970).
112. P. B. Hays, D. W. Rusch, R. G. Roble, and J. C. Walker, *Rev. Geophys. Space Phys.* **16**, 225 (1978).
113. E. C. Zipf, *J. Geophys. Res.* **85**, 4232 (1980).
114. C. S. Weller and M. A. Biondi, *Phys. Rev.* **172**, 198 (1968).
115. C. M. Huang, M. A. Biondi, and R. Johnsen, *Phys. Rev. A* **11**, 901 (1975).
116. M. R. Torr and D. G. Torr, *Planet. Space Sci.* **27**, 1233 (1979).
117. M. C. Sauer and W. A. Mulac, *J. Chem. Phys.* **56**, 4995 (1972).
118. M. Whitaker, M. A. Biondi, and R. Johnsen, *Phys. Rev. A* **24**, 743 (1981).
119. M. Fitaire, A. M. Pointu, and D. Stathopoulos, *J. Chem. Phys.* **81**, 1753 (1984).
120. L. V. Gurvich et al., *Thermodynamic Properties of Individual Substances* , Vol. 1 [in Russian], Nauka, Moscow (1978).
121. E. C. Zipf, *Geophys. Res. Lett.* **7**, 645 (1980).
122. A. Kh. Mnatsakanyan and G. V. Naidis, Institute of High Temperatures, Preprint No. 1–43, Moscow (1979).
123. D. R. Bates, *J. Phys. B* **14**, 3525 (1981).
124. V. L. Bychkov and A. V. Eletskii, *Plasma Chemistry*, Vol. 12 [in Russian], Énergoatomizdat, Moscow (1985), pp. 119–152.
125. D. Bates and A. Dalgarno, *Atomic and Molecular Processes* , Academic Press, New York (1962).
126. E. Kuffl, *Proc. Phys. Soc.* **74**, 237 (1959).
127. P. A. Chatterton and J. D. Craggs, *Proc. Phys. Soc.* **85**, 355 (1965).
128. H. Hessenauer, *Z. Phys.* **204**, 142 (1967).
129. K. H. Wagner, *Z. Phys.* **241**, 258 (1971).
130. H. Ryzko and E Aström, *J. Appl. Phys.* **38**, 328 (1967).
131. L. Frommhold, *Fortschr. Phys.* **12**, 597 (1964).
132. M. Ducos, P. Segur, and M. Yousfi, Proc. 7th. Int. Conf. on Gas Discharges and Their Applications, London (1982), pp. 335–336.
133. T. Taniguchi, K. Kawamura, S. Sakamoto, and H. Tagashira, *J. Phys. D* **15**, 1187 (1982).
134. K. Masek, *Czech J. Phys.* **B34**, 655 (1984).
135. R. S. Sigmond, Proc. 16th. Int. Conf. on Phenomena in Ionized Gases, Dusseldorf (1983), Invited Papers, p. 174–191.
136. B. M. Smirnov, *Negative Ions* [in Russian], Atomizdat, Moscow (1978).
137. N. L. Aleksandrov and A. M. Konchakov, *Teplofiz. Vys. Temp.* **22**, 254 (1984).
138. N. L. Aleksandrov, *Plasma Chemistry*, Vol. 8 [in Russian], Énergoizdat, pp. 90–122.

139. H. Massey, *Negative Ions* [Russian translation], Mir, Moscow (1979).
140. B. M. Smirnov, *Complex Ions* [in Russian], Nauka, Moscow (1983).
141. A. Kh. Mnatsakanyan, G. V. Naidis, and Yu. M. Solozobov, Abstracts of Talks at the 2nd All Union Seminar "Elementary Processes in Plasmas of Electronegative Gases" [in Russian], Erevan (1984), pp. 7–9.
142. F. Bastien, R. Haug, and M. Lecuiller, *J. Chem. Phys.* **72**, 105 (1975).
143. F. E. Niles, *J. Chem. Phys.* **52**, 408 (1970).
144. K. Smith and R. Thompson, *Computer Modelling of Gas Lasers*, Plenum Press, New York (1978).
145. M. McEwan and L. Phillips, *The Chemistry of the Atmosphere*, Halsted Press, New York (1975).
146. D. L. Albritton, *At. Data Nucl. Data Tables* **22**, 1–101 (1978).
147. D. L. Baulch, R. A. Cox, P. J. Crutzen, et al., *J. Phys. Chem. Ref. Data* **11**, 327 (1982).
148. A. Kh. Mnatsakanyan, G. V. Naidis, and Yu. M. Solozobov, *Teplofiz. Vys. Temp.* **24**, 1060 (1986).
149. D. L. Albritton, I. Dotan, W. Lindinger, et al., *J. Chem. Phys.* **66**, 410 (1977).
150. G. H. Wannier, *Bell. Syst. Tech. J.* **32**, 170 (1953).
151. H. W. Ellis, R. Y. Pay, E. W. McDaniel, et al., *At. Data Nucl. Data Tables* **17**, 178–214 (1976).
152. S. B. Woo and S. F. Wong, *J. Chem. Phys.* **55**, 3531 (1971).
153. F. Rebentrost, *Chem. Phys. Lett.* **17**, 486–488, 489–491 (1972); **21**, 368–371 (1973).
154. L. A. Viehland and E. A. Mason, *J. Chem. Phys.* **66**, 422 (1977).
155. I. Okada, Y. Sakai, H. Tagashira, and S. Sakamoto, *J. Phys. D* **11**, 1107 (1978).
156. S. L. Lin, J. N. Bardsley, I. Dotan, et al., *Int. J. Mass Spectrom. Ion Phys.* **34**, 113 (1980).
157. R. M. Snuggs, D. J. Volz, I. R. Gatland, et al., *Phys. Rev. A* **3**, 477–486, 487–494 (1971).
158. D. S. Burch and R. Geballe, *Phys. Rev.* **106**, 188 (1957).
159. J. L. Moruzzi and A. V. Phelps, *J. Chem. Phys.* **45**, 4616 (1966).
160. N. R. Varney, *Phys. Rev.* A **2**, 370 (1970).
161. L. G. McKnight, *Phys. Rev. A* **2**, 762 (1970).
162. P. R. Kinsman and J. A. Rees, *Int. J. Mass Spectrom. Ion Phys.* **5**, 71–79 (1970).
163. L. Harrison and J. L. Moruzzi, *J. Phys. D* **5**, 1239 (1972).
164. B. C. O'Neil and J. D. Craggs, *J. Phys. B* **6**, 2625 (1973).
165. D. A. Price, J. Lucas, and J. L. Moruzzi, *J. Phys. D* **5**, 1249 (1972).
166. J. Comer and G. J. Schulz, *J. Phys. B* **7**, L249 (1974).
167. S. W. Rayment and J. L. Moruzzi, *Int. J. Mass Spectrom. Ion Phys.* **26**, 321 (1978).
168. C. Doussot, F. Bastien, E. Marode, and J. L. Moruzzi, *J. Phys. D* **16**, 2451 (1982).
169. D. Rapp and D. D. Briglia, *J. Chem. Phys.* **43**, 1480 (1965).
170. D. A. Parkes, *Vacuum* **24**, 561 (1974).
171. J. L. Pack and A. P. Phelps, *J. Chem. Phys.* **44**, 1870 (1966).
172. D. W. Goodson, R. J. Corbin, and L. Frommhold, *Phys. Rev. A* **9**, 2049 (1974).
173. D. C. Conway and L. E. Nesbitt, *J. Chem. Phys.* **48**, 509 (1968).

# VIBRATIONAL RELAXATION IN ION–MOLECULE COLLISIONS

**G. V. Karachevtsev and V. L. Tal'roze**

## 1. INTRODUCTION

Molecular vibrational relaxation processes play an important role in nonequilibrium plasma kinetics. For example, they essentially determine the operating efficiency of molecular lasers [1]. Vibrational relaxation in collisions of neutral particles has been under intensive study for several decades by experimental and theoretical methods. Data obtained from these studies have been systematized in several recent reviews [2–4].

Until recently, vibrational relaxation in ion–molecule collisions has received considerably less attention. From a qualitative standpoint vibrational relaxation processes in ion–molecule collisions are in many ways analogous to such processes in neutral particle collisions. During collisions of a monatomic ion with a molecule or of a molecular ion with an atom, such as

$$Li^+ + H_2(v_1, j_1) \rightleftarrows Li^+ + H_2(v_2, j_2),$$

$$H_2^+(v_1, j_1) + He \leftrightarrows H_2^+(v_2, j_2) + He$$

the vibrational and rotational states of the molecular particle undergo changes. During collisions of the type

$$NO^+(v_1, j_1) + N_2(v_2, j_2) \rightleftarrows NO^+(v_1', j_1') + N_2(v_2', j_2')$$

between molecular particles, vibrational–vibrational exchange also occurs and may take place simultaneously with energy exchange in the other degrees of freedom. Specific to charged particles are charge exchange processes with vibrational energy exchange, such as

$$NO^+(v_1, j_1) + NO(v_2, j_2) \rightleftarrows NO(v_1', j_1') + NO^+(v_2', j_2'),$$
$$O_2^-(v_1, j_1) + O_2(v_2, j_2) \rightleftarrows O_2(v_1', j_1') + O_2^-(v_2', j_2').$$

Usually this is a very efficient channel for the vibrational relaxation of molecular ions.

Under the conditions of a nonequilibrium low-temperature plasma, relaxation processes of this type can have a substantial effect on the plasma characteristics. For example, in a number of cases the cross section for dissociative recombination of positive molecular ions with electrons depends strongly on the vibrational excitation of the ion [5, 6]. Thus, the time for recombinative decay of a plasma and electron density of this plasma in the steady state may depend strongly on the efficiency of vibrational relaxation processes among the ions. These processes must, presumably, be taken into account when creating exact models for the ionospheres of planets in order to calculate the electron density in the upper layers of the atmosphere [7, 8].

There is yet another channel through which vibrational relaxation of the ions affects the rate of recombinative decay of a plasma–ion-molecule reactions. These reactions are usually very fast in exothermic and thermally neutral channels and substantially determine the composition of the plasma ions [9–12], while different ions can have very different coefficients of recombination with electrons. The ion–molecule reactions themselves, however, can depend strongly on the vibrational excitation of the reagent ions. For example, it has been shown [13] that the rate constant for the exothermic ion–molecule reaction

$$NH_3^+ + D_2 = NH_3D^+ + D$$

increases with higher vibrational excitation of $NH_3^+$, while the rate constant for the analogous exothermic reaction

$$NH_3^+ + NH_3 = NH_4^+ + NH_2$$

decreases.

Basic information on the effect of vibrational excitation on the rates of ion–molecule reactions is obtained experimentally. For example, highly informative methods based on detection of coincidences between an ion and a threshold photoelectron have been developed [14, 15] and can be used to study ion–molecule reactions involving ions with strictly defined vibrational excitation.

Vibrational excitation of ions shows up in their absorption spectra [16, 17]. Photoabsorption on bound–free transitions of molecular ions of the inert gases is an important mechanism for radiative losses in excimer lasers [18, 19].

Vibrational relaxation of ions controls the energy of vibrational excitation of the ions extracted from ion sources [20]. Vibrational excitation has a substantial influence on the dissociation of ions in different types of collisions. Collisional dissociation of ions is widely used to determine the structure of complex ions [21]. This requires a careful distinction between the effects of structural factors and those of excitation energy on the processes in which the ions are broken up.

Translational–vibrational exchange leads to an enhanced vibrational temperature of ions drifting through a gas in an electric field [22, 23]. Inelastic collisions lower the drift velocity of ions in a gas [24]. These processes are important for the theory of such scientific and analytic apparatus as drift tubes [25] and plasma chromatographs [26].

Mention should also be made of the major contribution of research on energy exchange in ion–molecule collisions to solving the fundamental problems in atom–molecule gas phase kinetics which are common to ions and neutral particles. The great advantage of experimental studies of collisions involving ions over those involving only neutral particles lies in the simplicity of varying the kinetic energy of an ion with the aid of electric fields, in the ease of analyzing the mass of ions with the aid of different mass-spectroscopic methods, and in the high efficiency with which ions can be detected following acceleration in an electric field. This ensures a very high selectivity and sensitivity in the experiments and even makes it possible to study processes involving a single isolated ion.

Studies of energy exchange in ion–molecule collisions are conducted over a very wide range of kinetic energies, $10^{-2}$–$10^{4}$ eV. For kinetic energies greatly in excess of the vibrational energy, the method of crossed beams combined with analysis of the angular and energy distribution of the ions after collisions makes it possible to obtain detailed information on the mechanism of the collisions. This method has been used, for example, to study inelastic collisions of $Li^+ + CO_2$ and $Li^+ + N_2O$ (at $Li^+$ energies of 3–8 eV) [27], $Na^+ + CO_2$ ($Na^+$, 50–350 eV) [28], $Na^+ + CF_4$ ($Na^+$, 4–24 eV) [29], and $Na^+ + N_2$ and $Na^+ + CO$ ($Na^+$, 27–192 eV) [30].

The experiments of [27] are in good agreement with trajectory calculations based on a semiclassical model [31] in which the translational and rotational motions were treated as classical while the vibrational motion was treated as quantum mechanical. High-energy inelastic scattering processes involving charged and neutral particles are similar in many respects and have been examined in detail in a recent review by Leonas and Rodionov [32]. Here we limit ourselves to inelastic processes at the sub-electron volt energies characteristic of low-temperature plasmas.

## 2. COMPUTATIONAL METHODS

The theory of energy exchange processes in ion–molecule collisions does not now provide the accuracy attained in experiments and required for practical applications. In a number of cases, however, even approximate results can be useful. In the earliest work [33–35] a distorted wave approximation analogous to that for processes involving neutrals was used. The interaction potential between an ion and a molecule was given by the Lennard-Jones potential $U_{LJ}(r)$ plus an additional polarization potential,

$$U(r) = U_{LJ}(r) - \frac{\alpha e^2}{2r^4} = 4D\left[\left(\frac{\sigma}{r}\right)^{12} - \left(\frac{\sigma}{r}\right)^{6}\right] - \frac{\alpha e^2}{2r^4},$$

where $\sigma$ is determined from the equation $U_{LJ}(\sigma) = 0$, $D$ is the depth of the well in the potential $U_{LJ}(r)$, $\alpha$ is the polarizability of the molecule $M$, $e$ is the ionic charge, and $r$ is the distance between the centers of mass of the colliding ion and molecule. If we assume that the vibrational frequency $\omega$ of the molecular ion $ABC^+$ is equal to that of the neutral molecule ABC, then the ratio of the probabilities of vibrational deactivation of the ion ($P^+$) and molecule ($P$) in a single collision through the processes

$$ABC^{+*} + DE = ABC^+ + DE$$

and

$$ABC^* + DE = ABC + DE$$

is given by

$$P^+/P \approx \exp\left[0.254\left(\frac{\alpha e^2}{D^{1/3}\sigma^4}\right)\frac{\chi^{1/3}}{kT}\right],$$

where

$$\chi = \left[ \frac{\Gamma(19/12)\,\sqrt{2\pi\mu}\,(4D)^{1/12}\sigma\omega kT}{\Gamma(1/12)} \right]^{12/19},$$

$k$ is Boltzmann's constant, $T$ is the gas temperature, $\Gamma(x)$ is the gamma function, and $\mu$ is the reduced mass of the collision partners.

This formula shows that $P^+ > P$. In terms of the distorted wave approximation [33–35] this inequality follows from the fact that the polarization attraction causes an increase in the relative velocity of the colliding particles.

This approximation has been used for calculating the probabilities of vibrational–translational deactivation of ions in the following processes: $O_2{}^{+*} + O_2$, $N_2{}^{+*} + N_2$, $CO^{+*} + CO$, $H_2O^{+*} + H_2O$, $CH_4{}^{+*} + CH_4$, and $CO_2{}^{+*} + CO_2$. Analytic expressions for several other types of ion–molecule interaction potentials have also been given in [34, 35].

There are several shortcomings in this theory. In neutral particle collisions $M^* - M$, the vibrational relaxation of $M^*$ involves the conversion of vibrational energy into translational and rotational energy; however, in the case of ion–molecule collisions $M^{+*} - M$ there is another channel for vibrational relaxation, ion-vibrational–vibrational exchange $M{+}^* + M = M^+ + M^*$. For $CO^+$ and $N_2{}^+$ in $v = 1$ this is an endothermic process, while for $O_2{}^+$ it is exothermic. During vibrational deactivation of the ion in the process

$$O_2^{+*}(v = 1) + O_2 = O_2^+ + O_2^*\,(v = 1), \tag{1}$$

at low collision energies, a small fraction of the vibrational energy of the molecular ion goes into translational and rotational energy. Thus, the probability of process (1) will be greater than that of vibrational–translational relaxation in

$$O_2^{+*}(v = 1) + O_2 = O_2^+ + O_2,$$

calculated by Shin [34, 35]. In addition, vibrational exchange processes accompanied by charge exchange, such as

$$O_2^{+*}(v = 1) + O_2 = O_2 + O_2^+$$

and

$$O_2^{+*}(v = 1) + O_2 = O_2^*(v = 1) + O_2^+,$$

were neglected [33–35]. These processes can also lead to more rapid vibrational relaxation of the ions than the processes considered by Shin [33–35].

In the case of collisions involving polyatomic ions [35], fast ion–molecule reactions with rate constants on the order of $10^{-9}$ cm$^3$/sec, such as

$$H_2O^+ + H_2O = H_3O^+ + HO \quad \text{and} \quad CH_4^+ + CH_4 = CH_5^+ + CH_3$$

must be taken into account. Thus, in water vapor and $CH_4$, only the vibrational relaxation of $H_3O^+$ and $CH_5^+$ ions is of practical interest. The most efficient process for vibrational relaxation of the ions in these systems, at least at high levels of vibrational excitation, is probably proton exchange in which a substantial fraction of the excitation remains in the neutral molecular product,

$$H_3O^{+*} + H_2O = H_2O^* + H_3O^+.$$

The overall rate constant for proton exchange is about $10^{-9}$ cm$^3$/sec.

The theory discussed by Shin [33–35] should be applied to collisional processes of the type

$$M^{+*} + X = M^+ + X,$$

where X is an atom or molecule with which rapid exothermic or thermally neutral ion–molecular reactions cannot occur. Examples of such collisions include $O_2^{+*}$ ($v = 1$) + Ne, $H_2O^{+*}$ + Ne, and $CH_4^{+*}$ + Ne. Even in this case, however, the calculation will include an error associated with inaccuracies in the model approximations for the energy of interaction between an ion and an atom at the small distances of importance for energy exchange.

The statistical theory has been exceptionally useful in the study of bimolecular exchange reactions. It was first applied to ion–molecular reactions by Tal'roze and Firsov [36, 37] in a simplified version where the conservation of the total angular momentum was neglected and given a systematic, completed form by Nikitin [38] and Phechukas and Light [39]. In the statistical theory it is assumed that the first stage of a reaction involves the formation of an intermediate complex in which all the degrees of freedom interact strongly. Up to the moment of breakup, a complete statistical redistribution of the energy over all degrees of freedom takes place inside this complex. The second stage involves breakup with equal probability in arbitrary directions allowed by the conservation of the total energy and angular momentum, but independent of the other initial conditions under which the complex was formed.

This strong interaction is possible in an ion–molecule collisional complex because a positive ion is a strong electron acceptor and the collisional complex often corresponds to a stable ion with substantial chemical bonds owing to exchange forces. The statistical theory and the conditions under which it is applicable to ion–molecule reactions have been discussed in several reviews [40–44].

The statistical theory was first used to calculate the efficiency of vibrational–translational exchange in ion–molecule collisions by Karachevtsev [45].

The probability of formation of a molecular ion $AB^{*+}$ with vibrational energy $E$ during breakup of the complex $ABC^{+**}$,

$$AB^{+**} + C \rightarrow ABC^{+**} \rightarrow AB^{+*} + C,$$

has the form [45]

$$\gamma_v = \omega(v) \Big/ \sum_{v=0}^{[\varepsilon_I/\hbar\omega]} \omega(v), \tag{2}$$

where

$$\omega(v) = (\varepsilon_v I)^{1/2} \Big\{ J_T (x_2 - |x_1|) - ((2I\varepsilon_v)^{1/2}/2)(x_2^2 - x_1^2) - (8\alpha e^2\mu^2\varepsilon_v)^{1/4} \int_{|x_1|}^{x_2} (1 - x^2)^{1/4}\, dx \Big\};$$

$x_1$ and $x_2$ are the roots of the equation

$$\pm (1 - x^2)^{1/4} = J_T/(8\alpha e^2\mu^2\varepsilon_v)^{1/4} - (2I\varepsilon_v)^{1/2} x/(8\alpha e^2\mu^2\varepsilon_v)^{1/4},$$

which satisfy the conditions $1 \geq x_2 > x$; $\varepsilon_T = E_t' + E_r' + E_v' = E_t + E_r + E_v$; $E_t'$, $E_r'$, and $E_v'$ are the translational, rotational, and vibrational energies, respectively, of the colliding particles $AB^{+**}$ and C; $E_t$, $E_r$, and $E_v$ are the same for the decay products $AB^{+*}$ and C of the collisional complex; $I$ is the moment of inertia of the molecular ion $AB^+$; $\omega$ is the vibrational frequency of $AB^+$; $[\varepsilon_T/\hbar\omega]$ is the largest integer that does not exceed $\varepsilon_T/\hbar\omega$; $\alpha$ is the polarizability of the atom C; $e$ is the charge of the $AB^{+**}$ ion; $\mu$ is the reduced mass of the colliding particles; $\varepsilon_v = \varepsilon_T - \varepsilon_v$; $E_v = v\hbar\omega$; and $J_T$ is the total angular momentum of the system $AB^+ + C$.

As an example, a calculation was done for the deactivation of vibrationally excited $NO^{+*}$ ions in collisions with Xe atoms [45]. It was assumed that the ion is initially excited to the fourth vibrational level. The parameters used in the calculation are the following: $E_t' = E_r' = 0.025$ eV; $E_v' = 1.04$ eV; $\varepsilon_T = 1.09$ eV, $\alpha = 4\cdot10^{-24}$ cm$^3$, I = $1.4\cdot10^{-39}$ g·cm$^2$, and $J_T = J_{max} = (8e^2\mu^2\alpha E_t')^{1/4}$.

The total angular momentum was taken equal to the maximum orbital momentum at which the ion–molecule collisional complex can be formed through the polarization attraction forces between the ion and atom,

$$NO^{+*}(v=4) + Xe = NO^{+} + Xe, \tag{3}$$

since the bulk of the collisions occur at large impact parameters. Here for simplicity in the calculations, the intrinsic angular momentum of the molecule before the collision process was neglected compared to the orbital angular momentum of the ion–molecule system. The probabilities of being in vibrational states $v = 0$–$4$ after the collision are $\gamma_0 = 0.46$, $\gamma_1 = 0.30$, $\gamma_2 = 0.17$, $\gamma_3 = 0.07$, and $\gamma_4 = 0.004$, respectively.

As can be seen from this example, in cases where the statistical theory is applicable, we should expect highly efficient vibrational–translational deactivation in ion–molecule collisions, with a high probability of multi-quantum transitions. The formula given above for the probability of vibrational–translational exchange was obtained using the exact statistical theory with the conservation of both energy and angular momentum taken into account. If we use an approximate statistical theory [37, 46, 47] that includes only the conservation of the total energy, then it is easy to obtain a simple expression for the probability of vibrational–translational exchange:

$$\tilde{\gamma}_v = (\varepsilon_T - \hbar\omega v)^{3/2} \Big/ \sum_{v=0}^{[\varepsilon_T/\hbar\omega]} (\varepsilon_T - \hbar\omega v)^{3/2}. \tag{4}$$

Here the notation is as in Eq. (2). Calculations using this formula for reaction (3) yield $\gamma_0 = 0.45$, $\tilde{\gamma}_1 = 0.30$, $\tilde{\gamma}_2 = 0.17$, $\tilde{\gamma}_3 = 0.068$, and $\gamma_4 = 0.004$. Thus, in this case the results of Eqs. (2) and (4) were practically the same.

Now let the $NO^+$ ion in the example (3) considered above be excited to the first vibrational level. Then

$$\tilde{\gamma}_0 = \varepsilon_T^{3/2}/[\varepsilon_T^{3/2} + (\varepsilon_T - \hbar\omega)^{3/2}] = 0.945,$$

$$\tilde{\gamma}_1 = (\varepsilon_T - \hbar\omega)^{3/2}/[\varepsilon_T^{3/2} + (\varepsilon_T - \hbar\omega)^{3/2}] = 0.055.$$

In this approximation vibrational–translational exchange occurs in practically every ion–molecule collision. In the theory which takes only energy conservation into account, at low initial translational and rotational energies the cross section and rate coefficient for vibrational–translational relaxation are given by formulas similar to the Langevin formulas:

$$\sigma_{\Gamma} \approx 2\pi \left( \frac{\alpha e^2}{2E_t'} \right)^{1/2} \tilde{\gamma}_0 \tag{5}$$

and

$$K_{\Gamma} \approx 2\pi \left( \frac{\alpha e^2}{\mu} \right)^{1/2} \tilde{\gamma}_0. \tag{6}$$

Typical rate constants for vibrational–translational relaxation in ion–molecule collisions given by Eq. (6) are $10^{-9}$–$10^{-10}$ cm$^3$/sec.

It should, however, be noted that the assumptions on which the statistical theory is based are not always strictly satisfied. The statistical theory obviously gives an upper limit to the rate constants for vibrational–translational exchange.

We have been examining energy exchange in collisions of diatomic and monatomic particles in the framework of the statistical theory. There are generalizations to more complicated cases. For example, collisions of a spherically symmetric polyatomic molecular ion with an atom have been analyzed [45]. The molecular ion was modelled by a system of identical oscillators.

In collisions of molecular ions with molecules, vibrational–vibrational exchange is possible, as well as vibrational–translational exchange. It is usually assumed that the internal energy redistributes itself uniformly over all the internal degrees of freedom of the collisional complex [48, 49]. Thus, a statistical theory in which the rotational energy and the conservation of angular momentum are neglected has been used [49] to examine the vibrational deactivation of $D_3^{+*}$ ions in the process

$$D_3^{+**} + D_2 = D_3^{+*} + D_2^{*}.$$

A comparison of theory and experiment showed that the observed rate of deactivation of $D_3^{+**}$ is about 60% of the calculated value. As an example, when this correction is included the following probabilities $P(v)$ for transitions of the $D_3^{+**}$ ion from an initial vibrational state $v = 7$ to states $v$ are obtained:

$$P(0)=0.025,\quad P(1)=0.060,\quad P(2)=0.091,\quad P(3)=0.109,$$
$$P(4)=0.122,\quad P(5)=0.107,\quad P(6)=0.066.$$

Thus, even when vibrational–vibrational exchange occurs, the statistical theory yields excessive probabilities for the vibrational deactivation of ions. It has been pointed out [49] that for low vibrational excitation of $D_3^+$ ($H_3^+$), when the excitation energy is less than the vibrational quantum for $D_2$ ($H_2$), the probability of vibrational deactivation of the ion can decrease sharply because vibrational–vibrational exchange is no longer possible.

An attempt [50, 51] has been made to modify the statistical theory of energy transfer in ion–molecule collisions without assuming that the energy is distributed uniformly over all the internal degrees of freedom of the complex. This postulate is replaced by the assumption that the average lifetime of the collisional complex (equal to the difference between the flight times of the colliding particles with and without the mutual attraction taken into account) determines the number of internal degrees of freedom of the complex over which the energy can be redistributed with equal probability. The number of effective, active degrees of freedom is determined from the collision time with the aid of the relationship between the density of states and the lifetime for decay obtained from conventional statistical theory.

When the ion and molecule are treated as structureless particles, the collision time determined by the attractive part of the interaction potential is very short [$(2\text{–}3)\cdot 10^{-12}$ sec for nonpolar particles].

This means that in the process

$$A^{+\bullet} + M = A^{+} + M^{\bullet}$$

a polyatomic particle M will have only one active vibrational degree of freedom that is effective in redistributing the energy inside the collisional complex. This mechanism for redistributing the energy yields substantially lower rates than the conventional statistical theory. The average energy loss in $C_5H_9^{+*}$–M collisions with an initial excitation of 38 kcal/mole in the $C_5H_9^{+*}$ has been calculated [51]. It was found that the energy transfer depends very weakly on the number of degrees of freedom of M. According to the conventional statistical theory, the energy transfer should be proportional to the number of degrees of freedom of M. Unfortunately, so far there is little experimental data with which the theory could be tested [50, 51]. The weak point in this theory is the absence of any accounting for the rotational degrees of freedom of $A^+$ and M, which can lead to a substantial increase in the lifetime of the collisional complex. In addition, the lifetime of the complex was calculated neglecting the short-range exchange forces that determine the electronic structure of the ion and molecule. Thus, the statistical theory without extra assumptions gives an upper bound on the rate of vibrational relaxation. How close this limit is to the real value depends on the specific features of the ion–molecule system. With this, we end our discussion of the statistical theory of energy exchange in ion–molecule collisions and examine some other approaches.

Ivanov and Sukhomlinov [19] have calculated the probability of vibrational transitions in diatomic molecules during collisions with atoms using a model of a harmonic oscillator with an external force for values of the Massey parameter $\omega\tau_{coll} \simeq 1$, where $\omega$ is the vibrational frequency of the molecular ion and $\tau_{coll}$ is the ion–atom collision time.

The calculations were done with a model interaction potential of the form

$$V(r) \propto [(r_0/r)^{12} - (r_0/r)^4],$$

where $r$ is the distance between the centers of mass of the colliding particles. The rate constant for vibrational deactivation in

$$Ne_2^{+*}(v=1) + Ne = Ne_2^+ + Ne \tag{7}$$

(the vibrational quantum of $Ne_2^+$ is $\hbar\omega = 0.074$ eV) was found to be $K$ (300 K) = $5 \cdot 10^{-15}$ cm$^3$/sec and $K$ (600 K) = $2 \cdot 10^{-14}$ cm$^3$/sec. Analogous calculations were done for the process

$$He_2^{+*}(v=1) + He = He_2^+ + He. \tag{8}$$

Here we should make the same comment as before about the work of Shin [35]. In these systems rapid exchange ion–molecule reactions of the type

$$^{20}Ne\,^{20}Ne^+ + {}^{22}Ne = {}^{20}Ne + {}^{20}Ne\,^{22}Ne^+$$

are possible and vibrational relaxation may take place in the course of the exchange reactions.

Exchange reactions and vibrational–translational relaxation have been taken into simultaneous account by Karachevtsev [52] using a well-known simple approximation [53, 54]. The calculations require knowledge of the curvature of the reaction path, the profile of the reaction path, and the vibration frequency along an axis perpendicular to the reaction path. These dependences were determined using the coupling parameter-binding energy approximation [55–57]. The formula for the probability of the vibrational transition $n_1 \to n_2$ in the exchange reaction $X_2^{+*} + X = X + X_2^+$ has the form

$$P_{n_2 \leftarrow n_1} = J^2_{n_1 - n_2}(\gamma), \tag{9}$$

where

$$\gamma = 0.8\sqrt{2}\left\{(n_1 + n_2 + 1)\frac{1}{\hbar\omega}\left[E_k^0 + D(X_3^+) + (n_1 - n_2)\hbar\omega/2\right]\right\}^{1/2} \frac{\sinh\left(\frac{2}{3}\varphi\right)}{\sinh\varphi},$$

$$\varphi = \frac{\pi\omega\mu^{1/2}b}{\left\{2\left[E_k^0 + D(X_3^+) + (n_1 - n_2)\hbar\omega/2\right]\right\}^{1/2}};$$

$J_{n_1 - n_2}$ is the Bessel function of the first kind of the $(n_1 - n_2)$th order; $E_k^0$ is the initial kinetic energy; $\omega$ is the vibrational frequency of $X_2^+$; $D(H_3^+)$ is the bond breaking energy in $X_3^+$ ($X_3^+ = X_2^+ + X$); $\mu = 2m/3$; $m$ is the mass of an X atom; and $b$ is the parameter in the Pauling equation for this system [55–57].

The cross sections and rate constants for relaxation are given in this case by formulas of the type (5) and (6) with $\tilde{\gamma}$ replaced by $P_{n2 \leftarrow n1}$. These formulas have been used for relaxation of the molecular ions $X_2^+$ in a monatomic gas X, where X is an inert gas atom. As an example, for deactivation of a single vibrational quantum at a relative translational energy of 0.025 eV, $K = 2 \cdot 10^{-11}$ cm$^3$/sec. These calculations demonstrate the high efficiency of vibrational–translational deactivation in the course of exchange reactions. The theory [53, 54] is also applicable to asymmetric exchange reactions and to vibrational–translational exchange in nonreactive collisions involving polyatomic molecules. The solution of this problem reduces to obtaining reliable characteristic potential energy surfaces for the colliding systems. The coupling order–binding energy approximation [55–57] seems to provide sufficient accuracy, and a generalization of this method for symmetric exchange reactions of diatomic ions with atoms [52] to more complicated systems is of some interest.

A semiempirical potential for the triatomic system ABC obtained in the coupling order–binding energy approximation has recently been used to study the influence of intermolecular attraction on the efficiency of vibrational–translational exchange [58]. The expression for the potential mirrored an interaction of the Morse type for the diatomic pairs AB and BC and a Lennard-Jones type interaction for A and C.

Changing the parameters of the potential made it possible to vary the intermolecular attraction. Quasiclassical trajectory calculations on a computer showed that even a small attraction between AB and BC leads to a sharp increase in the efficiency of energy transfer [58].

This type of calculation has not been done for ion–molecule collisions. By analogy with the above papers [33–35] a transition to ion–molecule collisions from neutral collisions could lead in the first approximation to adding a polarization term $V_1 \propto 1/r^4$ to the potential (10). Even without additional calculations, however, it is clear from qualitative considerations that the transition to ion–molecule systems generally leads to increased attraction between the colliding particles, because of both polarization and exchange forces, and, therefore, in accordance with calculations [58], to more efficient energy transfer.

It is often useful to treat the process of energy exchange as occurring in two stages: formation of a complex between a vibrationally excited molecule and a molecule (or atom) in the ground state and the decay or vibrational predissociation of the complex. The second phase of the collision, a "semicollision" corresponding to vibrational predissociation of the excited complex, is subject to direct experimental investigation through the excitation and ionization of van der Waals molecules [59–62].

The dynamics of decaying charged clusters of inert gas atoms, formed during ionization of the corresponding neutral van der Waals clusters, has been modelled on a computer. The charged clusters are excited because of the relatively high bond energy in the dimer ions of inert gases. It has been shown that the clusters $Ar_6^+$ and $Kr_6^+$ break up into the dimer ions $Ar_2^+$ and $Kr_2^+$ after roughly $10^{-5}$ sec. The computations are in agreement with an experiment [61] in which complete vaporization of light $Ar_3^+$ – $Ar_6^+$ clusters was observed.

It has been shown theoretically [59] that the lifetime of the trimer $Ar_3^+$ is anomalously long with respect to vibrational dissociation. In trajectory calculations many of the trajectories ended in the triatomic ion up to times of $2 \cdot 10^{-6}$ sec, although under these conditions the vibrational energy of the $Ar_2^+$ dimer corresponded to thousands of degrees, while the binding energy of $Ar_2^+ \ldots Ar$ corresponds to 820 K. It should be noted that the potential energy surface for the triatomic ion of an inert gas used in [59] corresponded to a toroidal well positioned symmetrically around the axis of the $Ar_2^+$ dimer. This position of the atoms should inhibit predissociation, since the directions in which the dimer vibrates are perpendicular to the escape directions for neutral atoms. At the same time, it has been shown [63] that the most stable configuration for the triatomic ions of the inert gases is linear. When the atoms are positioned in this way the efficiency of vibrational predissociation should be higher.

In recent years the role of van der Waals molecules in vibrational relaxation processes has been discussed widely [64–69]. In the trimolecular collision

$$M^* + M + M = M^*M + M$$

the vibrationally excited molecule M* forms an excited van der Waals molecule M*M which then undergoes vibrational predissociation

$$M^*M = M + M.$$

This mechanism is especially efficient at low gas temperatures for dipole molecules, when the intermolecular interaction is relatively large. The vibrational predissociation time for the M*M complex has been calculated [65] by examining crossings of the potential curves of the M – M interaction for different vibrational–rotational states of M. These crossings develop because of the anisotropy of the intermolecular potential and because of the dependence of the potential on the rotational state. The probability of a transition at the crossing point was calculated using the Landau–Zener model. For the HF* ($\nu = 1$) – HF complex at room temperature a vibrational predissociation lifetime of $\tau \simeq 10^{-10}$ sec was obtained [65], in good agreement with the experimental value, $5 \cdot 10^{-10} < \tau < 3 \cdot 10^{-9}$ sec [67].

As mentioned above, ion–molecule complexes have very much higher bond energies than those of van der Waals complexes of neutral particles. Thus, relaxation through the ion–molecule complex should play a major role in the vibrational relaxation of ions that is analogous to the relaxation of neutral molecules through a van der Waals complex. So far there is only one experimental paper [70] which reported the vibrational relaxation of ions through vibrational predissociation of ion–molecule complexes which had been partially stabilized in termolecular collisions. The interaction in the ion–molecule complex has a strong anisotropy, which according to the theory [65] should result in a short time for vibrational predissociation of the ion–molecule complex.

A model [71] has been proposed for vibrational relaxation through a metastable complex which is formed in bimolecular collisions and decays monomolecularly through different channels:

$$M^* + M \rightarrow [M^* - M]^* \rightarrow M + M,$$
$$[M^* - M]^* + M \rightarrow M^* - M,$$
$$M^* - M \rightarrow M + M.$$

The statistical theory was used in order to evaluate the probabilities of the different decay channels, while the cross section for formation of the complex was determined from the cross section for capture owing to the long-range attractive forces. This model is one of the approximations in the above-mentioned statistical theory of energy exchange.

A similar model has been used in a semiempirical theory of vibrational relaxation in ion–molecule collisions [8]. In this theory, in accordance with the equation

$$AB^{+*} + C \rightarrow ABC^{+*} \rightarrow AB^+ + C,$$

the expression

$$K_r = \begin{cases} K_c K_p \tau & \text{for } K_p < \tau^{-1}, \\ K_c & \text{for } K_p > \tau^{-1} \end{cases}$$

is obtained for the rate constant $K_r$ for vibrational relaxation of the molecular ion $AB^{+*}$, where $K_c$ is the rate constant for formation of the Langevin complex, $K_p$ is the rate constant for vibrational predissociation of the complex, which an analysis of experimental data shows is roughly equal to $10^{-9}$ sec, and $\tau$ is the lifetime of the complex. The lifetime of the complex was estimated in accordance with the trimolecular rate constant for formation of the complexes

$$AB^+ + C + X \xrightarrow{K_3} ABC^+ + X$$

from the equation

$$K_3 = K_c K_{\text{coll}}\, \tau,$$

where $K_{\text{coll}}$ is the rate constant for collisional stabilization of the complex, which is assumed equal to the Langevin rate constant $K_L$ for $AB^+ + X$ collisions with heavy X and to $K_L/4$ for X = He.

One effective mechanism for vibrational relaxation of molecules is a nonadiabatic mechanism which shows up when electron–vibrational levels cross [4]. Analogous processes should also be efficient for ion–molecule collisions as well. A lot of work remains to be done to transfer the relatively well developed theory of vibrational relaxation of molecules [2–4] to vibrational relaxation of ions, the theory of which has thus far received relatively little attention.

Charge exchange accompanied by energy exchange is specific to energy exchange processes in ion–molecule systems. For example, the cross section for the process

$$H_2^+(v') + H_2 \rightarrow H_2 + H_2^+(v'')$$

with different values of $v'$ and $v''$ and different kinetic energies of collision has been calculated [72] with exact account being taken of the adiabatic interaction between the ion and molecule.

## 3. EXPERIMENTAL TECHNIQUES

The techniques for investigating the vibrational relaxation of ions are in many respects analogous to those used for studying ion–molecule reactions [10, 11]. We shall, therefore, not dwell on the details of these methods, but examine only the general principles and some special features. These methods usually include the following stages: production of vibrationally excited ions, organizing collisions between the ions and molecules, and diagnosing the state of the molecular ions after collisions.

Vibrationally excited ions are obtained during ionization of molecules by electron or photon impact, as well as during collisions of ions with neutral particles as a result of exothermic ion–molecule reactions and of translational–vibrational exchange.

Collisions of excited ions with atoms and molecules can occur at the place where ions are created when an ion beam collides with a gas target, in a gas flow, as ions drift in an electric field, and during collisions of ion and molecule beams.

Diagnostics of the vibrational states of ions is based on using ion–molecule processes leading to dissociation, charge exchange, and energy exchange. For example, of the two charge exchange reactions

$$N_2^+(v = 1) + Ar = N_2 + Ar^+$$

and

$$N_2^+(v = 0) + Ar = N_2 + Ar^+,$$

the first is exothermic and has a large collision cross section at low kinetic energies, while the second is endothermic and has a very small cross section under the same conditions. These processes have been used for measuring the rates of vibrational relaxation of $N_2^{+*}$ [73].

Vibrational relaxation of ions was first observed through the dependence of the dissociation cross section in

$$D_3^+ + X = D^+ + D_2 + X \tag{10}$$

on the gas pressure in an ion source with a $D_3^+$ zone [74]. The apparatus was a double, tandem mass spectrometer. In the ion source, deuterium molecules were ionized by electron impact,

$$e + D_2 = D_2^{+*} + 2e.$$

At high pressures the primary ions in the ionization chamber could participate in a fast exothermic reaction with deuterium,

$$D_2^{+*} + D_2 = D_3^{+**} + D.$$

Subsequently, the ions were extracted from the ion source, accelerated, and passed through a mass analyzer which separated the $D_3^{+**}$ ions, which then entered a collision chamber (a gas target) where dissociation took place through reaction (10). The dissociation was detected by the formation of $D^+$ ions which were separated from the $D_3^+ + D_2^+ + D^+$ beam by means of a second mass analyzer. From the dependence of the dissociation cross section on the gas pressure in the ion source it was possible to determine the rate constant for vibrational relaxation

$$D_3^{+**} + D_2 = D_3^+ + D_2.$$

We now consider the simplest kinetic model as an example. For the excited $D_3^{+**}$ ions in the ionization chamber we can write the kinetic equation

$$d\,[D_3^{+**}]/dt = -K_r\,[D_2]\,[D_3^{+**}],$$

where $[D_3^{+**}]$ is the density of $D_3^{+**}$, $[D_2]$ is the density of $D_2$, and $K_r$ is the rate constant for vibrational relaxation.

The solution of this equation is

$$[D_3^{+**}] = [D_3^{+**}]\,e^{-K_r[D_2]\,\tau}.$$

where $\tau$ is the lifetime of $D_3^+$ ions in the ion source.

Let the cross section for dissociation of $D_3^{+**}$ ions be $\sigma_1$ and that for $D_3^+$ ions be $\sigma_2$. Then the effective cross section for dissociation of the ions leaving the ion source is

$$\bar{\sigma} = \sigma_1 e^{-K_r [D_2] \tau} + \sigma_2 (1 - e^{-K_r [D_2] \tau}).$$

By measuring the values of $\bar{\sigma}$ for different values of $n$ and $\tau$, it is possible to determine $K_r$. It was shown that deactivation of vibrational excitation takes place in three to four ion–molecule collisions in the ion source [74].

The analogous process of vibrational deactivation of $H_3^{+**}$ in collisions with $H_2$ molecules has also been studied [75–77, 79–81]. An ion cyclotron resonance (ICR) mass spectrometer [78] has been used for this purpose [76, 77]. This method makes it possible to record the time dependence of the densities of ions of different types in the ionization region after an ionizing pulse. To illustrate this method, let us examine a case in which the ion–molecule reaction

$$A^{+*} + B \xrightarrow{K_r} C^+ + D$$

involves only excited ions $A^{+*}$, while the $A^+$ ions can no longer produce the product $C^+$ after vibrational relaxation in the reaction

$$A^{+*} + B \xrightarrow{K_r} A^+ + B$$

The kinetic equations for this case have the form

$$d[A^{+*}]/dt = -(K + K_r)[B][A^{+*}],$$
$$d[A^+]/dt = -K_r [B][A^{+*}],$$
$$d[C^+]/dt = K[B][A^{+*}].$$

The solution of this system is

$$[A^{+*}] = [A_0^{+*}] \exp[-[B](K_r + K)t],$$

$$[A^+] = \frac{K_r [B][A_0^{+*}]}{K + K_r} [1 - \exp[-[B](K_r + K)t]],$$

$$[C^+] = \frac{K[B][A_0^{+*}]}{K + K_r} [1 - \exp[-[B](K_r + K)t]].$$

In the mass spectrometry experiment the time ($t$) dependence of the densities $[C^+]$ and $[A^{+*}] + [A^+]$ are recorded. When [B] is known, these data can be used to determine $K$ and $K_r$. The rate constant for relaxation in

$$H_3^{+*} + H_2 = H_3^+ + H_2$$

has been found to be $K_r = (2.7 \pm 0.6)\cdot 10^{-10}$ cm$^3$/sec using the ICR method [77].

The relaxation of $H_3^{+**}$ ions has been detected through the dependence of the yield of reaction products of the reactions of $H_3^+$ with $N_2$, CO, $N_2O$, and $CO_2$ on the residence time of the ions in the ionization region by Ryan [79]. In this case the ionization region was an ion trap which made it possible to vary the residence time of the ions in the ionization region over a wide range. The primary and secondary ions were mass analyzed with the aid of a conventional magnetic analyzer following the ionization pulse, a delay time, and pulsed extraction of ions from the trap. The pulsed method was first used for studying the kinetics of ion–molecule processes by Tal'roze and Frankevich [82] and an ion trap was used by Tal'roze and Karachevtsev [83].

Colliding ion and atomic beams have been used [80, 81] to demonstrate that the endothermic reaction

$$H_3^+ + Ar = ArH^+ + H_2$$

proceeds with a significant cross section at low collision kinetic energies, even at pressures of 0.5 Torr in an ion source. Calculations based on these experimental data show that the average energy of excitation of $H_3^+$ is equal to $0.6 \pm 0.15$ eV at 0.1 Torr and to $0.3 \pm 0.15$ eV at 0.5 Torr. In the kinetic analysis of these data, account was taken of the fact that the rate constant for vibrational relaxation depends on the vibrational excitation of the ion [80, 81]. For example, it follows from general considerations that the highly excited states of $H_3^{+**}$ can relax through rapid vibrational–vibrational exchange, while the weakly excited states of $H_3^{+*}$, which are not in good resonance with the vibrational levels of $H_2$, can only relax through relatively slow vibrational–translational exchange. The experimental data [80, 81] can be described if we assume that the rate constant for relaxation of $H_3^{+*}$ in collisions with $H_2$ is roughly equal to $10^{-11}$ cm$^3$/sec for $H_3^{+*}$ in the third vibrational level and to $10^{-12}$ cm$^3$/sec for relaxation of the lower excited states (in the first or second excited vibrational level).

Relatively slow relaxation of $D_3^{+*}$ into $D_2$ has been observed in the reaction of $D_3^+$ with Ar [84]. In this experiment $D_3^{+*}$ was formed and underwent relaxation in the ionization chamber of a magnetic mass spectrometer, and then a beam of $D_3^+$ ions was directed into the analysis chamber of an ICR spectrometer where the effective cross section for the reaction of $D_3^+$ with Ar at low kinetic energies was measured. It appeared that even after 14 collisions in the ionization chamber, the cross section for the reaction with Ar fell by only 8%.

The discrepancies between the estimates for the rates of relaxation of $H_3^+$ and $D_3^+$ ions found in the preceding experiment and earlier work was explained [84] by the different methods used for identifying excited states. For example, in one earlier paper [85] the reaction

$$D_3^+ + CH_4 = CH_3^+ + D_2 + HD - 0.6 \text{ eV}$$

was used to determine whether $D_3^+$ was excited. Strictly speaking, it is not true that the cross section for this reaction drops to zero at exciting energies $\leq 0.6$ eV. The cross section may fall sharply even at considerably higher exciting energies, since a rapid exothermic competing channel exists for this reaction,

$$D_3^+ + CH_4 = CH_4D^+ + D_2 + 1.1 \text{ eV}.$$

When the reaction

$$D_3^+ + Ar = ArD^+ + D_2$$

is used, there are no such competing channels and the determination of the excited state becomes more reliable.

An experimental arrangement similar to that employed in [84] has been used to study vibrational relaxation involving complex ions [48]. A beam of $C_3H_5^+$ ions was formed when propane was ionized by electron impact in an ion source, separated by a magnetic mass analyzer, and injected into the reaction chamber of an ICR spectrometer which contained ethylene and a buffer gas M. In this system, an exchange reaction and stabilization of the complex in collisions with the molecule M are possible:

$$C_3H_5^+ + C_2H_4 \rightleftarrows C_5H_9^{+} \begin{cases} \nearrow C_5H_7^+ + H_2 \\ \xrightarrow{K[M]} C_5H_9^+ + M. \end{cases}$$

In the experiment the densities of $C_5H_9^+$ and $C_5H_7^+$ ions were measured as functions of the pressure of the buffer gas, whose molecules do not enter the reaction but only participate in energy exchange.

A kinetic analysis of the experimental data makes it possible to determine the rate constants for stabilization of the $C_5H_9^{+*}$ complex by various gas molecules. It is assumed that the rate constant for the exchange reaction with formation of $C_5H_7^+$ does not depend on the energy excess above threshold. It turns out that the rate constants for stabilization are about three orders of magnitude lower than the Langevin rate constant for ion–molecule collisions. We note, however, that in this case the rate constant for stabilization corresponds to transfer of a relatively large portion of the energy, after which the reaction leading to formation of $C_5H_7^+$ becomes impossible.

The stabilization efficiency usually rises as the mass and the number of degrees of freedom of the neutral particle are increased. The experimental data are in qualitative agreement with similar data obtained previously [86] using pulsed mass spectrometry and with theoretical calculations based on a modification of the statistical theory [50]. Because of the complexity of these processes, we shall not examine the large amount of data on the efficiency of stabilization of ion–molecule complexes obtained from the rate constants for trimolecular reactions.

Pulsed mass spectrometry has been used to study the vibrational relaxation of $SF_5^{+*}$ molecules [70]. This method is based on investigations of the kinetics of trimolecular ion–molecule association reactions. In studies of ion–molecule reactions in a $SF_6/H_2O$ mixture, it was found that the ion current of complex ions $SF_5^+\cdot H_2O$ begins to increase at short reaction times with a delay which is especially noticeable for low densities of $SF_6$ and $H_2O$ in the ion source. This behavior has been interpreted with the aid of the following simplified kinetic scheme:

$$SF_5^{+\bullet} + H_2O + SF_6 \overset{K}{=} SF_5^{+\bullet}\cdot H_2O + SF_6,$$

$$SF_5^{+\bullet}\cdot H_2O \xrightarrow{K'} SF_5^+ + H_2O,$$

$$SF_5^+ + H_2O + SF_6 \overset{K}{=} SF_5^+\cdot H_2O + SF_6.$$

If the rate constant for vibrational predissociation of the $SF_5^+\cdot H_2O$ complex ion $K'$ is small, then the observed current of complex ions varies as

$$J\,(SF_5^+\cdot H_2O) = J_0\left(1 - e^{-kt}\right).$$

If, on the other hand, $K'$ is large, then the kinetic scheme reduces to the two sequential reactions and

$$J\,(SF_5^+\cdot H_2O) = J_0\,(1 - e^{-kt} - kte^{-kt}).$$

At early times the first formula gives

$$J\,(SF_5^+\cdot H_2O) \simeq J_0kt,$$

while the second gives

$$J\,(SF_5^+\cdot H_2O) \simeq J_0k^2t^2/2.$$

The ratio of these currents is $t/2$. Clearly, for large $K'$ the current $J(SF_5^+\cdot H_2O)$ is small compared to the current when $K'$ is small.

An examination of the kinetics of the ion–molecule processes taking place in the ion source and a comparison of the data from kinetic calculations with the experiments show that the rate constant for vibrational predissociation of $SF_5^{+*}\cdot H_2O$ is $K' \geq 1.5\cdot 10^6$ sec$^{-1}$.

The rate constant for relaxation in collisions with molecules,

$$SF_5^{+*} + SF_6 \xrightarrow{K_r} SF_5^+ + SF_6,$$

obtained from this type of experiment [70] lies in the interval $3\cdot 10^{-11} \leq K_r \leq 1.2\cdot 10^{-10}$ cm$^3$/sec. A comparison of these limits to $K_r$ with the rate constant for Langevin collisions of $SF_5^+$ and $SF_6$ shows that 5–20 collisions are required to deactivate $SF_5^{+*}$. The ion–molecule system examined in [70] is very complicated, so that a large number of ion–molecule reactions which have not been mentioned in the present review take place in it. There is some interest in determining the times for vibrational predissociation in simpler cases where there are no complicating reactions, such as in the $Ar/H_2O/NO^+$ system:

$$NO^{+*} + H_2O + Ar \rightarrow NO^{+*}\cdot H_2O + Ar;$$
$$NO^{+*}\cdot H_2O \rightarrow NO^+ + H_2O,$$
$$NO^+ + H_2O + Ar \rightarrow NO^+\cdot H_2O.$$

The maximum time for vibrational predissociation determined by this method is shorter than the lifetime of the ions in the ion source and the vibrational relaxation time in bimolecular collisions. On the other hand, only an upper bound estimate can be obtained for short vibrational predissociation times.

A method has been proposed and tested for determining the rate of vibrational–translational relaxation based on measuring the dependence of the kinetic energy of molecular ions in an equipotential region on the number of ion–molecule collisions [7, 12, 87]. The rate constant for vibrational–translational exchange in ion–molecule collisions was measured for the first time by this method [7].

Let us take as an example the process

$$NO^{+*} + Xe \xrightarrow{K_1} \overrightarrow{NO}^{+} + Xe.$$

When vibrational energy is converted to translational energy rapidly enough, heating of the translational degrees of freedom of the ions through vibrational–translational exchange should be observed. At the same time, translational–translational exchange occurs and reduces the density of fast ions:

$$\overrightarrow{NO}^{+} + Xe \xrightarrow{K_2} NO^{+} + Xe.$$

This simple model yields the following equation for the fast ion density:

$$[NO^{+}] = [NO^{+*}]_0 \frac{K_1}{K_2 - K_1} (e^{-K_1 [Xe] t} - e^{-K_2 [Xe] t}),$$

where $[NO^{+*}]_0$ is the initial density of $NO^{+*}$. Thus, if we increase the density of Xe atoms in the ion source while raising the number of ion–molecule collisions during the time $t$ spent by the ions in the source, then the density of fast ions initially rises, reaches a maximum when [Xe] = $\ln(K_2/K_1)/(K_2 - K_1)$, and then falls.

The average kinetic energy of the $NO^{+}$ ions was determined by the deflection method, which was first used in [88] for measuring the kinetic energies of the products of ion–molecule reactions. An NO/Ar/Xe gas mixture was admitted into the ion source of a mass spectrometer. The influx of NO + Ar is relatively small and the pressure of this mixture in the ion source was about $10^{-3}$ Torr. The pressure of Xe was varied over the range $(1–8)\cdot 10^{-2}$ Torr. Ionization was by means of electrons with energies of 60 eV.

The velocity distributions of the $NO^{+}$ and $Ar^{+}$ ions were studied. The $Ar^{+}$ ions served as a standard for a thermal distribution. The average energy of the ions in a direction perpendicular to the accelerating field was determined from the inverse slope of the logarithm of the ion current plotted against the square of the deflection voltage. The inverse slope is proportional to the average kinetic energy for a Maxwellian distribution. These dependences were used at each pressure of Xe to obtain the ratio of the kinetic energies of the $NO^{+}$ and $Ar^{+}$ ions. It turned out that as the Xe pressure was raised, this ratio increased very little. This means that deactivation of vibrational excitation in the $NO^+$ ions through collisions with Xe atoms occurs with a probability considerably less than unity.

In order to carry out a kinetic analysis of the experimental data, it was necessary to determine the number of collisions experienced by an $NO^+$ ion before it leaves the ion source. This was determined from the variation with the Xe pressure in the average kinetic energy of the $^{35}Cl^+$ ions formed from $CH_3Cl$ by electron impact. The data were analyzed under the assumption that the distribution over the energy states of $NO^+$ after vibrational–translational exchange corresponds to the statistical theory [45]. The rate constant $K_1$ and the probability $W$ of vibrational–translational exchange in a single collision were found to be [7]

$$4 \cdot 10^{-10} > K_1 > 3.8 \cdot 10^{-12} \text{ cm}^3/\text{sec}$$

and

$$1/20 \gg W \gg 1/200.$$

A similar method has been used to study the vibrational–translational relaxation of hydrogen and deuterium molecular ions in collisions with He and Ne atoms [12, 87]. In this case the $H_2^{+*}$ and $D_2^{+*}$ ions were obtained through ionization by the resonance line of He with a photon energy of 21.2 eV. The probabilities of the following collisional vibrational–translational relaxation processes were evaluated:

$$H_2^{+*} + He \rightarrow \vec{H}_2^{+} + \vec{He}, \qquad (11)$$

$$H_2^{+*} + Ne \rightarrow \vec{H}_2^{+} + \vec{Ne}, \qquad (12)$$

$$D_2^{+*} + Ne \rightarrow \vec{D}_2^{+} + \vec{Ne}. \qquad (13)$$

As in the studies of $NO^{+*}$ relaxation, the dependence of the average kinetic energy of the ions on the inert gas pressure was measured. Under conditions such that an ion experiences only one collision with an inert gas atom during the time it is in the ion source, the relative increase in the kinetic energy of the ion above the thermal energy was $1.3 \pm 0.5$ for $H_2^+$, He and $H_2^+$, Ne, and $1.38 \pm 0.05$ for $D_2^+$ + Ne.

A kinetic analysis of these data yields $0.55 \pm 0.2$, $0.63 \pm 0.2$, and $0.65 \pm 0.2$, respectively, for the probabilities of deactivation in a single collision through processes (11)–(13).

Process (11) has been studied in terms of the dependence of the products of the endothermic reaction

$$H_2^{+\bullet} + He = HeH^+ + H$$

on the gas pressure and on the time using an ion cyclotron resonance technique [89].

The probability of deactivation of $H_2^{+*}$ ($v \geq 3$) in a single collision with He was found to be 0.3, in agreement with the data of [87].

The high efficiency of deactivation of vibrational excitation in these systems is probably related to some feature of the potential energy surfaces. In the region where the particles approach one another at atomic distances, there is a potential well corresponding to formation of a stable $HeH_2^+$ ion with an equilibrium H–H separation which differs greatly from the equilibrium distance in $H_2^+$ [90].

In slow collisions an ion may spend a considerable time in the region of this potential energy minimum. This favors efficient redistribution of the energy over the internal degrees of freedom.

Note that for the $He^+ + H_2^*$ system, the potential energy surface does not have a potential well and is repulsive in character [43]. From this we may conclude that vibrational relaxation involving the $He^+$ ion,

$$H_2^\bullet + He^+ = \vec{H}_2 + \vec{He}^+,$$

will be slow compared to reaction (11).

Among the inadequacies of this method for studying vibrational–translational relaxation [7, 12] we must include the complexity of separating the fast (relaxed) and slow (unrelaxed) ions when the deactivation probability is low. Increasing the gas pressure in the relaxation region may lead to a rapid drop in the excess (beyond the thermal energy) kinetic energy of the relaxed ions because of ion–molecule collisions and does not increase the sensitivity of the measurements. In addition, when the gas pressures are higher in the ion source region, scattering in the zone where the ions are extracted and accelerated leads to an increase in the component of the momentum perpendicular to the ion beam. This is equivalent to an effective increase in the kinetic energy of the ions as measured by the deflection method.

Molecular beam techniques can probably increase the sensitivity of the method substantially [7, 12]. For example, suppose we have a molecular beam of Xe with an NO additive. At some cross section of the beam, let photoionization of NO take place. The $NO^{+*}$ ions and Xe atoms continue to move in the beam, experiencing infrequent mutual collisions. At some distance from the ionization site, the ions are extracted perpendicular to the beam direction and the ions are analyzed in terms of their kinetic energy.

In this case, the ions will undergo considerably less scattering at the time they are accelerated than when they are being extracted from the region with a higher gas pressure, so that the accuracy of the measurement of the kinetic energy of the relaxed ions will be improved.

Several attempts have been made to study vibrational relaxation of ions directly under gas discharge plasma conditions. In [91] excited $He_2^{+*}$ molecular ions have been obtained in a decaying He plasma through associative ionization, undergone relaxation through collisions with He atoms, and recombined with electrons. The recombination radiation from the atoms and molecules was detected. A kinetic analysis of the processes during decay of the complicated recombining plasma did not yield reliable quantitative data. It could only be stated qualitatively that vibrational–translational exchange was rapid.

Analogous experiments have been carried out on Ne plasmas [92]. The time variation in the intensities of atomic line emission from neon in the decaying plasma was measured in these experiments. It was assumed that the emission takes place through dissociative recombination of $Ne_2^+$ with electrons. A kinetic analysis of the observed dependences made it possible to determine the rate constant for vibrational relaxation of $Ne_2^{+*}$ in collisions with Ne. For example, at a temperature of 300 K, the rate constant for deactivation of $Ne_2^+$ ($v = 1$) is $5 \cdot 10^{-15}$ cm$^3$/sec, and at 600 K, $2 \cdot 10^{-14}$ cm$^3$/sec.

It has been pointed out [93] that the kinetic scheme used to analyze these experiments [92] should include processes involving metastable atoms. Taking such processes into account leads to a large uncertainty in the deactivation rate constants derived from the data of [92].

Methods based on the interaction of photons with molecular ions are more reliable optical methods for recording the vibrational relaxation of ions. It has been found [22] that the absorption spectra of the $Ne_2^+$ and $Ar_2^+$ ions depend on the drift velocity of these ions in an electric field. This is because the ions are excited vibrationally as they drift through the gas.

Absorption in vibrational–vibrational transitions in the infrared has been used to determine the vibrational state of relaxing $NO^+$ ions in gaseous $N_2$ and NO. The following values of the rate constants for deactivation of $NO^{+*}$ at $T = 297$ K have been obtained: $K(N_2) = (2–3) \cdot 10^{-12}$ cm$^3$/sec and $K(NO) \simeq 10^{-13}$ cm$^3$/sec [94]. The small value for deactivation by NO molecules is probably associated with some errors in the measurements, since in this case, the cross section for resonant charge exchange is large and this usually leads to rapid loss of vibrational energy by the ion. Rapid relaxation of the vibrational energy of the ions in a $NO^+/NO$ system was, in fact, observed subsequently [8] by another method.

Laser-induced fluorescence has been used for detecting the vibrational states of molecular ions [95]. Ion–molecule reactions which produced vibrationally excited product molecular ions were produced in a device with a flowing afterglow. Primary ions were formed in an electrical discharge at some cross section of a gas flow of He with a small amount of added $N_2$. CO molecules were introduced into the gas flow downstream of this region and reacted with $N^+$. As the vibrational populations of the products of the ion–molecule reaction

$$N^+ + CO = CO^+ (v = 0,\ 1,\ 2) + N$$

were being measured, it turned out that the initial populations were strongly distorted by vibrational relaxation and Penning ionization

$$He^* + CO = CO^{+*} + He + e.$$

These data show that the rate constant for relaxation of $CO^+$ ($v = 1$) is large for collisions with both CO and $N_2$ and is roughly 0.1–1 times the Langevin value. In the case of collisions with CO the speed of the reaction is explained by resonant charge exchange, but a nearly resonant vibrational–vibrational exchange probably occurs in collisions with $N_2$. It has been pointed out [95] that if the method of ion production were improved, for example by injecting preselected ions into the reactor in order to avoid processes involving neutral particles, laser-induced fluorescence could offer still greater promise.

The photodetachment of electrons has been used for diagnostics of vibrational excitation of negative ions [96]. Excited ions were produced in the ion trap of an ICR spectrometer through the reaction

$$O^- (O_2^-) + NO_2 = NO_2^{-*} + O\,(O_2).$$

The resulting $NO_2^{-*}$ ions were irradiated by photons whose energy was sufficient to detach an electron from the vibrationally excited state of the ion, but not enough to detach an electron from its ground state. Mass spectrometry was used to measure the change in the ion density owing to photodetachment processes as a function of the pressure of the deactivating molecules. The following values of the deactivation rate constant were obtained: $10^{-12}$ cm$^3$/sec ($O_2$), $8 \cdot 10^{-10}$ cm$^3$/sec ($NO_2$), and

$10^{-11}$–$10^{-12}$ cm$^3$/sec ($CO_2$). As with the positive ion reactions examined above, the large deactivation rate constant in the symmetric case ($NO_2^{-*}$ + $NO_2$) is probably the result of charge exchange.

Vibrational deactivation of negative ions has also been studied by photodissociation [97, 98]. Excited ions were created in the exothermic electron attachment reaction

$$e + CF_3OOCF_3 \rightarrow CF_3O^{-*} + CF_3O.$$

For diagnostics of the vibrational excitation, the laser light was tuned so that multiphoton dissociation was possible only for the excited ions,

$$nh\nu + CF_3O^{-*} = F^- + CF_2O.$$

The ICR method was used to measure the change in the ion density owing to photodissociation as a function of the pressure of the deactivating molecules. The following upper limits on the rate constant for vibrational deactivation were obtained (in units of $10^{-10}$ cm$^3$/sec): 0.19 ± 0.02 ($N_2$), 0.46 ± 0.03 ($CH_4$), 1.4 ± 0.2 ($CF_2O$), and 1.6 ± 0.1 ($n$-$C_5H_{12}$) [98].

A similar method was used [98] to study the deactivation of the $CH_3OHF^-$ ion in collisions with $HCO_2CH_3$. It was shown that 40–80 collisions are needed for vibrational deactivation.

The long-range interaction of negative ions with molecules is completely analogous to the interaction of positive ions with molecules, but the short-range forces may be quite different. The short-range interaction between a positive ion and a molecule is usually stronger than the analogous interaction for a negative ion. Thus, the efficiency of vibrational deactivation by negative ions should be lower than that by positive ions and should approach that by neutral molecules.

Photodissociation of ions has also been used to study the deactivation of positive ions [99]. $C_6H_5Br^+$ ions in the cell of an ICR mass spectrometer were irradiated with photons whose energy was such that photodissociation occurred only through the absorption of two photons. If an ion loses an energy of 1.3 eV as a result of collisions after it has absorbed one photon, then it can no longer be dissociated by a second photon. The following rate constants for deactivation were obtained from the dependence of the dissociation efficiency on the pressure of the deactivating gases (in units of $10^{-9}$ cm$^3$/sec): 1.5 ± 0.3 ($C_6H_5Br$), ≤0.05 ($C_6H_5NO_2$), 0.25 ± 0.08 ($C_6H_6$), 0.22 ± 0.06 ($C_6D_6$), 0.46 ± 0.09 ($C_6H_5F$), ≤0.05 ($CO_2$), ≤0.02 ($CH_4$), ≤0.05 ($C_3H_8$), and 0.074 ± 0.02 ($SF_6$).

The relative complexity of this deactivation process is noteworthy. From the above data [99] it cannot be stated whether the measured rate constants correspond to an energy transfer of 1.3 eV in a single collision or whether they are effective rate constants for deactivation of the same energy by small steps in a sequence of collisions.

The greatest amount of reliable experimental data on vibrational relaxation of ions has so far been obtained by the method of relaxation in a gas flow with detection of vibrational excitation from the kinetics of ion–molecule reactions. Ions leaving an ion source are mass selected in a vacuum and injected into a reactor region with an elevated pressure where they move in a gas flow, relax, react, and then enter a vacuum region, where they are mass selected and detected. An electric field in the reactor can be used to change the kinetic energy of the ions and vary their drift velocity through the gas. A large amount of experimental data obtained in this way has been reported in a review by Ferguson [100].

A rate constant of $1.3 \cdot 10^{-11}$ $cm^3/sec$ for relaxation through the process

$$ArH^{+*} + Ar \rightarrow ArH^{+} + Ar$$

has been obtained [101] from the dependence of the rate constant for the reaction

$$ArH^{+} + H_2 = H_3^{+} + Ar$$

on the pressure of the inert gas. Relaxation in an analogous system,

$$N_2H^{+*} + N_2 \rightarrow N_2H^{+} + N_2\,,$$

has also been studied [102]. Identification of the excited states was based on the exothermicity and, therefore, on the high rate of the reaction

$$N_2H^{+*}\,(v = 2) + Kr = KrH^{+} + N_2$$

and on the endothermicity of the analogous reactions for $N_2H^{+*}$ ($v = 1$) and $N_2H^{+}$ ($v = 0$). The rate constants for vibrational deactivation of the $N_2H^{+}$ ion by $N_2$ molecules and inert gas atoms obtained by this method are (in units of $10^{-10}$ $cm^3/sec$): 3 ($N_2$), $\leq 10^{-3}$ (He), $\leq 5 \cdot 10^{-1}$ (Ar), and $\leq 5 \cdot 10^{-1}$ (Kr).

The rates of relaxation of ions in a gas flow have been determined from the kinetics of ion–molecule reactions for $O_2^{+}$ [103–106], $N_2^{+}$ [107–

111], and $NO^+$ [8]. Most of the data have been obtained for low translational temperatures of the ions ($E \leq 0.1$ eV).

Vibrational–translational exchange at higher ion temperatures have been studied in detail by Federer et al. [23] Mass-selected ions were injected into a drift region A with a fast gas flow and then entered a gas flow region B without an electric field. In region A the ions gained and lost vibrational energy in the electric field, while in region B the vibrational energy remained approximately unchanged. The vibrational excitation of the ions was established from the progress of endothermic ion–molecule reactions with a reagent gas that was specially added to flow region B. For $N_2^+$ the reaction $N_2^+$ + Ar was used and for $O_2^+$, the reaction $O_2^+ + CH_4$ was used. Secondary ions were detected by a quadrupole mass spectrometer after the ions passed from the high-pressure region. In these experiments the vibrational temperatures $T_v$ of $N_2^+$ and $O_2^+$ ions drifting in He were determined quantitatively for the first time for different ratios of the electric field strength to the He density (*E*/*N*). Kinetics experiments were also carried out. For example, excited $N_2^+[\alpha = N_2^+(v \geq 1)/N_2^+(v \geq 0) > 0.4]$ ions were injected into the drift region and their cooling in region A was observed for different values of *E*/*N* as $\alpha$ was measured in region B. In another variant, the length *L* of region A was varied for constant *E*/*N* and the dependence of $\alpha$ on *L* was measured.

It was found that the rate constants for vibrational excitation and deactivation rise rapidly with increasing *E*/*N*. For both ions $T_v$ reaches ~3000 K when $E/N = 150$ Td. Over the range of $E/N = 50$–150 Td, the temperature increases approximately linearly from 1000 to 3000 K.

When $E/N = 90$ Td and $p$(He) = 0.25 Torr, the "equilibrium" value of $T_v$ is established over a drift length $L = 50$ cm.

Using these data and the formulas for the Langevin mobility of ions in gases and for the kinetic energy with which drifting ions collide with He atoms, we find, for example, that for $O_2^+$ and $N_2^+$ with energies of ~0.45 eV the rate constant for vibrational deactivation of these ions is $\sim 1.5 \cdot 10^{-12}$ $cm^3$/sec. The observation that the rate constants for vibrational excitation and deactivation increase rapidly with the kinetic energy of ions drifting in He indicates that earlier data must be checked. Previously, the rate constants for vibrational deactivation were measured in an electric field without taking the possibility of vibrational excitation into account. Failure to take that into account yields lower values for the deactivation rate constant.

Inelastic ion–molecule collisions of $NO_3^-$, $NO_2^-$, $NO^+$, and Cl in $N_2$ have been studied in a drift tube [112]. The difference between the measured ion mobility and the theoretical value for elastic collisions was used to evaluate the ratios of the inelastic energy losses to the energy loss during elastic ion–molecule collisions.

## 4. CONCLUSION

Vibrational relaxation of ions has an important influence on ion–molecule exchange and association reactions, dissociative recombination of positive ions with electrons, and several other processes in low-temperature plasmas. In recent years significant progress has been made in both theoretical and experimental research on vibrational relaxation.

On the theoretical side there is a tendency toward detailed calculations of the characteristics of the potential energy surfaces of the colliding particles and numerical modelling of the dynamics of energy exchange [58, 11]. For example, the classical trajectory approach has been used to evaluate the lifetime of the collision complex $H^+ + H_2$ [113]. It was found that the lifetime is given roughly by the formula obtained from the statistical theory,

$$\tau = \tau_0 \, [E/(E + D)]^S,$$

where $\tau_0$ is the vibrational period of the complex (~$14 \cdot 10^{-15}$ sec), $E$ is the total energy, $D$ is the depth of the potential well, and $S$ = 1.9–2.0. Changing the collision energy over 2–20% of the depth of the potential well causes the average lifetime to vary over the range (2000–20)$\tau_0$.

A rather substantial time (~20$\tau_0$) is required for the particles to "forget" the initial conditions. They are found to depend on the total angular momentum.

It would be interesting to carry out analogous calculations for energy exchange as well. Since the lifetime of the complex is the same according to dynamical calculations and to the statistical theory, however, we might expect that the dynamical calculations in the approximation of classical trajectories would yield a high value of the rate constant for vibrational–translational relaxation in the reaction

$$H_2^{+*} + H = H + H_2^+.$$

Although there are no experimental data on this process, it is probably fast because of the large binding energy in the $H_3^+$ ion and the absence of a potential barrier along the path for formation and decay of the complex.

Such calculations are also of interest for relaxation processes during which complexes with relatively weak bonds are formed. Here we should expect substantial differences from the rate constants calculated using the statistical theory, as has been observed, for example, in the relaxation of $NO^{+*}$ on Xe [7, 8, 45].

In the experimental area, improvements are possible in all the known techniques, ranging from methods based on stabilization of excited complexes [114, 115] to drift tubes with double selection of the ions [8, 23]. For example, there is some interest in extending the method of ion–electron coincidences used for studying ion–molecule reactions with a fixed excitation energy for the molecular ion [14, 15] to the vibrational relaxation of ions.

New methods based on laser spectroscopy are being developed. For example, a method employing overlapping ion and laser beams with modulation of the ion-beam velocity has been used to obtain the vibrational spectrum of $H_3^+$. From the measured width of the spectral lines it has been shown that the lifetimes of the predissociating levels lie in the range above $10^{-7}$–$10^{-9}$ sec. The existence of such metastable levels is of interest for the theory of energy exchange.

To conclude, let us briefly discuss some questions about energy exchange among ions which have not been dealt with in this review, but are closely related to vibrational relaxation in ions [7]. As mentioned above, an efficient mechanism for vibrational deactivation through vibrational–rotational transitions has recently been pointed out [65].

Vibrational–translational relaxation is related to the processes through which ions are dissociated in collisions with atoms and molecules at intermediate kinetic energies. It has been shown that these processes take place through vibrational, rather than electronic, excitation [117–119]. This, in turn, provides the prerequisites for explaining the mechanism of an important analytic technique, collisional ion spectroscopy [21]. The foundations of this technique were laid by several Soviet researchers [120–122].

Increasing attention is being devoted to vibrational relaxation of ions in excited electronic states. For example, it has been shown that rapid relaxation in the reaction

$$CO^{+**}(A,\ v') + He = CO^{+*} + He$$

proceeds by an electronic–vibrational mechanism through the ground state with high vibrational excitation of the $CO^{+**}(X, v'')$ with $v'' \gg v'$ [123]. It has been pointed out [124] that over the lifetime of the A-state (≃4 μsec), vibrational and rotational relaxation of the $HBr^{+*}(A)$ and $DBr^{+*}(A)$ ions cannot occur in Ar at a pressure of 1 Torr and a temperature of 300 K. The rate of vibrational deactivation of $HCl^{+*}(A)$ in Ne is low. On the other hand, $HBr^{+*}(A)$ ions undergo rapid deactivation in He and Ne, as do $HCl^{+*}(A)$ ions in He.

In plasmas with a high density of excited neutrals, collisions of ions with excited particles are important. It has been shown [125] that relaxation involving the metastable singlet state of oxygen,

$$NO^+ (v = 1) + O_2 (^1\Delta_g) = NO^+ (v = 0) + O_2$$

proceeds with a rate constant of $(3 \pm 2)\cdot 10^{-10}$ $cm^3/sec$. This is considerably higher than the rate constant for relaxation of this ion by oxygen in the ground electronic state and illustrates the role of short-range exchange forces in energy exchange processes.

In highly ionized, dense plasmas, ion–ion interactions can become important in energy exchange [126–128].

In low-density plasmas, however, collisionless radiative vibrational relaxation plays an important role [129–131]. For example, the radiative lifetimes of vibrationally excited $SH^{+*}$ and $SH^{-*}$ ions have been shown [129] to be considerably shorter than the radiative lifetime of the excited neutral molecule $SH^*$. As in the case of collisional deactivation, the presence of charge greatly accelerates the energy conversion process.

Deactivation of the vibrational energy of negative ions can proceed by collisionless spontaneous electron detachment. The autodeionization time of $LiH^{-*}$ has been found [132] to be $10^{-9}$–$10^{-10}$ sec, and that of $OH^{-*}$, to be $10^{-5}$–$10^{-6}$ sec.

Positive and negative polyatomic ions with high vibrational excitation can undergo spontaneous dissociation at low pressures [133] with the vibrational energy being expended in the breaking of chemical bonds.

## REFERENCES


1. B. F. Gordiets, A. I. Osipov, and L. A. Shelepin, *Kinetic Processes in Gases and Molecular Lasers* [in Russian], Nauka, Moscow (1980).
2. E. A. Andreev and E. E. Nikitin, in: *Plasma Chemistry*, B. M. Smirnov (ed.), Vol. 3 [in Russian], Atomizdat, Moscow (1976), pp. 28–94.
3. E. E. Nikitin and A. I. Osipov, *Vibrational Relaxation in Gases* [in Russian], VINITI, Moscow, (1977).
4. E. E. Nikitin and S. Ya. Umanskii, *Theoretical Problems in Chemical Physics* [in Russian], Nauka, Moscow (1982), pp. 34–51.

5. A. V. Eletskii and B. M. Smirnov, *Usp. Fiz. Nauk* **136**, 25–59 (1982).
6. N. G. Adams, D. Smith, and E. Alge, *Chem. Phys.* **81**, 1778–1784 (1984).
7. P. S. Vinogradov, G. V. Karachevtsev, A. Z. Marutkin, et al., *Ann. Geophys.* **28**, 859–852 (1972).
8. W. Federer, W. Dobler, F. Howorka, et al., *J. Chem. Phys.* **83**, 1032–1039 (1985).
9. V. L. Tal'roze and A. K. Lyubimova, *Dokl. Akad. Nauk SSSR* **86**, 909–912 (1952).
10. L. I. Virin, R. V. Dzhagatspanyan, G. V. Karachevtsev, et al., *Ion–Molecule Reactions in Gases* [in Russian], Nauka, Moscow (1979).
11. M. Venugopalan (ed.), *Reactions under Plasma Conditions*, Vol. 2, Wiley Interscience, New York (1971).
12. V. L. Talrose, P. S. Vinogradov, and I. K. Larin, in: *Gas-Phase Ion Chemistry*, M. T. Bowers (ed.), Vol. 1, Academic Press., New York (1979), pp. 305–347.
13. G. V. Karachevtsev, V. M. Matyuk, V. K. Potapov, and A. A. Prokof'ev, *Khim. Vys. Énerg.* **14**, 81–83 (1980).
14. K. Tanaka, T. Kato, P. M. Guyon, and I. Koyono, *J. Chem. Phys.* **79**, 4302–4305 (1983).
15. D. van Pijkeren, J. Vaneck, and A. Niehaus, *Chem. Phys.* **95**, 449–457 (1985).
16. A. Carrington and R. A. Kennedy, *J. Chem. Phys.* **81**, 91–112 (1984).
17. R. Kacoschke, V. Boesl, J. Herman, and E. W. Schlag, *Chem. Phys. Lett.* **119**, 467–472 (1985).
18. W. R. Wadt, *J. Chem. Phys.* **73**, 3915–3926 (1980).
19. V. A. Ivanov and V. Sukhomlinov, *Khim Fiz.* **3**, 1646–1650 (1984).
20. R. K. Boyd, F. M. Harris, and J. H. Beynon, *Int. J. Mass Spectrom. Ion Phys.* **66**, 185–194 (1985).
21. R. G. Cooks (ed.), *Collision Spectroscopy*, Plenum, New York (1978).
22. L. C. Lee and G. P. Smith, *Phys. Rev.* **A19**, 2329–2334 (1979).
23. W. Federer, H. Ramler, H. Villinger, and W. Lindinger, *Phys. Rev. Lett.* **54**, 540–543 (1985).
24. L. A. Viehland and D. W. Fahey, *J. Chem. Phys.* **78**, 435–441 (1983).
25. W. Lindinger, T. D. Mark, and F. Howorka, *Swarms of Ions and Electrons in Gases*, Springer, Berlin (1984).
26. T. W. Carr (ed.), *Plasma Chromatography*, Plenum, New York–London (1984).
27. W. Eastes, V. Ross, and J. P. Toennies, *J. Chem. Phys.* **66**, 1919–1928 (1977).
28. A. Kohlhase, T. Hasegawa, S. Kita, and H. Inouye, *Chem. Phys. Lett.* **117**, 555–560 (1985).
29. V. Gierz, J. P. Toennies, and M. Wild, *Chem. Phys. Lett.* **95**, 517–519 (1983).
30. T. Hasegawa, S. Kita, M. Izawa, and H. Inaye, *J. Phys. B, At. Mol. Phys.* **18**, 3775–3782 (1985).
31. G. D. Billing, *Chem. Phys.* **60**, 199–213 (1981).
32. V. B. Leonas and I. D. Rodionov, *Usp. Fiz. Nauk* **146**, 7–34 (1985).
33. H. Shin, *J. Chem. Phys.* **41**, 2864–2868 (1964).
34. H. Shin, *J. Chem. Phys.* **42**, 1739–1743 (1965).
35. H. Shin, in: *Ion–Molecule Reactions in the Gas Phase*, Am. Chem. Soc., Washington (1966), pp. 44–62.
36. V. L. Tal'roze, *Izv. Akad. Nauk SSSR, Ser. Fiz.* **24**, 1001–1005 (1960).
37. O. B. Firsov, *Zh. Éksp. Teor. Fiz.* **42**, 1307–1310 (1962).
38. E. E. Nikitin, *Teor. Éksp. Khim.* **1**, 428–435 (1965).
39. Ph. Phechukas and J. C. Light, *J. Chem. Phys.* **42**, 3281–3291 (1965).

40. E. E. Nikitin and S. Ya. Umanskii, in: *Plasma Chemistry*, B. M. Smirnov (ed.), Vol. 3 [in Russian], Atomizdat, Moscow (1974), pp. 8–66.
41. W. J. Chesnavich and M. T. Bowers, in: *Gas-Phase Ion Chemistry*, M. T. Bowers (ed.), Vol. 1, Academic Press, New York (1979), pp. 119–149.
42. S. Ya. Umanskii, in: *Theoretical Problems in Chemical Physics*, N. M. Kuznetsov, E. E. Nikitin, and N. D. Sokolov (eds.) [in Russian], Nauka, Moscow (1982), pp. 52–70.
43. G. V. Karachevtsev, A. Z. Marutkin, and V. V. Savkin, *Usp. Khim.* **51**, 1849–1874 (1982).
44. E. E. Nikitin, *Dynamics of Molecular Collisions. Progress in Science and Technology, Series on Kinetics and Catalysis*, Vol 11 [in Russian], VINITI, Akad. Nauk SSSR (1983).
45. G. V. Karachevtsev, *Khim. Vys. Énerg.* **4**, 387–393 (1970).
46. G. E. Spezhakova, N. N. Tunitskii, and M. V. Tikhomirov, *Zh. Fiz. Khim.* **39**, 2002–2007 (1965).
47. N. N. Tunitskii, G. E. Spezhakova, and M. V. Tikhomirov, *Zh. Fiz. Khim.* **40**, 1634–1637 (1966).
48. R. Hourief and J. H. Futrell, *Adv. Mass Spectrom.* **7A**, 335–341 (1977).
49. V. G. Anicich and J. H. Futrell, *Int. J. Mass Spectrom. Ion Phys.* **55**, 189–215 (1984).
50. R. C. Bhattacharje and W. Forst, *Adv. Mass Spectrom.* **7A**, 229–233 (1977).
51. W. Forst and R. C. Bhattacharje, *Chem. Phys.* **37**, 343–353 (1979).
52. G. V. Karachevtsev, *Khim. Fiz.* **5**, 875–877 (1986).
53. W. H. Miller and S. Shi, *Chem. Phys.* **75**, 2258–2264 (1981).
54. W. H. Miller, *J. Phys. Chem.* **87**, 3811–3818 (1983).
55. H. S. Johnston, *Gas-Phase Reaction Rate Theory*, The Ronald Press, New York (1966).
56. R. L. Brown, *J. Res. Natl. Bur. Stand.* **86**, 605 (1981).
57. G. V. Karachevtsev and V. V. Savkin, *Khim. Fiz.* **1**, 928–932 (1982).
58. M. K. Osborn and I. W. M. Smith, *Chem. Phys.* **91**, 13–26 (1984).
59. J. J. Saenz, J. M. Soler, and N. Garcia, *Surf. Sci.* **156**, 121–125 (1985).
60. B. M. Smirnov, *Usp. Fiz. Nauk* **142**, 31–60 (1984).
61. V. Buck and H. Meyer, *Phys. Rev. Lett.* **52**, 109–112 (1984).
62. J. M. Soler, J. J. Saenz, N. Garcia, and O. Echt, *Chem. Phys. Lett.* **109**, 71–75 (1984).
63. W. R. Wadt, *Appl. Phys. Lett.* **38**, 1030–1032 (1981).
64. G. E. Ewing, *Chem. Phys.* **29**, 253–270 (1978).
65. G. E. Ewing, *Chem. Phys.* **63**, 411–418 (1981).
66. K. Range, J. Chesnow, and D. Ricard, *Chem. Phys.* **67**, 347–353 (1982).
67. A. S. Pine and W. J. Lafferty, *J. Chem. Phys.* **78**, 2154–2162 (1983).
68. J. Chesnoy, *J. Chem. Phys.* **79**, 2793–2798 (1983).
69. S. K. Gray and S. A. Rice, *J. Chem. Phys.* **83**, 2818–2825 (1985).
70. G. V. Karachevtsev and V. V. Savkin, *Khim. Fiz.* **2**, 849–850 (1983).
71. R. J. Gordon, *J. Chem. Phys.* **74**, 1676–1681 (1981).
72. C. Y. Lee and A. E. De Pristo, *J. Chem. Phys.* **80**, 1116–1126 (1984).
73. P. R. Kemper and M. T. Bowers, *J. Chem. Phys.* **81**, 2634–2638 (1984).
74. J. J. Levental and L. Friedman, *J. Chem. Phys.* **49**, 1974–1976 (1968).
75. J. J. Levental and L. Friedman, *J. Chem. Phys.* **50**, 2928–2931 (1969).
76. W. T. Huntress and M. T. Bowers, *Int. J. Mass Spectrom. Ion Phys.* **12**, 1–18 (1973).

77. J. K. Kim, L. Theard, and W. T. Huntress, *Int. J. Mass Spectrom. Ion. Phys.* **15**, 223–244 (1974).
78. E. N. Nikolaev, *Zh. Khim. Ova. im. D. I. Mendeleeva*, **30**, No. 2, 136–142 (1985).
79. K. R. Ryan, *J. Chem. Phys.* **61**, 1559–1570 (1974).
80. M. L. Vestal, C. R. Blakley, K. R. Ryan, and J. H. Futrell, *Adv. Mass. Spectrom.* **7A**, 234–243 (1977).
81. C. R. Blakley, M. L. Vestal, and J. H. Futrell, *J. Chem. Phys.* **66**, 2392–2399 (1977).
82. V. L. Tal'roze and E. L. Frankevich, *Zh. Fiz. Khim.* **34**, 2709–2718 (1960).
83. V. L. Talrose and G. V. Karachevtsev, *Adv. Mass. Spectrom* **3**, 211–233 (1966).
84. R. D. Smith and J. H. Futrell, *Int. J. Mass Spectrom. Ion. Phys.* **20**, 33–41 (1976).
85. D. L. Smith and J. H. Futrell, *J. Phys. B* **8**, 803–815 (1975).
86. P. G. Miasek and A. G. Harrison, *J. Am. Chem. Soc.* **97**, 714–721 (1975).
87. G. V. Karachevtsev, V. L. Talrose, and P. S. Vinogradov, in: Proc. 12th Int. Conf. on Phenom. in Ionized Gases, J.G.A. Hölscher and D.C. Schram (eds.), Am. Elsevier, New York (1975), p. 44.
88. V. L. Talrose, *Chemical Effects of Nuclear Transformations*, Int. Atomic Energy Agency, Vienna (1961), pp. 103–112.
89. L. P. Theard and W. T. Huntress, Jr., *J. Chem. Phys.* **60**, 2840–2848 (1974).
90. G. V. Karachevtsev, A. Z. Marutkin, V. L. Tal'roze, and S. P. Tkachenko, *Khim. Vys. Énerg.* **13**, 11–18 (1979).
91. F. Robben, *Phys. Rev.* **5A**, 1516–1522 (1972).
92. V. A. Ivanov and V. S. Sukhomlinov, *Zh. Tekh. Fiz.* **53**, 843–853 (1983).
93. G. V. Karachevtsev and V. L. Tal'roze, *Khim. Fiz.* **5**, 415–417 (1986).
94. F. Bien, *J. Chem. Phys.* **69**, 2631–2638 (1978).
95. C. E. Hamilton, M. A. Duncan, T. S. Zwier, et al., *Chem. Phys. Lett.* **94**, 4–9 (1983).
96. B. A. Huber, R. C. Cosby, J. R. Peterson, and J. T. Mosely, *J. Chem. Phys.* **66**, 4520–4526 (1977).
97. J. M. Jasinski and J. I. Brauman, *J. Chem. Phys.* **73**, 6191–6195 (1980).
98. R. N. Rosenfeld, J. M. Jasinski and J. I. Brauman, *J. Am. Chem. Soc.* **104**, 658–667 (1982).
99. M. S. Kim and R. C. Dunber, *Chem. Phys. Lett.* **60**, 247–250 (1979).
100. E. E. Ferguson, *J. Phys. Chem.* **90**, 731–738 (1986).
101. A. B. Rakskit and P. Warneck, *J. Chem. Phys.* **74**, 2853–2859 (1981).
102. H. Villinger, J. H. Futrell, A. Saxer, et al., *J. Chem. Phys.* **80**, 2543–2547 (1984).
103. H. Böhringer, M. Durup-Ferguson, E. E. Ferguson, and D. W. Fahey, *Planet. Space Sci.* **31**, 483–487 (1983).
104. H. Böhringer, M. Durup-Ferguson, D. W. Fahey, et al., *J. Chem. Phys.* **79**, 4201–4213 (1983).
105. E. E. Ferguson, N. G. Adams, D. Smith, and E. Alge, *J. Chem. Phys.* **80**, 6095–6098 (1984).
106. M. Durup-Ferguson, H. Böhringer, D.W. Fahey, et al., *J. Chem. Phys.* **81**, 2657–2666 (1984).
107. D. Smith and N. G. Adams, *Phys. Rev.* **A23**, 2327–2330 (1981).
108. W. Lindinger, F. Howorka, P. Lukac, et al., *Phys. Rev.* **A23**, 2319–2326 (1981).

109. W. Dobler, F. Howorka, and W. Lindinger, *Plasma Chem. Plasma Process.* **2**, 353–359 (1982).
110. W. Dobler, H. Ramler, H. Villinger, et al., *Chem. Phys. Lett.* **97**, 553–556 (1983).
111. W. Dobler, H. Villinger, F. Howorka, and W. Lindinger, *Int. J. Mass Spectrom. Ion Phys.* **47**, 171–174 (1983).
112. L. A. Viehland and D. W. Fahey, *J. Chem. Phys.* **78**, 435–441 (1983).
113. Ch. Schlier and V. Vix, *Chem. Phys.* **95**, 401–409 (1985).
114. P. Kebarle and R. M. Haynes, in: *Ion–Molecule Reactions in the Gas Phase*, Am. Chem. Soc., Washington (1986), pp. 210–242.
115. R. C. Dunbar, Abstracts of papers presented at the 10th Int. Mass Spectrom. Conf., Swansea, Sept. 1985, No. 254.
116. A. Carrington and R. A. Kennedy, *J. Chem. Phys.* **81**, 91–112 (1984).
117. Z. Herman, and J. H. Futrell and E. Fridrich, *Int. J. Mass Spectrom. Ion Phys.* **58**, 181–199 (1984).
118. J. Los, P. G. Kistemaker, S. A. McLuckey, W. J. van der Zande, and D. P. De Bruijn, *Int. J. Mass Spectrom. Ion Phys.* **66**, 161–170 (1985).
119. J. Los and D. P. de Bruijn, in: Abstracts of papers presented at the 10th Int. Mass Spectrom. Conf., Swansea, Sept. 1985, No. 13.
120. N. N. Tunitskii, M. V. Tikhomirov, S. E. Kupriyanov, et al., *Problems of Physical Chemistry*, Vol. 1 [in Russian], p. 122 (1958).
121. S. E. Kupriayanov and A. A. Perov, *Zh. Fiz. Khim.* **42**, 857–859 (1968).
122. S. E. Kupriayanov, *Elementary Processes in High-Energy Chemistry* [in Russian], Nauka, Moscow (1965), pp. 23, 28.
123. D. H. Katayama and J. A. Welsh, *Chem. Phys. Lett.* **106**, 74–78 (1984).
124. H. Obase, M. Tsuji, and Y. Nishimura, *Chem. Phys.* **99**, 111–119 (1985).
125. I. Dotan, S. E. Barlow, and E. E. Ferguson, *Chem. Phys. Lett.* **121**, 38–10 (1985).
126. F. Bonillard and J. McGowan, *Physics of Ion–Ion and Electron–Ion Collisions*, Plenum, New York (1983).
127. V. A. Belyaev, B. G. Brechnev, and E. M. Erastov, in: W. Botticher, H. Wenk and E. Schulz-Gulde (eds.), 16th Int. Conf. Phenom. in Ionized Gases, Contributed Papers (1983), No. 9, pp. 4364–4373.
128. M. R. Flannery, *J. Phys. B, At. Mol. Phys.* **18**, L531–L537 (1985).
129. J. Senekowitsch, K. R. Lykke, T. Andersen, and W. C. Lineberger, *J. Chem. Phys.* **83**, 4364–4373 (1985).
130. M. Okumura, J.-C. L. Jeh, D. Normad, et al., *Tetrahedron* **41**, 1423–1426 (1985).
131. J. P. Honovich, R. C. Dunbar, and J. Lehman, *J. Phys. Chem.* **89**, 2513–2516 (1985).
132. P. K. Acharya, R. A. Kendall, and J. Simons, *J. Am. Chem. Soc.* **18**, 5389 (1985).
133. K. Lifschitz, *J. Phys. Chem.* **87**, 2304–2313 (1983).
134. G. V. Karachevtsev, in: *Quantitative Data on Vibrational Relaxation in Ion–Molecule Collisions. Physicochemical Kinetics in Gas Dynamics*, S. A. Losev and O. P. Shatalov (eds.) [in Russian], Izd. MGU, Moscow (1986), pp. 67–79.